SINA

ADVANCES IN

METAL-ORGANIC CHEMISTRY

Volume 2 • 1991

ADVANCES IN METAL-ORGANIC CHEMISTRY

A Research Annual

Editor: LANNY S. LIEBESKIND
Department of Chemistry
Emory University

VOLUME 2 • 1991

 JAI PRESS LTD

London, England   Greenwich, Connecticut

JAI PRESS LTD
118 Pentonville Road
London N1 9JN, England

JAI PRESS INC.
55 Old Post Road No. 2
Greenwich, Connecticut 06836-1678

ISBN: 0-89232-948-3

Printed in the United States of America

CONTENTS

LIST OF CONTRIBUTORS

Steven J. Coote

Dyson Perrins Laboratory
University of Oxford
Oxford, England

William E. Crowe

Central Research and Development
E.I. du Pont de Nemours &
 Company
Wilmington, Det., U.S.A.

G. Doyle Daves, Jr.

Dean, School of Science
Rensselaer Polytechnic
 Institute
Troy, N.Y., U.S.A.

Stephen G. Davies

Dyson Perrins Laboratory
University of Oxford
Oxford, England

William A. Donaldson

Department of Chemistry
Marquette University
Milwaukee, Wis., U.S.A.

Craig L. Goodfellow

Dyson Perrins Laboratory
University of Oxford
Oxford, England

Paul Helquist

Department of Chemistry
University of Notre Dame
Notre Dame, Ind., U.S.A.

Koichiro Oshima

Department of Industrial
 Chemistry
Kyoto University
Kyoto, Japan

Stuart L. Schreiber

Department of Chemistry
Harvard University
Cambridge, Mass., U.S.A.

Motokazu Uemura

Faculty of Science
Osaka City University
Osaka, Japan

INTRODUCTION

Volume 2 of "Advances in Metal–Organic Chemistry" continues in the same spirit as Volume 1, published approximately two years ago. Authors have been encouraged to write detailed, informal accounts of their research efforts in the field of metal-oriented organic chemistry. Although authors were given guidelines in an attempt to maintain some formatting continuity between the various chapters, I have chosen to minimize editorial interference in order to allow each author to maximize the information presented according to his own style.

Topics included in Volume 2 have been selected to emphasize the virtues of metal-oriented organic chemistry utilizing stoichiometric as well as catalytic reagents. In addition to processes of value for the synthesis of generally useful organic structures (Chapter 3, "Transition Metal Catalyzed Silylmetallation of Acetylenes and Et_3B-Induced Radical Addition of Ph_3SnH to Acetylenes" by Koichiro Oshima; Chapter 4, "Development of Carbene Complexes of Iron as New Reagents for Synthetic Organic Chemistry" by Paul Helquist; and Chapter 7, "Palladium-Mediated Methylenecyclopropane Ring Opening: Applications to Organic Synthesis" by William A. Donaldson), a topic of relevance to the synthesis of the pharmaceutically interesting C-glycosides is included (Chapter 2, "Palladium-Mediated Arylation of Enol Ethers" by G. Doyle Daves, Jr.). The last few years have witnessed a resurgence of interest in synthetic applications of arene complexes of chromiumtricarbonyl and two chapters are included within Volume 2 (Chapter 1, "Synthetic Applications of Chromium Tricarbonyl Stabilized Benzylic Carbanions" by Stephen G. Davies, Steven J. Coote and Craig L. Goodfellow and Chapter 5, "Tricarbonyl (η^6-Arene) Chromium Complexes in Organic Synthesis" by Motokazu Uemura). Chapter 6, "π-Bond Hybridization in Transition Metal Complexes: A Stereoelectronic Model for Conformational Analysis" by William E. Crowe and Stuart L. Schreiber, addresses the origins of the interesting conformational properties of

organometallic complexes. It is an important first step to the rational application of organometallic complexes to stereoselective organic synthesis.

A survey of the chapter titles in both Volumes 1 and 2 will show an obvious
emphasis on transition metal chemistry; however, it is my intent to begin to
expand the scope of chapters published in forthcoming volumes to include
metals from all regions of the periodic table.

Atlanta, Georgia   *Lanny S. Liebeskind*
January 1991   Samuel Candler Dobbs
                                                  Professor of Chemistry

SYNTHETIC APPLICATIONS OF CHROMIUM TRICARBONYL STABILIZED BENZYLIC CARBANIONS

Stephen G. Davies, Steven J. Coote and

Craig L. Goodfellow

OUTLINE

Advances in Metal-Organic Chemistry, Volume 2, pages 1–57
Copyright © 1991 JAI Press Ltd
All rights of reproduction in any form reserved
ISBN: 0-89232-948-3

I. INTRODUCTION

The ease of preparation and wide range of chemical and stereochemical properties imparted to arenes on complexation to chromium tricarbonyl has resulted in numerous studies of their synthetic applications.[1] This review will deal with one aspect of the chemistry of (arene)chromium tricarbonyl complexes, namely the synthetic applications of chromium tricarbonyl stabilized benzylic carbanions. However, a very brief outline of all the general chemical properties of these complexes is given as an introduction.

(Arene)chromium tricarbonyl complexes are bright yellow to red in colour. The complexes are generally air sensitive in solution; although as solids, whilst they should be stored under an inert atmosphere, they may be handled and weighed in air.

The X-ray crystal structure of (benzene)chromium tricarbonyl **1** is shown in Figure 1.[2] The 12 atoms which comprise the benzene unit are essentially coplanar with the chromium lying under one face, equidistant from all the carbon atoms. The chromium–arene carbon bond lengths are 2.23 Å and the chromium to the centroid of the benzene ring distance is 1.73 Å. In solution there is rapid rotation of the chromium tricarbonyl fragment about the chromium to benzene centroid axis. The carbon monoxide ligands thus provide an effective steric block to the whole face of the benzene to which the chromium tricarbonyl fragment is bound. The geometry about the chromium atom is pseudo-octahedral with the benzene occupying three of the coordination sites.

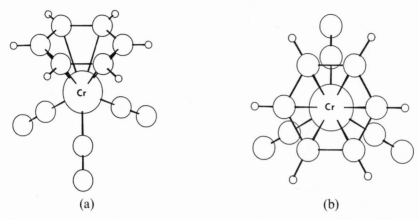

(a)   (b)

Figure 1. X-ray crystal structure of (benzene)chromium tricarbonyl **1**: (a) side view and (b) Newman projection from the benzene centroid to the chromium.

The above structural features are common to all (arene)chromium tricarbonyl complexes although some perturbation from planarity of the arene occurs when substituents possess lone pairs or are very bulky.[3] Complexation of arenes to chromium tricarbonyl causes an upfield shift of about 2 ppm in the ^{1}H-NMR spectrum of the aryl hydrogens. For example, the ^{1}H-NMR spectrum of (benzene)chromium tricarbonyl is a singlet at δ5.31 compared to free benzene at δ7.37 in deuteriochloroform as solvent.

Coordination of arenes to chromium tricarbonyl increases the acidity of the aryl protons by stabilizing, via induction, the conjugate base, an aryl anion. This may be illustrated by the ready fluoride-mediated desilylation of (phenyltrimethylsilane)chromium tricarbonyl **2** under conditions where phenyltrimethylsilane itself is completely inert.[4]

Arenes bound to chromium tricarbonyl are susceptible to nucleophilic addition reactions. Thus (chlorobenzene)chromium tricarbonyl **3** is converted to (anisole)chromium tricarbonyl **4** on treatment with methoxide.[5]

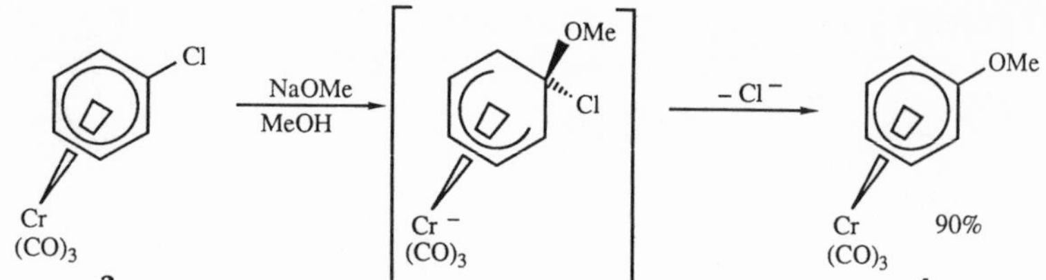

Complexes of arenes possessing benzylic leaving groups exhibit enhanced rates of S_N1 solvolysis when the leaving group can adopt an orientation antiperiplanar to the chromium to arene centroid axis. This enhanced rate of solvolysis results from neighbouring group participation by a lone pair on the chromium assisting the ionization process to form the corresponding resonance-stabilized carbenium ion. Such neighbouring group participation also accounts for the conversion of (+)-(S)-α-methylbenzyl alcohol)chromium tricarbonyl 5 under Ritter reaction conditions to (−)-6 with complete retention of configuration.[6]

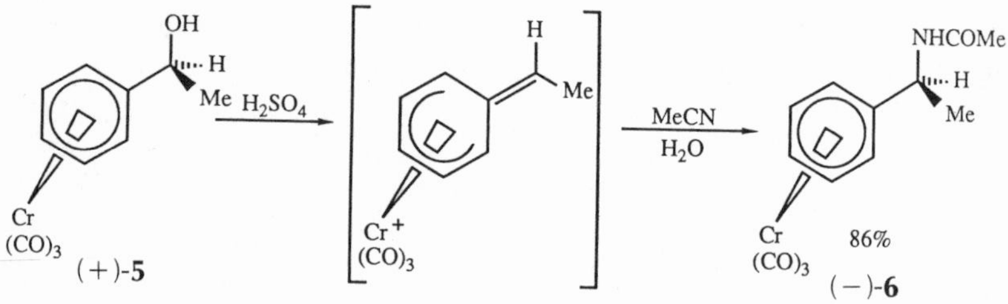

Complexation of arenes to chromium tricarbonyl also enhances the kinetic acidity of benzylic protons in the conformation which places the benzylic C–H bond antiperiplanar to the chromium–arene centroid axis. The resulting benzylic carbanions are also stabilized relative to their uncomplexed analogues by delocalization of the negative charge onto the chromium. These effects may be illustrated by the ready desilylation of (benzyltrimethyl-silane)chromium tricarbonyl 7 under conditions where benzyltrimethylsilane itself is inert.[7]

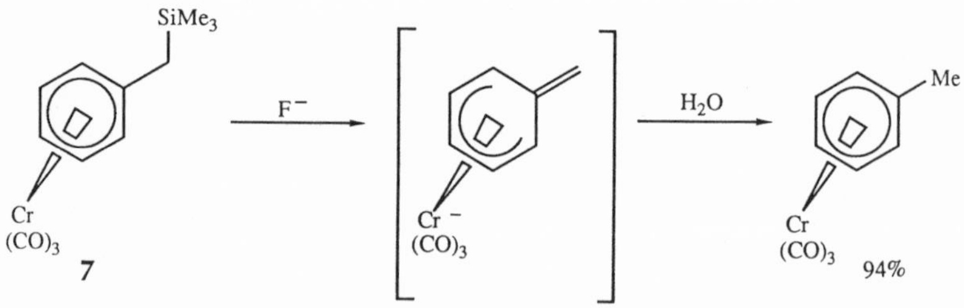

The faces of unsymmetrically substituted arenes are prochiral and hence complexation of, for example, *o*-methoxyethylbenzene **8** results in the formation of complex **9** as a racemate. Whereas the benzylic methylene protons of arene **8** are prochiral, in complex **9** they are diastereotopic.

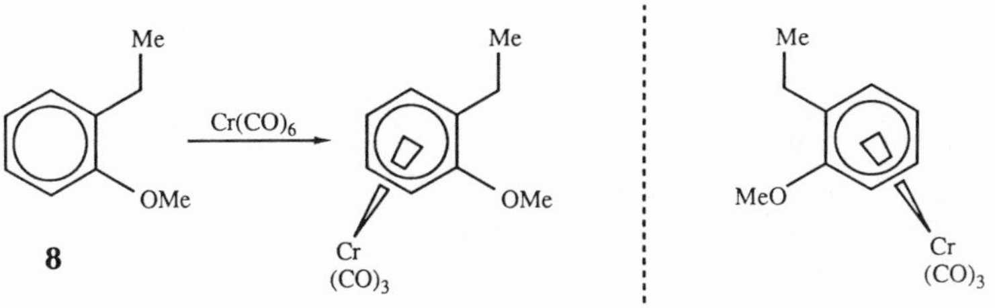

Me

Cr(CO)₆

OMe

8

Me

OMe

Cr
(CO)₃

Me

MeO

Cr
(CO)₃

9

The two *ortho* and the two *meta* protons of a simple monosubstituted (arene)chromium tricarbonyl complex are enantiotopic but are diastereotopic if the substituent is chiral.

II. PREPARATION OF ARENE CHROMIUM TRICARBONYL COMPLEXES

Coordination of the chromium tricarbonyl unit to an arene may be accomplished in a variety of ways, the simplest and most widely used of which is direct ligand exchange with $Cr(CO)_6$.[8] The preparation of all complexed starting materials referred to in this review may be carried out in this manner. Other methods for complexation include thermolysis of $Cr(CO)_3L_3$ [L = NH_3,[9] RCN (R = Me, Et),[10] or pyridine[11]] with the arene or arene exchange with (naphthalene)chromium tricarbonyl.[12]

(Phenyltrimethylsilane)Cr(CO)₃ (**2**).[4] All manipulations of (arene)chromium tricarbonyl complexes were carried out under an atmosphere of nitrogen and all solvents were rigorously degassed prior to use. A stirred solution of $Cr(CO)_6$ (steam distilled and dried, 13.0 g, 59.1 mmol) and phenyltrimethylsilane (7.4 g, 49.3 mmol) in dibutyl ether (Na dried, filtered through active alumina and distilled from CaH_2 under nitrogen, 100 mL) and THF (distilled free of stabilizer and distilled from benzophenone ketyl under nitrogen, 15 mL) was heated at reflux under an atmosphere of nitrogen (60 h). The initially colourless solution rapidly became yellow. The cooled solution was filtered (celite) and evaporated to leave a brown solid. Column chromatography (Al_2O_3, Grade V, 1:1 Et_2O/hexane) gave, after evaporation and recrystallization from CH_2Cl_2/petrol, the title compound **2** as yellow granular crystals (12.1 g, 86%).

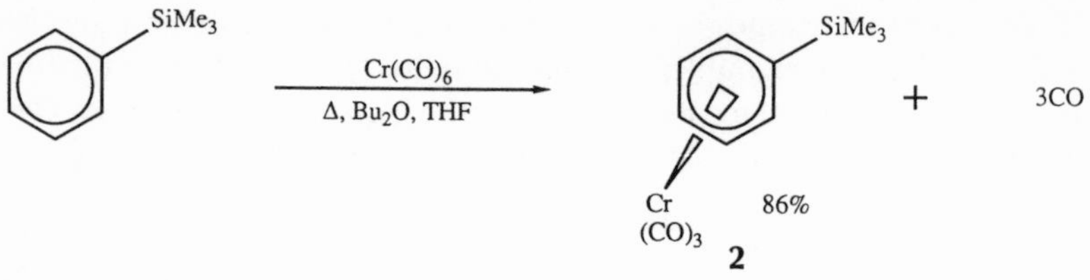

In principle most arenes may be coordinated to the chromium tricarbonyl unit, however, in practice certain functional groups are incompatible (e.g. NO_2 and CN)[5] whilst π-electron withdrawing groups (e.g. CHO, CO_2H, CO_2R, etc.) retard complexation and are best protected prior to complexation. Electron donating arene substituents accelerate the rate of complexation. Thus, complexation of a substrate containing more than one aromatic ring generally occurs preferentially at the most π-electron rich ring,[5,13] although where the face of an arene is sterically hindered the electronic directing effects of the ring substituents may be overriden. Face selectivity may be influenced by the presence of both steric and/or electronic factors. For example, in the case of 1-tetralol **10** stereoselective complexation with chromium hexacarbonyl occurs exclusively onto the *endo* face via chelation of the alcohol oxygen atom with the incoming metal unit to give *endo*-(1-tetralol)chromium tricarbonyl **11** as a single diastereoisomer.[14]

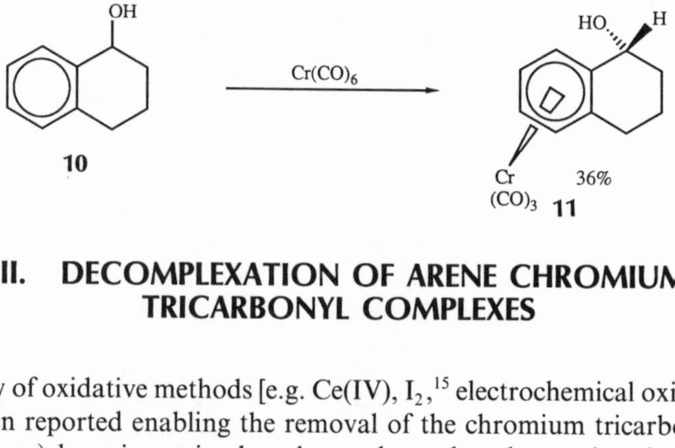

III. DECOMPLEXATION OF ARENE CHROMIUM TRICARBONYL COMPLEXES

A variety of oxidative methods [e.g. Ce(IV), I_2,[15] electrochemical oxidation[16]] have been reported enabling the removal of the chromium tricarbonyl unit from (arene)chromium tricarbonyl complexes, but decomplexation is most conveniently accomplished by aerial oxidation in the presence of sunlight.[17] Alternatively, (arene)chromium tricarbonyl complexes may be heated in pyridine at reflux to liberate the free arene.

α-(o-Tolyl)benzyl alcohol (**12**).[4] A yellow solution of [α-(*o*-tolyl)benzyl alcohol]-Cr(CO)$_3$ **13** (260 mg, 0.78 mmol) in Et$_2$O (20 mL) was exposed to air and sunlight until a colourless solution resulted (72 h). The precipitated chromium residues were filtered off (celite) and the filtrate evaporated to give a white solid.

Recrystallization from Et$_2$O/petrol gave the title compound **12** as white crystals (134 mg, 87%).

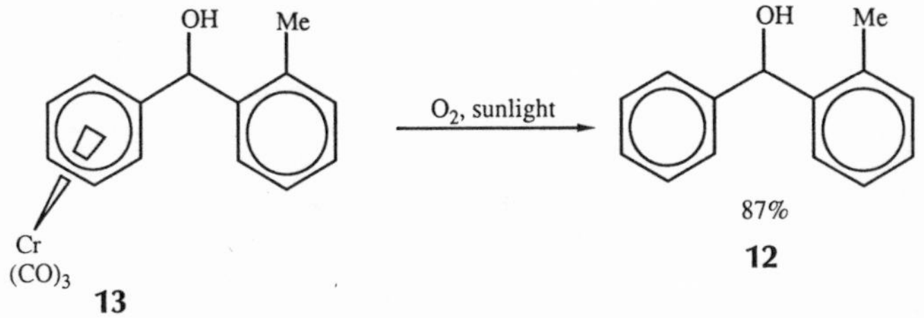

IV. BENZYLIC CARBANIONS DERIVED FROM (C$_6$H$_5$R)CHROMIUM TRICARBONYL COMPLEXES

(Toluene)chromium tricarbonyl **14** undergoes exclusive benzylic deprotonation to generate the benzylic carbanion **15** in the presence of bases such as potassium t-butoxide in dimethylsulphoxide or potassium hydride/18-crown-6 in tetrahydrofuran.[18-20] In contrast, free toluene is inert under the same conditions. Deprotonation of (toluene)chromium tricarbonyl **14** at low temperatures with strong bases such as butyllithium leads to preferential abstraction of a ring proton,[21,22] however, with a deficiency of base and warming, equilibration occurs to generate the thermodynamically more stable benzylic carbanion **15**.[23] Deprotonation of free toluene with butyllithium in the presence of tetramethylethylenediamine requires forcing conditions and yields α-lithiotoluene in good yield.[24]

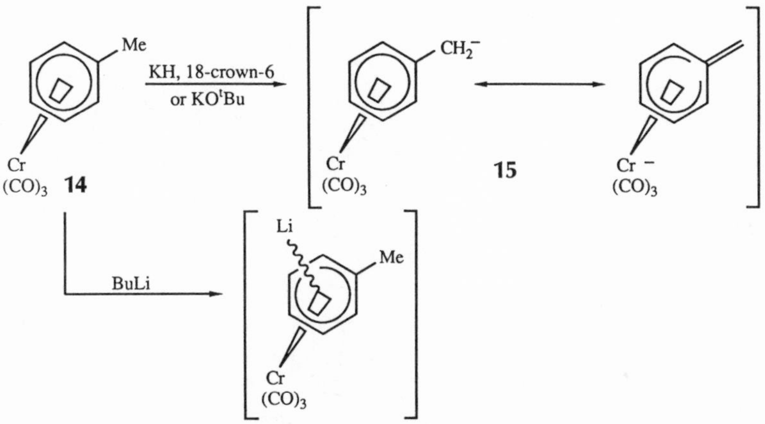

Benzylic carbanions complexed to chromium tricarbonyl are an intense red colour,[18] which is discharged back to yellow when they are quenched with

Table 1.   Electrophilic additions to carbanion **15**

Electrophile (E$^+$)	**16** (E)	Yield (%)	Ref.
D$_2$O	D	—	18
MeI	Me	34	18,21,25
PhCH$_2$Br	PhCH$_2$	56	25
HCHO	CH$_2$OH	16	19,26
PhCHO	CH(Ph)OH	86	19,20,26,27
CO$_2$	COOH	61	18,23,28

electrophiles. Anion **15** reacts with a variety of electrophiles such as alkyl halides, aldehydes, esters, etc., to generate the corresponding α-substituted complexes **16** (Table 1).

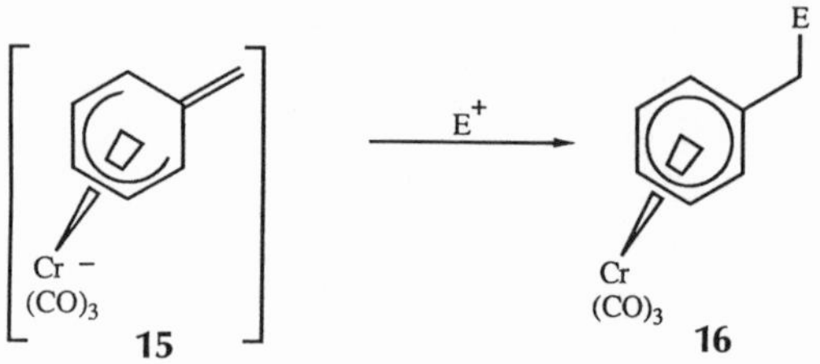

The carbanion **15** also reacts with diethyl oxalate[20,29] and t-butylnitrite[30] to yield the enol **17** and the oxime **18**, respectively. The X-ray crystal structure of **17** has been determined.[31]

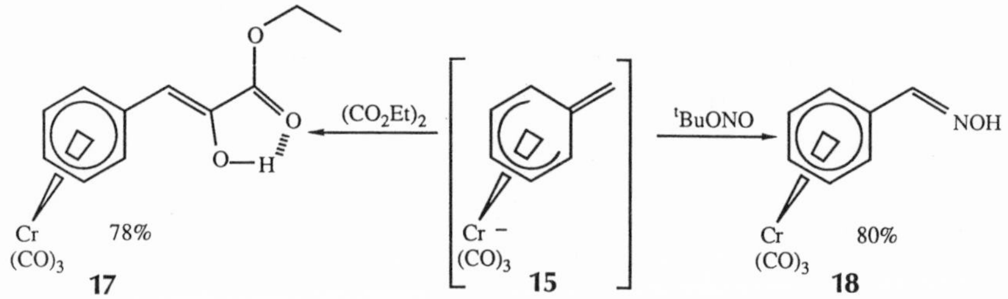

As a result of proton exchange reactions some disubstitution products are produced in the alkylation reactions of **15**. Thus in the methylation of **15**, (i-propylbenzene)chromium tricarbonyl **20** is produced in addition to (ethylbenzene)chromium tricarbonyl **19**.[18,25] Similarly the methylation of (ethylbenzene)chromium tricarbonyl **19** with potassium t-butoxide and methyl iodide gave a small quantity of (t-butylbenzene)chromium tricarbonyl **21** in addition to the major product **20**.[25,32]

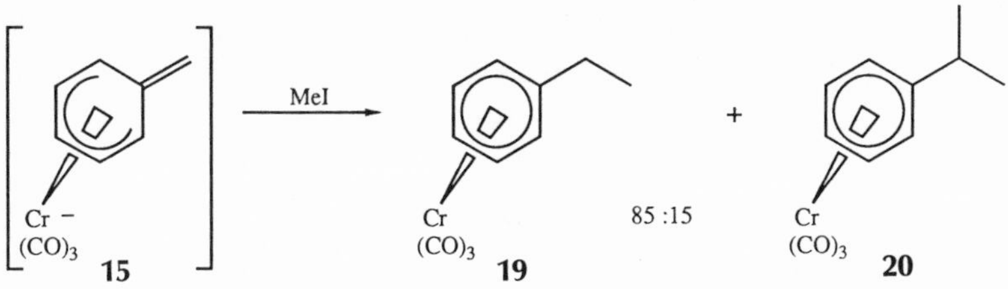

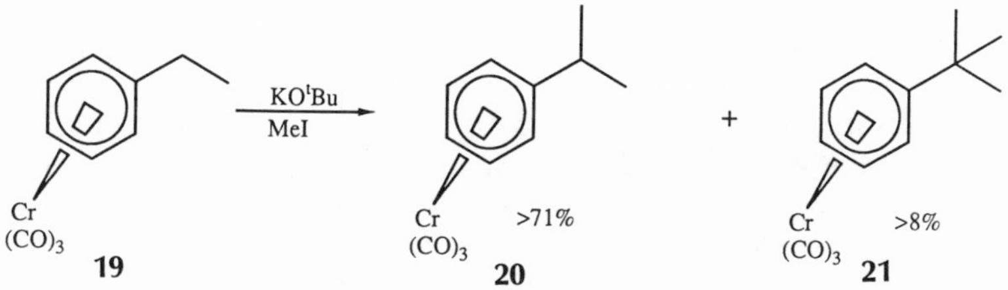

The benzylic carbanion derived from (ethylbenzene)chromium tricarbonyl **19** has been trapped by a variety of electrophiles, which are listed in Table 2, to give the benzylically substituted derivatives **22**.

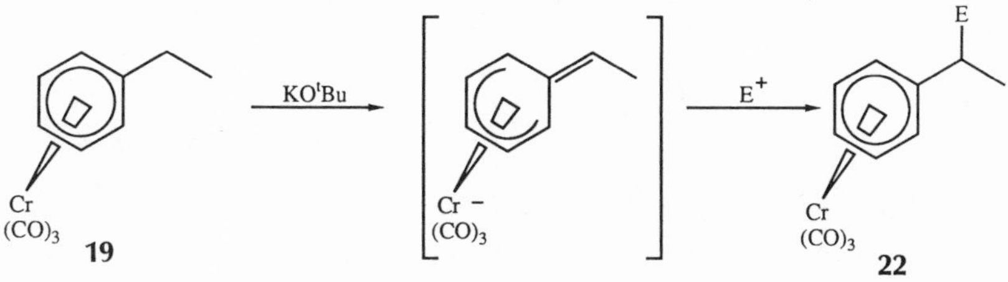

Table 2. α-Substitution reactions of (ethylbenzene)chromium tricarbonyl **19**

Electrophile (E$^+$)	**22** (E)	Yield (%)	Ref.
(CD$_3$)$_2$CO	D	55	19,26
MeI	Me	49	25,32
PhCH$_2$Br	PhCH$_2$	73	25
HCHO	CH$_2$OH	28	19,26
PhCHO	CH(Ph)OH	72	19,20,26,27
Furfural	(C$_4$H$_3$O)CHOH	40	27
m-(MeO)C$_6$H$_4$CHO	*m*-(MeO)C$_6$H$_4$CO	40	27
p-(Me)C$_6$H$_4$CHO	*p*-(Me)C$_6$H$_4$CHOH	37	27
(COOEt)$_2$	COCOOEt	62	29
t-BuONO	= NOH	49	30

The benzylic anion generated from (i-propylbenzene)chromium tricarbonyl **20** with potassium t-butoxide has been trapped by benzaldehyde.[27]

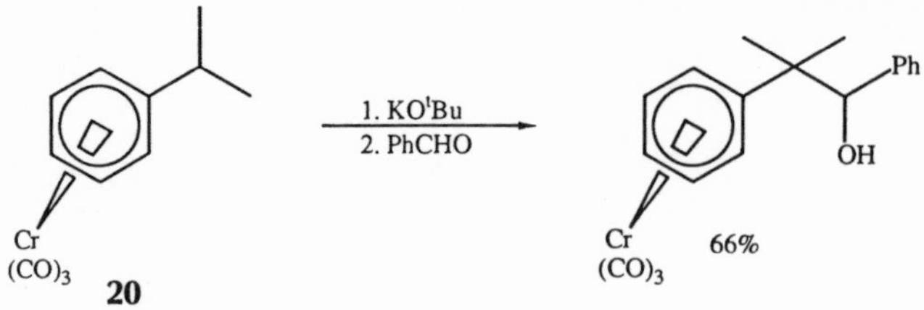

Other (alkylbenzene)chromium tricarbonyl complexes also deprotonate smoothly at the benzylic position. Thus (propylbenzene)chromium tricarbonyl **23** reacts with potassium t-butoxide and t-butylnitrite or benzaldehyde to produce the oxime **24**[30] and alcohol **25**,[20,27] respectively.

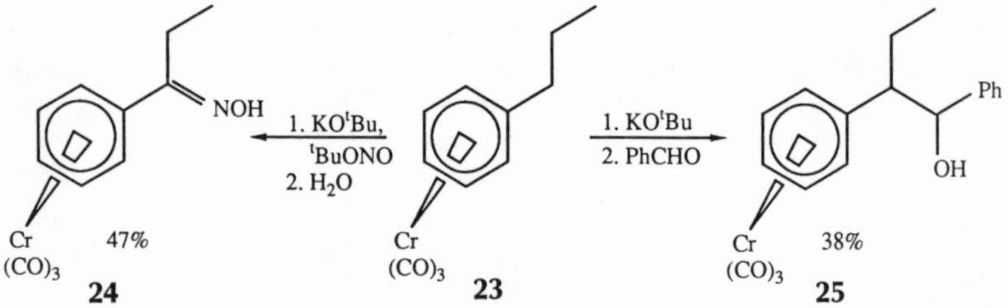

An NMR spectroscopic analysis of carbanions such as **26** has shown that the *ortho* hydrogens and carbons are magnetically inequivalent.[18] This is consistent with restricted rotation about the α-carbon to *ipso* carbon bond, i.e. the structure of the anions is best considered as possessing significant exocyclic double bond character with the negative charge delocalized onto the chromium tricarbonyl fragment.

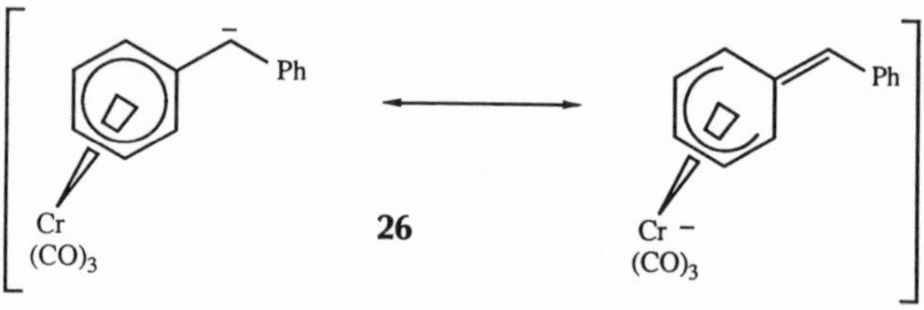

The enhancement of benzylic proton acidity upon complexation to chromium tricarbonyl is evident from the reactions of the mono-(chromium tricarbonyl) complexes of diphenylalkanes. Complexes **27** undergo H/D exchange on treatment with base in deuteriodimethylsulphoxide exclusively alpha to the complexed arene to generate **28**.[32-34]

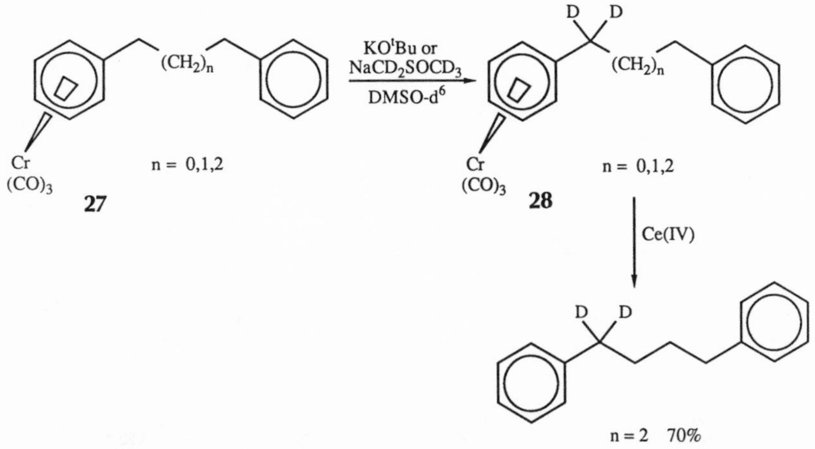

In the presence of potassium t-butoxide complex **27** ($n = 0$) could be monomethylated to give **29**[25] while complex **27** ($n = 1$) underwent disubstitution with formaldehyde to give **30**.[19,26]

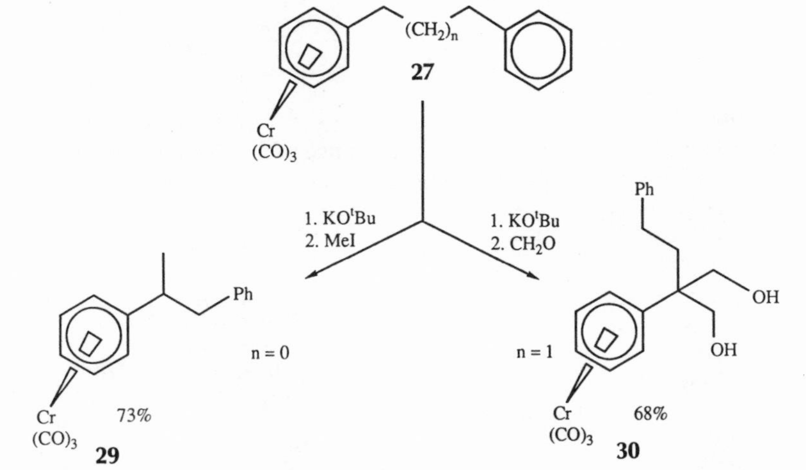

Phenylacetic esters complexed to chromium tricarbonyl undergo base-induced mono- and dialkylation reactions. For example, treatment of (methyl phenylacetate)chromium tricarbonyl **31** with excess sodium hydride and alkyl halides generated the corresponding dialkylated derivatives **32–34**.[25,35]

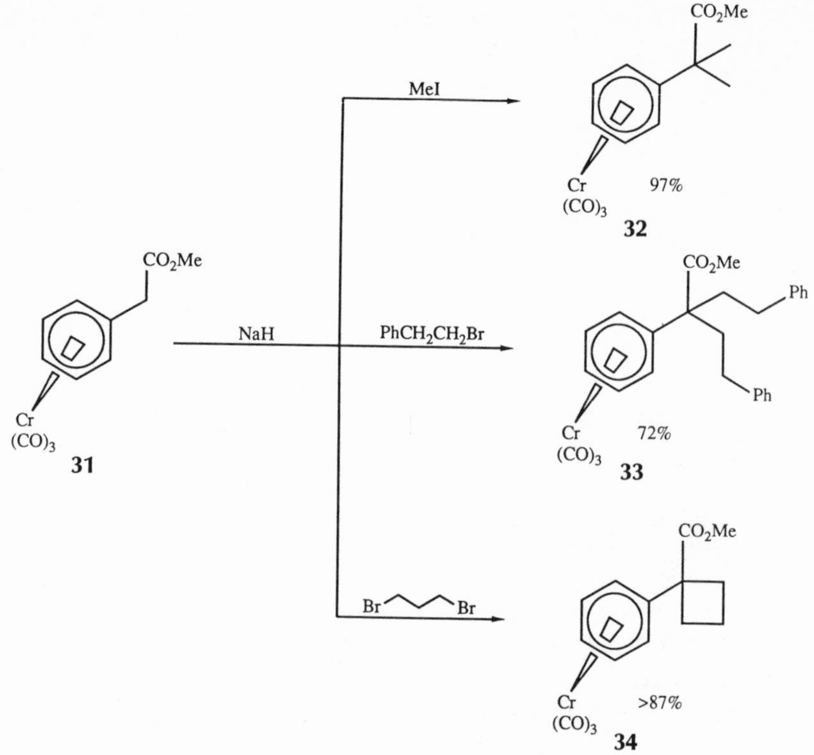

Correspondingly, deprotonation of (t-butyl phenylacetate)chromium tricarbonyl **35** with sodium hydride in the presence of 1,4-dibromopentane gave a 3:1 mixture of the cyclopentane diastereoisomers **36**; a reaction which fails for the uncomplexed arene.[36]

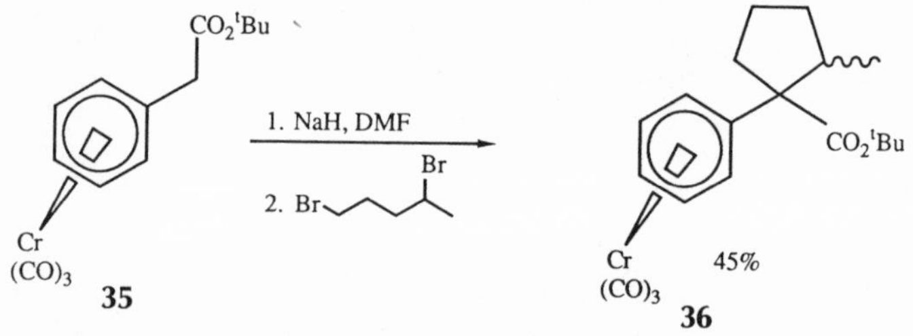

The α-substituted methyl phenylacetate complexes **37–39** have been successfully quaternized by treatment with base in the presence of an alkyl halide. Attempts to achieve the same reactions on the free arenes under phase transfer conditions resulted in ester hydrolysis.[36]

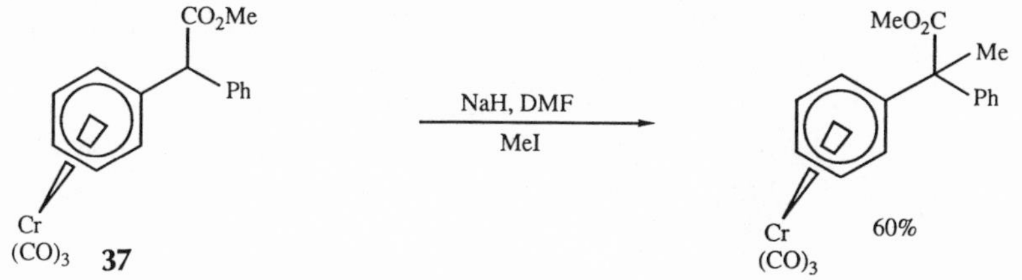

CO2Me
Ph
Cr
(CO)3
NaH, DMF
MeI
MeO2C
Me
Ph
Cr
(CO)3
60%

37

CO2Me
Cr
(CO)3
Cr
(CO)3
NaH, DMF
MeI
Me
CO2Me
Cr
(CO)3
100%
Cr
(CO)3

38

CO2Me
Cr
(CO)3
NaH, DMF
RX
RX = MeI
PhCH2Br
CH2=CHCH2Br
HC≡CCH2Br
MeO2C
R
Cr
(CO)3
R = Me 70%
PhCH2 100%
CH2=CHCH2 100%
HC≡CCH2 100%

39

The related γ-lactone complex **40** underwent analogous benzylic alkylation reactions to give complexes **41**.[36]

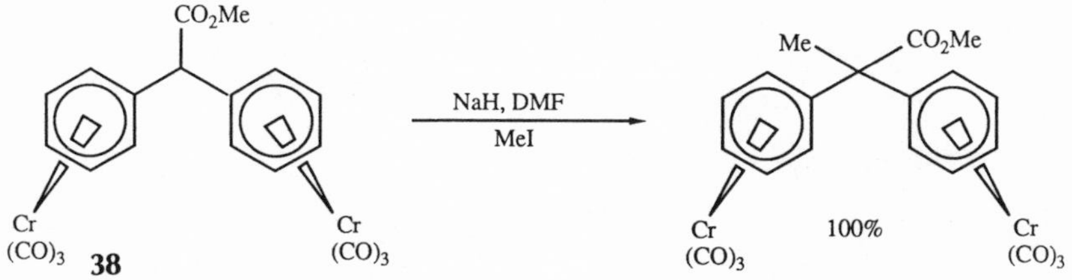

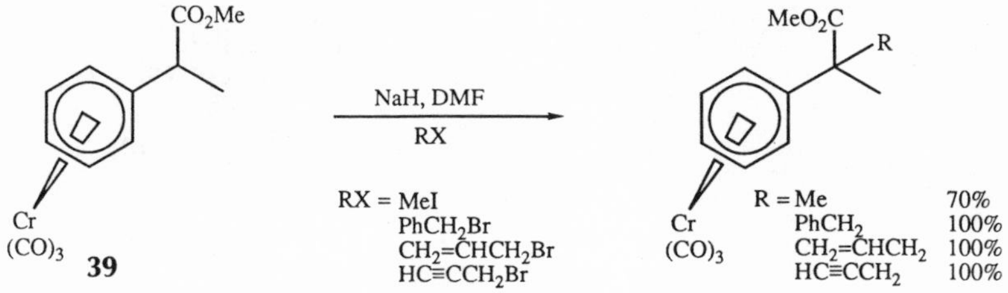

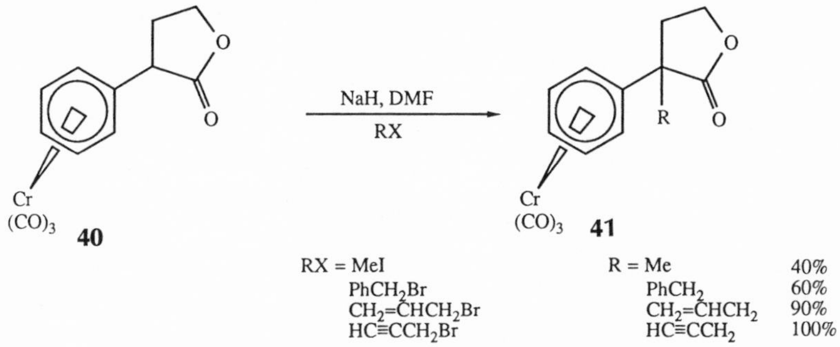

O
O
Cr
(CO)3
NaH, DMF
RX
O
R
O
Cr
(CO)3
RX = MeI
PhCH2Br
CH2=CHCH2Br
HC≡CCH2Br
R = Me 40%
PhCH2 60%
CH2=CHCH2 90%
HC≡CCH2 100%

40 **41**

V. BENZYLIC CARBANIONS DERIVED FROM THE CHROMIUM TRICARBONYL COMPLEXES OF XYLENES, INDANES AND TETRALINS

Although relatively little information is available concerning the parent (xylene)chromium tricarbonyl complexes they do exhibit reactivities analogous to (toluene)chromium tricarbonyl **14**. Treatment of (xylene)chromium tricarbonyl complexes sequentially with potassium t-butoxide and electrophiles led to α-monosubstitution and subsequent α,α′-disubstituted products. This trend is illustrated below by the reactions of (*m*-xylene)chromium tricarbonyl **42**[29,37] and (*p*-xylene)chromium tricarbonyl **43**.[19,26,30,37]

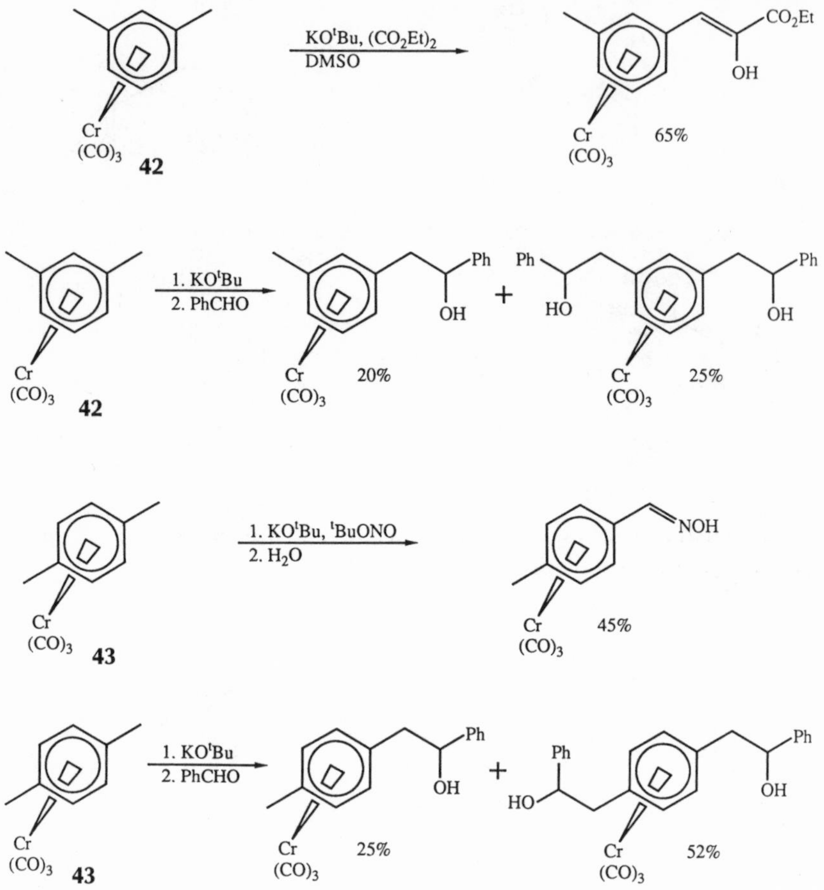

Whereas the use of potassium t-butoxide as base leads to exclusive benzylic deprotonation, kinetic deprotonation with a strong base such as butyllithium leads to a mixture of arene and benzylic deprotonation. Sequential treatment of complexes **42**, **43**, or (*o*-xylene)chromium tricarbonyl **44** with butyllithium and methyl iodide gave a mixture of products resulting from both benzylic and ring methylation; for complexes **42** and **44** some preference for ring methylation at the least hindered position was observed.[21]

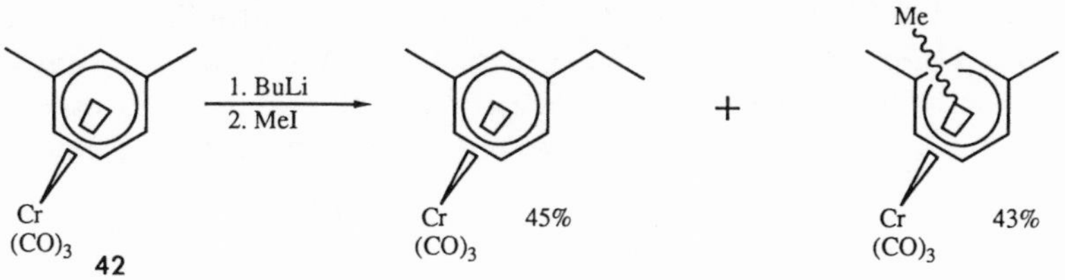

The enhancement of the acidity of benzylic protons on complexation to chromium tricarbonyl is apparent from the deuteriation of complex **45**, which led to the incorporation of five deuterium atoms alpha to the complexed arene, but none into the methyl group attached to the uncomplexed arene.[34]

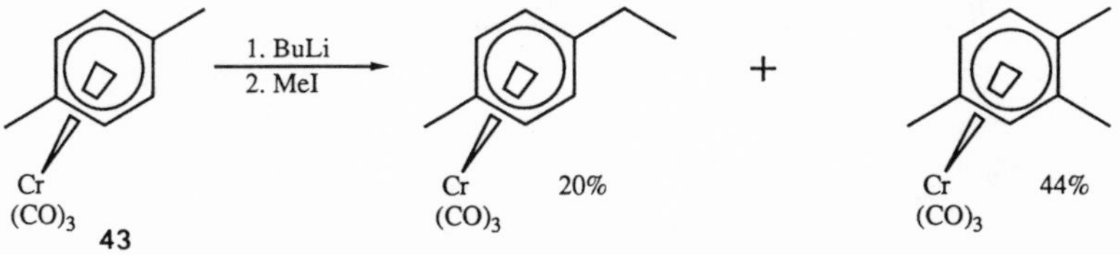

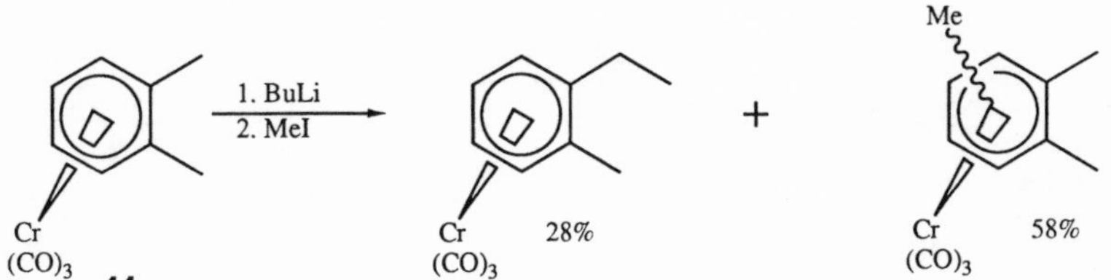

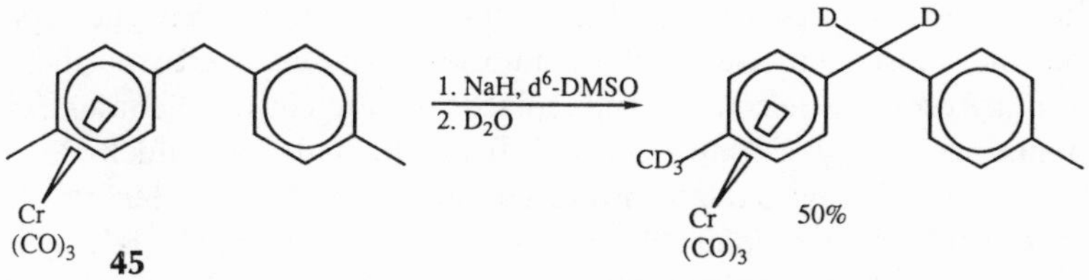

In (arene)chromium tricarbonyl complexes the enhancement of the acidity of benzylic protons is only apparent in complexes which can adopt conformations which place the benzylic C–H bond close to antiperiplanar to the chromium–arene centroid axis. In (indane)chromium tricarbonyl **46** and (tetralin)chromium tricarbonyl only the benzylic protons *exo* to the chromium tricarbonyl, i.e. on the uncoordinated face, can achieve this conformation. Deuteriation of (indane)chromium tricarbonyl **46** by treatment with potassium t-butoxide in deuteriodimethylsulphoxide resulted in the stereoselective incorporation of only two deuterium atoms exclusively in the

1-*exo* and 3-*exo* positions. The *endo*-(1-methylindane)chromium tricarbonyl complex **47** possesses two *exo*-benzylic protons whereas the corresponding *exo*-(1-methylindane)chromium tricarbonyl complex **48** has only one. Thus on deuteriation, complexes **47** and **48** stereoselectively incorporate two and one deuterium atoms, respectively.[33]

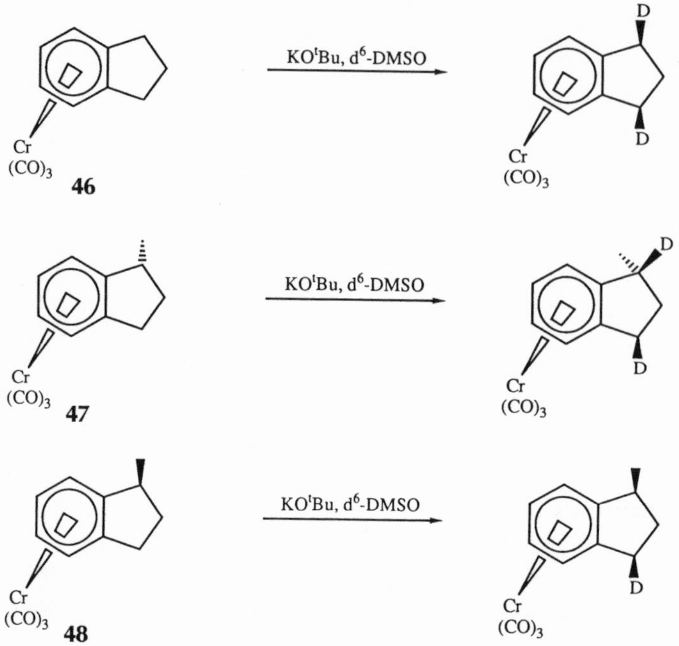

The above deuteriation results demonstrate that both the benzylic deprotonation and subsequent electrophilic addition steps are stereospecific. *exo*-Protons are removed selectively and replaced with retention of configuration. Trapping the benzylic anion derived from **46** with formaldehyde, benzaldehyde and diethyl oxalate produced complexes **49**,[19,26,38] **50**[19,20,27,38] and **51**,[20,29] respectively. Complex **49** was converted to complex **48** via tosylation and reduction.[19,26] The reaction with benzaldehyde also produced some of the 1-*exo*-ketone **52** presumably via an Oppenauer–Woodward type oxidation of the primary product **50**.[27,38]

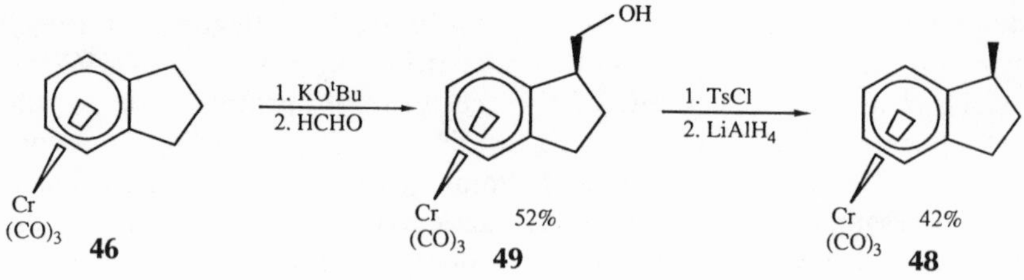

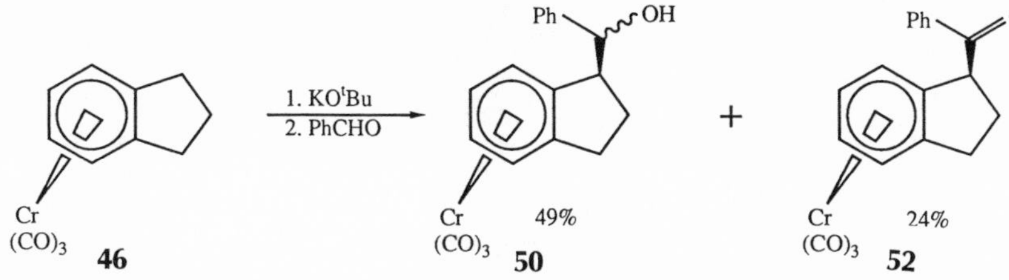

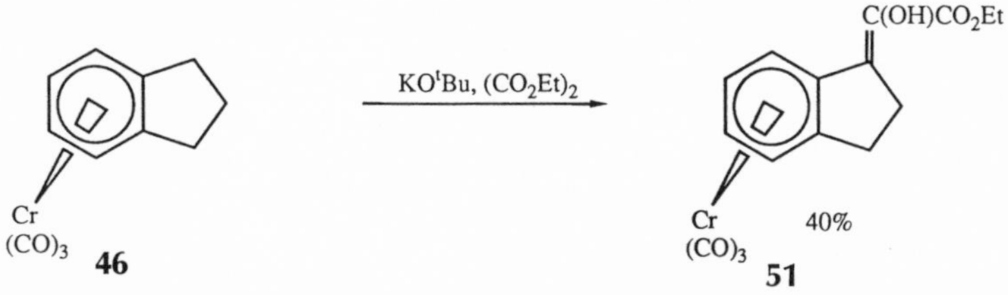

Protons alpha to esters are removed in preference to those alpha to (arene)chromium tricarbonyl moieties.[25] Thus both the *exo-* and *endo-*(1-methoxycarbonyl indane)chromium tricarbonyl complexes **53** and **54** are deprotonated in the 1-position to give the same stabilized carbanion **55**.[36] In the *exo* complex **53** the 1-*endo*-proton is activated by the ester group to a greater extent than the 3-*exo*-proton is activated by the (arene)chromium tricarbonyl fragment. The 1-*exo*-proton in the *endo* complex **54** is activated by both the ester and (arene)chromium tricarbonyl groups. Subsequent alkylations of the intermediate anion **55** are completely stereoselective giving the 1-*exo*-alkyl derivatives **56**, which result from approach of the electrophile from the uncoordinated face. This stereoselectivity results from a combination of stereoelectronic and steric effects.

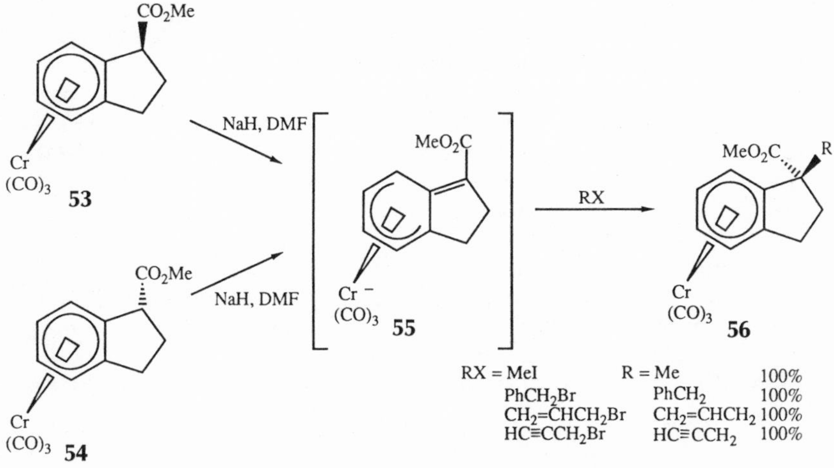

RX = MeI	R = Me	100%
PhCH₂Br	PhCH₂	100%
CH₂=CHCH₂Br	CH₂=CHCH₂	100%
HC≡CCH₂Br	HC≡CCH₂	100%

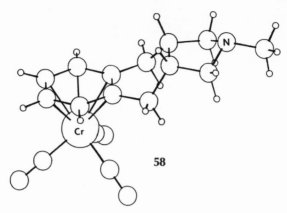

58

Figure 2. X-ray crystal structure of *endo*-(N-methyltetrahydrobenzisoindoline)chromium tricarbonyl **58**.

Complexation of N-methyltetrahydrobenzisoindoline **57** yields a 1.4:1 mixture of the *endo*- and *exo*-chromium tricarbonyl complexes **58** and **59**.[39] Figure 2 shows the X-ray crystal structure of the major product, *endo*-(N-methyltetrahydrobenzisoindoline)chromium tricarbonyl **58**.[39]

Sequential treatment of **58** with butyllithium and methyl iodide resulted in exclusive methylation of the arene ring with a mixture of the 6- and 7-methyl derivatives being formed. Clean benzylic methylation of **58** could be achieved using sodium hexamethyldisilazide as the base to yield, stereoselectively, the 5-*exo*-methyl derivatives **60**, which on decomplexation gave 5α-methyl-N-methyltetrahydrobenzisoindoline **61**. Stereoselective 5-*exo*-methylation under similar conditions of the complex **59** produced complex **62** as a single diastereoisomer which yielded **63** after decomplexation.[39]

5α-Methyl-4α,10α-N-methyltetrahydrobenz[f]isoindoline (**61**).[39] A solution of *endo*-(N-methyltetrahydrobenz[f]isoindoline)Cr(CO)₃ **58** (250 mg, 0.77 mmol) in DMF (distilled from calcium sulphate, 3 mL) was added dropwise via a cannula to a suspension of sodium hexamethyldisilazide (made from freshly prepared sodamide and hexamethyldisilazine, 560 mg, 3.06 mmol) in DMF (3 mL) at 0°C. The resulting red solution was stirred (3 h, 20°C), cooled to 0°C and treated with methyl iodide (dried over 4 Å molecular sieves, 0.05 mL, 0.85 mmol). After stirring (2 h, 20°C), water (50 mL) was added and the mixture extracted (Et₂O, 3 × 40 mL). The combined extracts were concentrated and chromatographed (Al₂O₃, Grade V, CH₂Cl₂) to give an orange oil which was dissolved in Et₂O and allowed to stand in air and sunlight until colourless. Filtration (celite) gave a clear oil. Flash chromatography (SiO₂, 78:20:2 toluene/EtOH/aq. NH₃) gave the title compound **61** as a white solid (51 mg, 33%).

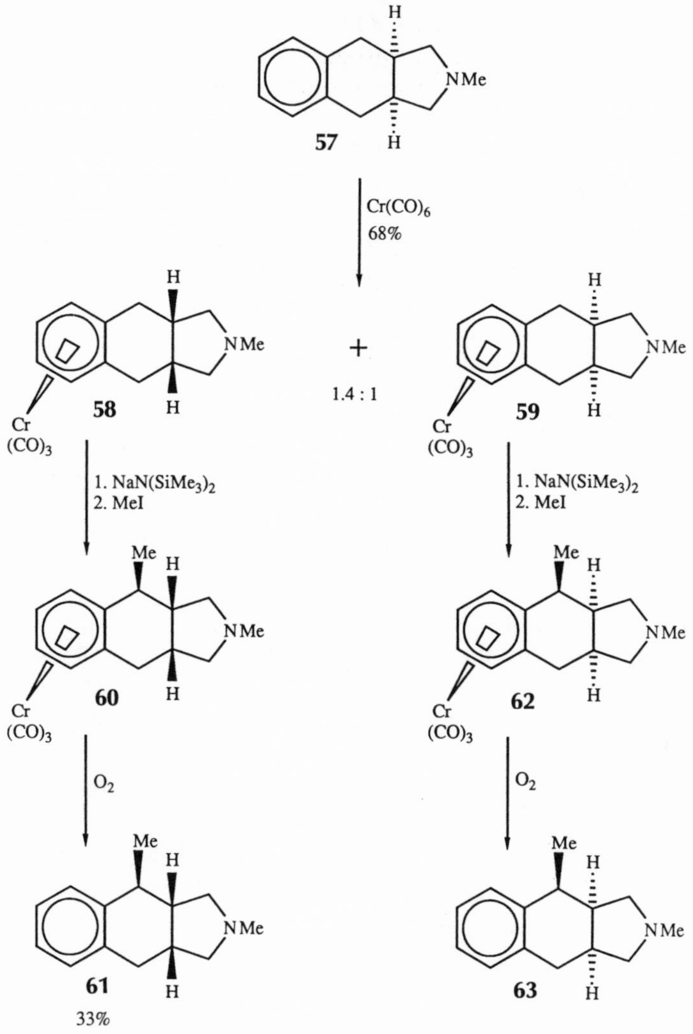

VI. INFLUENCE OF *META* AND *PARA* SUBSTITUENTS ON THE BENZYLIC DEPROTONATION OF (ARENE)CHROMIUM TRICARBONYL COMPLEXES

Meta substituents in (arene)chromium tricarbonyl complexes are not expected to have any significant effect on the acidity of benzylic protons. However, the acidity of benzylic protons will be significantly increased by *para*-π-electron acceptor substituents (CO_2R, $SiMe_3$) but decreased by *para*-

π-electron donating substituents (OR, NR_2, Cl). For example, in 4-substituted (*o*-xylene)chromium tricarbonyl complexes the regioselectivities observed in deprotonation–alkylation reactions are dependent on the nature of the substituent. If the 4-substituent is π-electron withdrawing then α-substitution occurs exclusively at the C-1 methyl; the substituent is increasing the acidity of the C-1 methyl protons relative to those of the C-2 methyl group. If the 4-substituent is π-electron donating then exclusive α-substitution at the C-2 methyl is observed; the substituent is decreasing the acidity of the C-1 methyl protons relative to the C-2 methyl protons. If the 4-substituent is neither π-electron donating nor withdrawing (H, ^{t}Bu) then random α-substitution occurs.

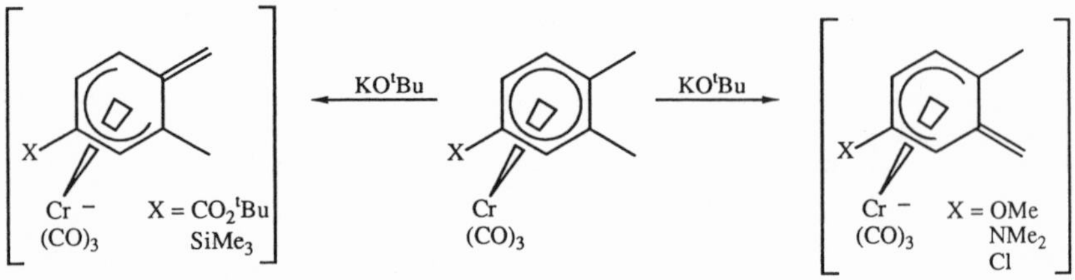

Treatment of (4-t-butoxycarbonyl-*o*-xylene)chromium tricarbonyl **64** with potassium t-butoxide in the presence of deuterioacetone results in exclusive deuteriation of the C-1 methyl group.[40] Treatment of complex **64** with potassium t-butoxide in the presence of aldehydes leads only to benzylic substitution at the C-1 methyl group.[37,38,40]

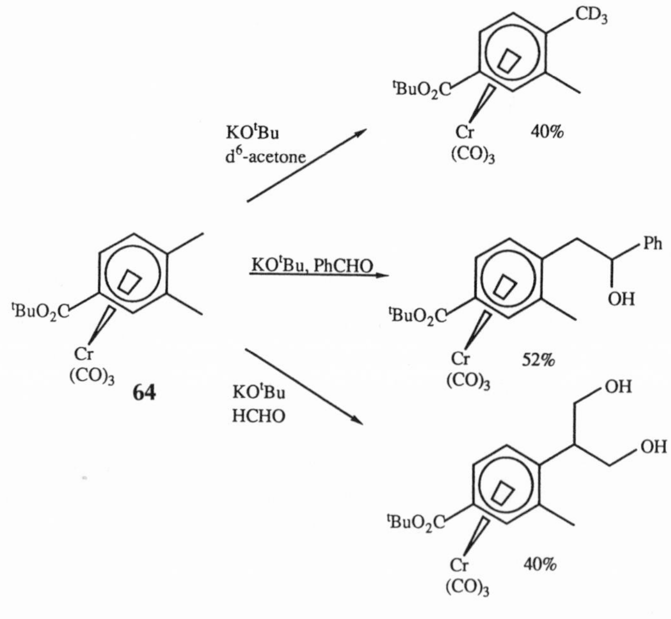

The trimethylsilyl substituent in (4-trimethylsilyl-*o*-xylene)chromium tricarbonyl **65** also activates the C-1 methyl protons. Addition of butyllithium followed by methyl iodide to complex **65** results in exclusive methylation of the C-1 methyl group to give (2-methyl-4-trimethylsilylethylbenzene)chromium tricarbonyl **66**.[7]

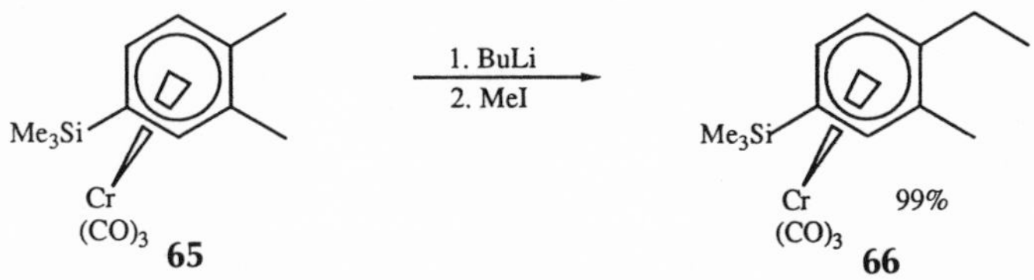

In contrast to the above, (4-methoxy-*o*-xylene)chromium tricarbonyl **67** undergoes deprotonation by potassium t-butoxide exclusively at the C-2 methyl group leading to selective deuteriation and alkylation reactions at this position.[37,38,40,41] This is consistent with the electron donating methoxyl group disfavouring C-1 methyl deprotonation, but not having any significant effect on the C-2 methyl proton acidity. Strong bases such as butyllithium are expected to deprotonate complex **67** on the arene adjacent to the methoxyl group.[42]

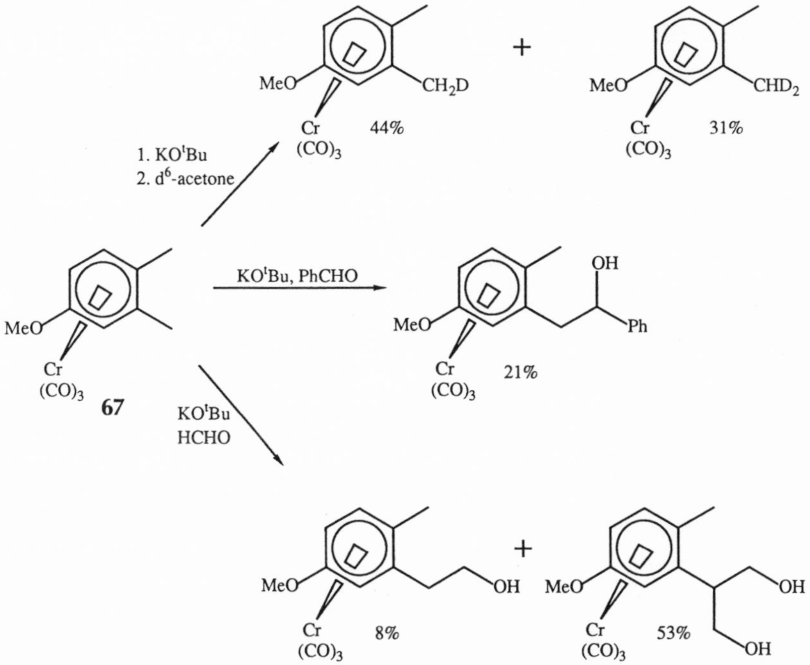

The dimethylamino group is also strongly electron donating and consequently (4-dimethylamino-*o*-xylene)chromium tricarbonyl **68** undergoes benzylic deprotonation–alkylation at the C-2 methyl position.[37,38]

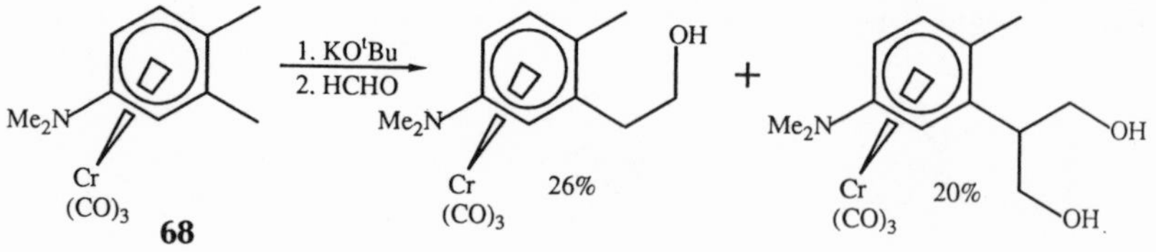

Similarly, the dimethylamino group in (1,2-diethyl-4-dimethylaminobenzene)chromium tricarbonyl **69** deactivates the methylene protons of the C-1 ethyl group relative to the methylene protons of the C-2 ethyl group. Thus, treatment of complex **69** with potassium t-butoxide and formaldehyde yields complex **70** regioselectively. Furthermore, complex **70** is formed stereoselectively.[43] Although the relative configuration in **70** has not been established, the most reasonable explanation for the observed stereoselective substitution is that the most favourable conformation for deprotonation of the C-2 ethyl group, which must place a C–H bond antiperiplanar to the chromium–arene centroid axis, places the methyl groups of the two ethyl substituents *anti* to each other. Deprotonation of **69** in this conformation would generate stereoselectively the anion **71**, which on alkylation from the uncoordinated face, will produce **70** stereoselectively.

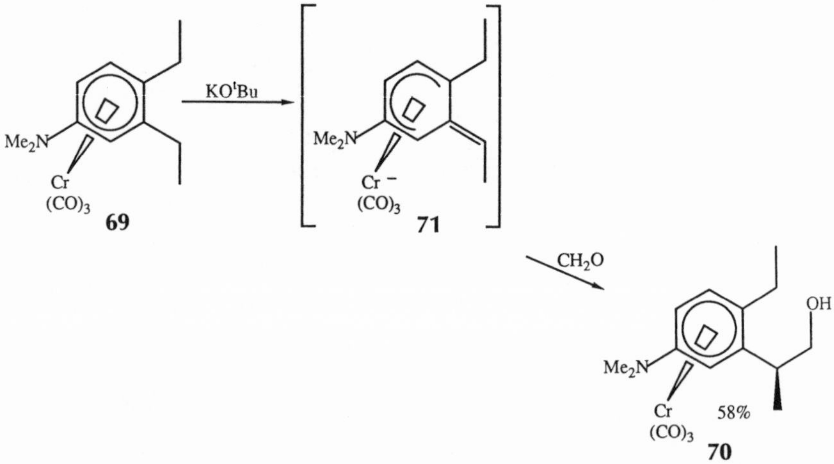

Benzylic deprotonation-alkylation reactions of (4-chloro-*o*-xylene)chromium tricarbonyl **72** occur regioselectively at the C-2 methyl position.[37,38]

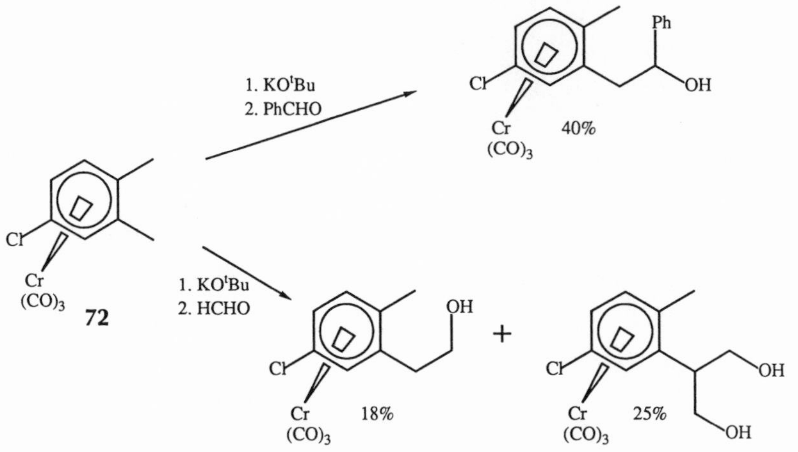

Treatment of (4-t-butyl-*o*-xylene)chromium tricarbonyl **73** with potassium t-butoxide and benzaldehyde results in an approximately equimolar mixture of the two possible α-substituted and the α,α'-disubstituted products, thus demonstrating the inoccuous nature of the t-butyl group.[37]

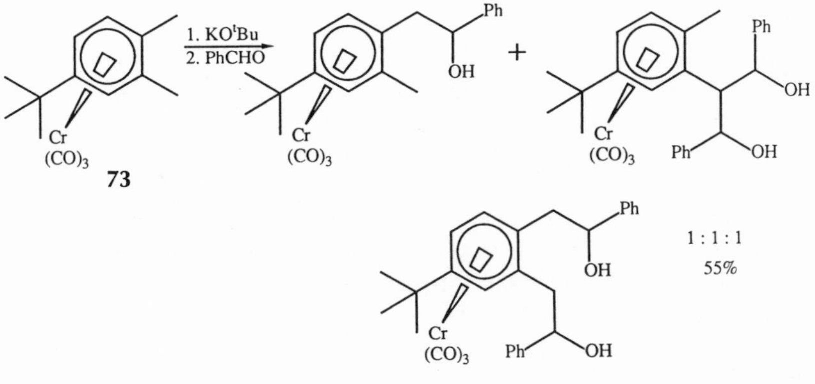

In 3-substituted (toluene)chromium tricarbonyl complexes the substituent exerts no influence on the benzylic proton acidity and the reactions of these complexes parallel those of the analogous unsubstituted complexes. The known reactions of such complexes are illustrated below. Of particular note is that even the presence of an electron donating 3-methoxyl group does not temper the benzylic reactivity. Thus (3-methoxytoluene)chromium tricarbonyl **74**, (3-methoxy-ethylbenzene)chromium tricarbonyl **75** and (3-methoxy-i-propylbenzene)chromium tricarbonyl **76** all undergo benzylic deprotonation in the presence of potassium t-butoxide and yield the corresponding α-substituted products.[27,30,37,38,40,41]

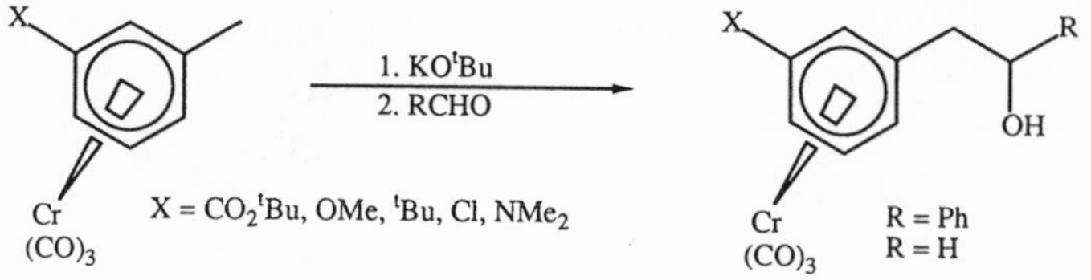

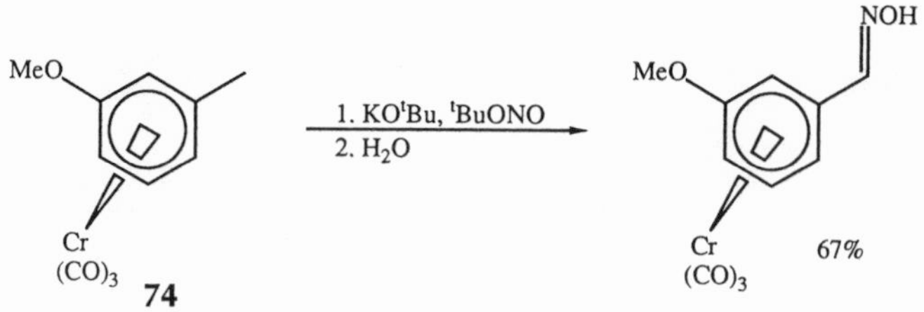

In 4-substituted (toluene)chromium tricarbonyl complexes the substituent does influence the acidity of the methyl protons. The π-electron withdrawing group, t-butoxycarbonyl, facilitates benzylic substitution reactions whereas the electron donating methoxyl, dimethylamino and chloro substituents retard the benzylic substitution reactions. In the latter cases lower yields are obtained.[27,29,30,31,37,38,40,41]

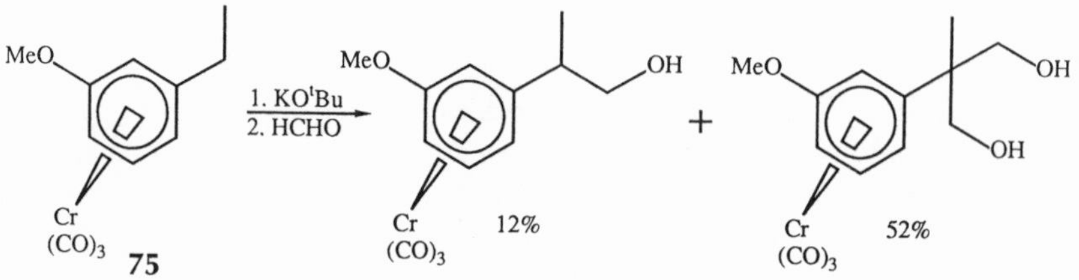

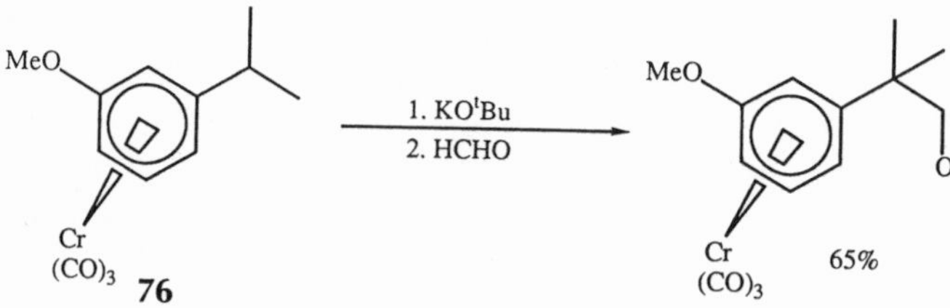

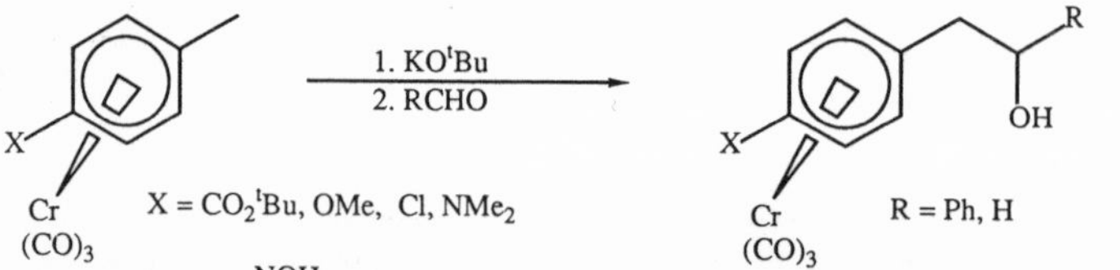

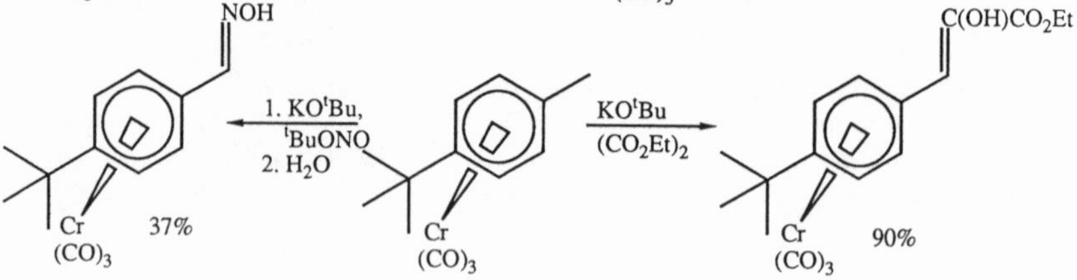

(4-Methoxy-ethylbenzene)chromium tricarbonyl **77** and (4-methoxy-i-propylbenzene)chromium tricarbonyl **78** are inert to benzylic deprotonation–substitution under conditions where (4-t-butoxycarbonylethylbenzene)chromium tricarbonyl **79** and the analogous 3-methoxy complexes **75** and **76** are reactive.[40,41]

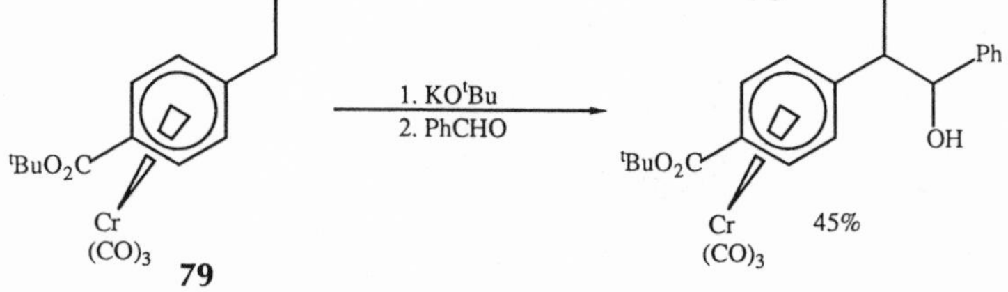

The introduction of the electron donating methoxy and dimethylamino substituents into the 5-position of (indane)chromium tricarbonyl **46** deactivates the 1-*exo*-proton. Treatment of these complexes with potassium t-butoxide and aldehydes resulted in the exclusive formation of the corresponding 3-*exo*-derivatives.[38,41]

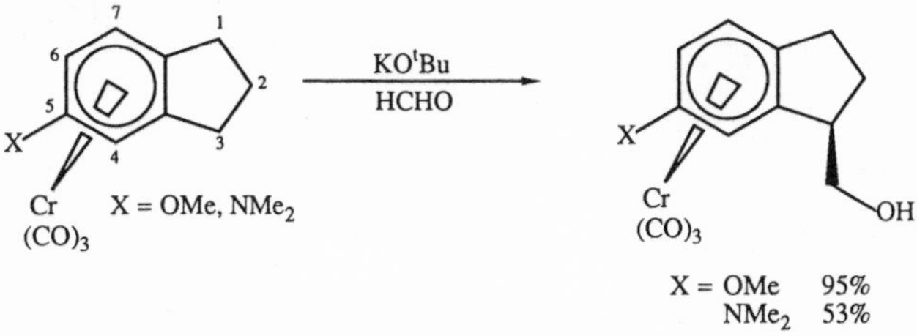

Treatment of (6-methoxytetralin)chromium tricarbonyl **80** with benzaldehyde and potassium t-butoxide produced the expected 4-*exo* derivative **81** together with its oxidation product **82**.[20]

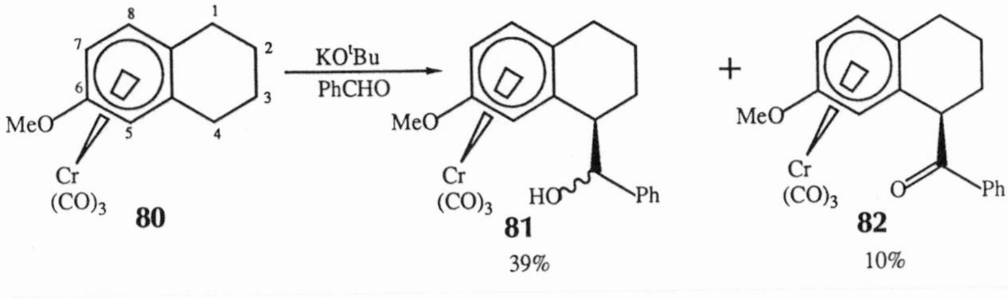

Complexation of estradiol derivatives generally occurs non-stereoselectively producing an approximately equimolar mixture of α- and β-chromium tricarbonyl complexes, which are normally separable. In the α-complex **83** the only benzylic proton which is *exo* to the chromium tricarbonyl moiety is the C-6β proton. Hence on deprotonation–alkylation complex **83** regio- and stereoselectively yields the corresponding 6β derivatives **84**. In the β-complex **85** both the C-9α and the C-6α protons are *exo* to the chromium tricarbonyl group. Deprotonation of the C-9α proton is, however, disfavoured by the oxygen substituent at C-3. Deprotonation–alkylation of complex **85** therefore occurs regio- and stereoselectively to generate the corresponding C-6α derivatives **86**. Decomplexation and deprotection of **84** and **86** yields the corresponding 6β and 6α derivatives of estradiol, respectively.[41,44–46]

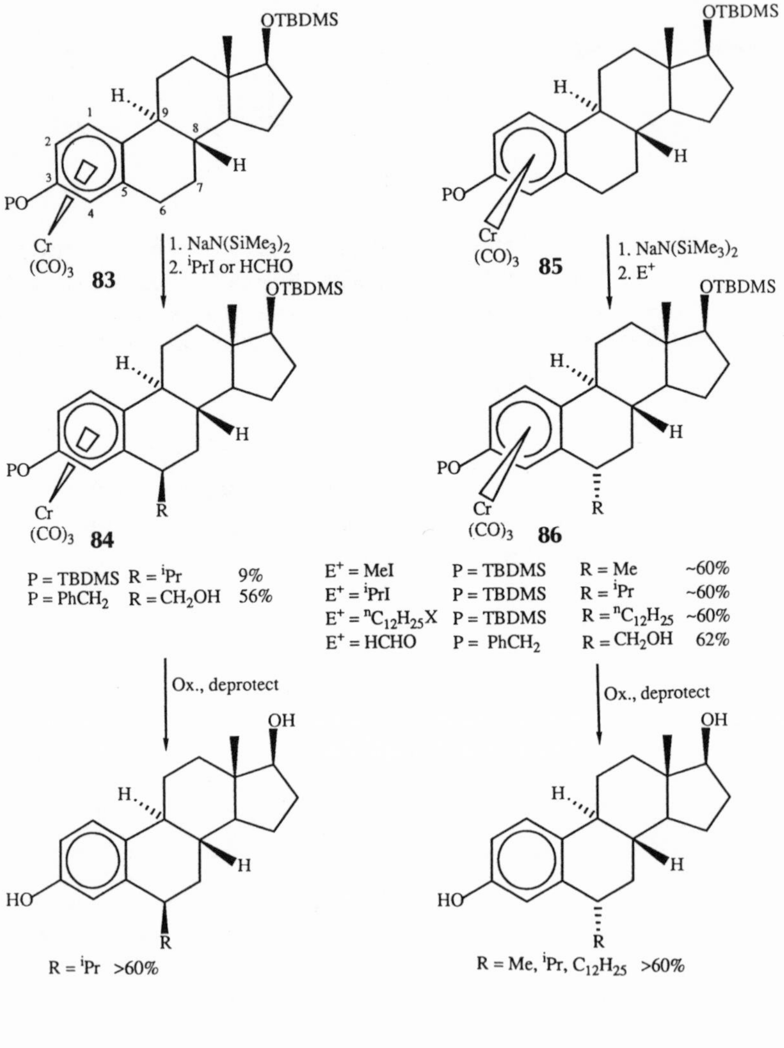

OTBDMS
H
PO
Cr
(CO)3
83
1. NaN(SiMe3)2
2. iPrI or HCHO
84
R
85
1. NaN(SiMe3)2
2. E+
86
P = TBDMS R = iPr 9%
P = PhCH2 R = CH2OH 56%
E+ = MeI P = TBDMS R = Me ~60%
E+ = iPrI P = TBDMS R = iPr ~60%
E+ = nC12H25X P = TBDMS R = nC12H25 ~60%
E+ = HCHO P = PhCH2 R = CH2OH 62%
Ox., deprotect
OH
HO
R = iPr >60%
R = Me, iPr, C12H25 >60%

VII. INFLUENCE OF *ORTHO* SUBSTITUENTS ON THE BENZYLIC DEPROTONATION OF (ARENE)CHROMIUM TRICARBONYL COMPLEXES

Ortho substituents in (arene)chromium tricarbonyl complexes influence the acidity of adjacent benzylic protons in the same way as they would in the *para* position: The acidity is decreased by π-electron donors and increased by π-electron withdrawers. More interestingly the close proximity of an *ortho* substituent to the benzylic group has important stereochemical consequences. The benzylic protons in such a complex are diastereotopic and high stereoselectivities are observed in many reactions

Treatment of the complexes **87** with potassium t-butoxide followed by formaldehyde resulted in benzylic hydroxymethylation, which produced the corresponding complexes **88**. For the complexes **87** where the *ortho* substituent is methoxyl, i-propoxy and diethylamino the substitution reactions are completely stereoselective.[43] In the case of an *ortho* dimethylamino substituent the reaction is also highly diastereoselective, but a small amount of the other diastereoisomer is observed.[43] The stereoselectivity is consistent with complexes **87** adopting the sterically least congested conformation with the methyl of the ethyl group *anti* to the *ortho* substituent. Removal of the *exo*-proton to generate the configurationally stable anion **89** and replacement with the electrophile from the same face would then generate the observed diastereoisomer **88**.

XR = OMe	60%
O^iPr	45%
NMe$_2$	82%
NEt$_2$	88%

Alkylation of the benzylic carbanion derived from treatment of (methyl *o*-methoxyphenylacetate)chromium tricarbonyl **90** with sodium hydride and methyl iodide gave an 82:18 mixture of the two possible distereoisomers **91** and **92**, whereas with benzyl bromide **93** was formed to the exclusion of diastereoisomer **94**. Deprotonation–protonation epimerized **91** into **92** and converted **93** into a 72:28 mixture of itself with **94**. Further deprotonation–alkylation of each of the complexes **91**, **92** (benzyl bromide), **93** and **94** (methyl iodide) gave the same complex **95** completely stereoselectively in each case.[36] The stereoselectivities observed in the reactions of complex **90** and derivatives indicates that the methoxycarbonyl substituent facilitates equilibration of the two geometric isomers of the derived benzylic carbanions **96**; alkylation occurring onto the *exo* face of the thermodynamically most stable isomer **96a** (R = H, Me) or **96b** (R = CH$_2$Ph) with the larger group *anti* to the *ortho* methoxyl substituent.

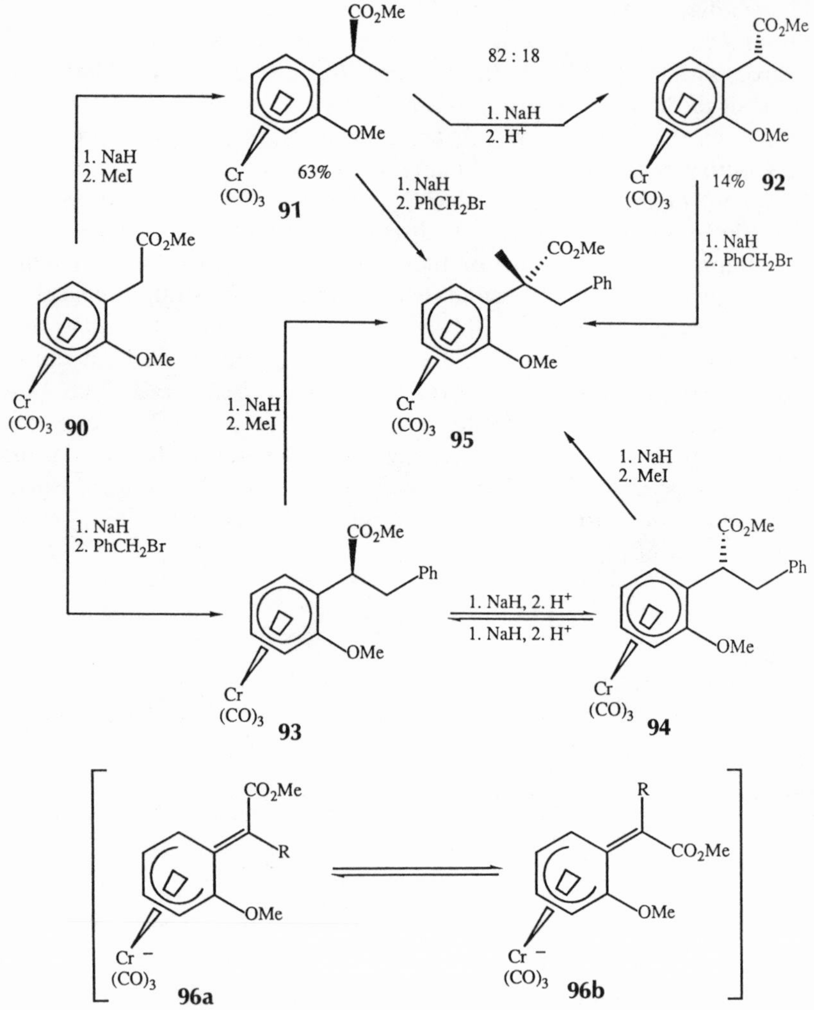

Kinetic deprotonation of (chroman)chromium tricarbonyl **97** with butyllithium occurs regioselectively at the 8-position.[47] However, the 4-benzylic carbanion may be generated exclusively by treatment of **97** with sodium hexamethyldisilazide. Alkylation of the 4-carbanion gave the 4-*exo* substituted products **98** completely stereoselectively while methylation of the

8-carbanion generated (8-methylchroman)chromium tricarbonyl **99**. Further methylation of complex **99** with sodium hexamethyldisilazide and methyl iodide occurred completely regio- and stereoselectively to produce *exo*-(4,8-dimethylchroman)chromium tricarbonyl **100**.[47] The regioselectivity can be accounted for by the secondary anion in the 4-position being thermodynamically preferred to the primary anion derived from deprotonation of the 8-methyl group consistent with these benzylic anions having a considerable amount of double bond character.

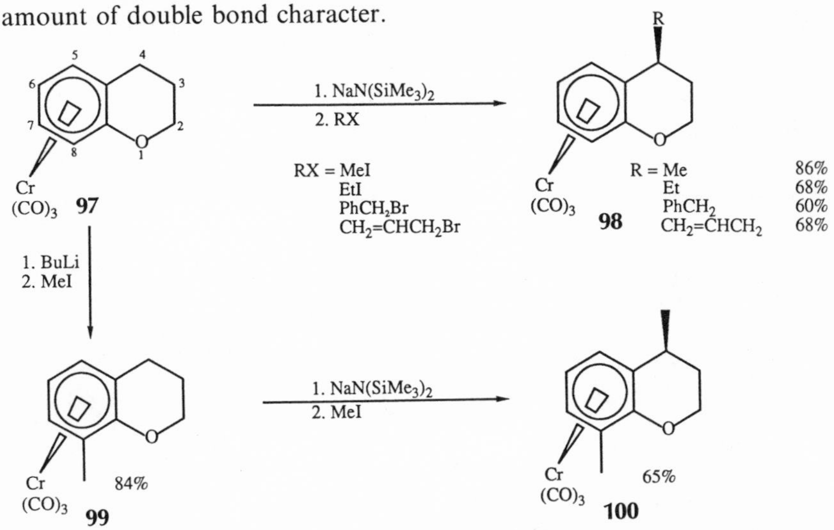

exo-(4-Allylchroman)Cr(CO)₃ (**98**) *(R = allyl)*.[47] Sodium hexamethyldisilazide (270 mg, 1.5 mmol) was added to a solution of (chroman)Cr(CO)₃ **97** (200 mg, 0.75 mmol) in DMF at room temperature. The resultant orange solution was stirred (10 min), allyl bromide (0.2 mL, 2.3 mmol) added and stirring continued (10 min). Methanol was added followed by water (50 mL) and the mixture extracted (Et₂O, 3 × 30 mL). The combined organic fractions were evaporated. Column chromatography (Al₂O₃, Grade V, 1:1 Et₂O/petrol) gave the title compound **98** (R = allyl) as a yellow powder (156 mg, 58%).

Exposure of complex **97** to excess sodium hexamethyldisilazide and dimethyldisulphide gave the 4,4-*gem* disubstituted product **101**; the second deprotonation presumably being promoted by the initially introduced thiomethyl substituent rather than by the chromium tricarbonyl moiety.[47]

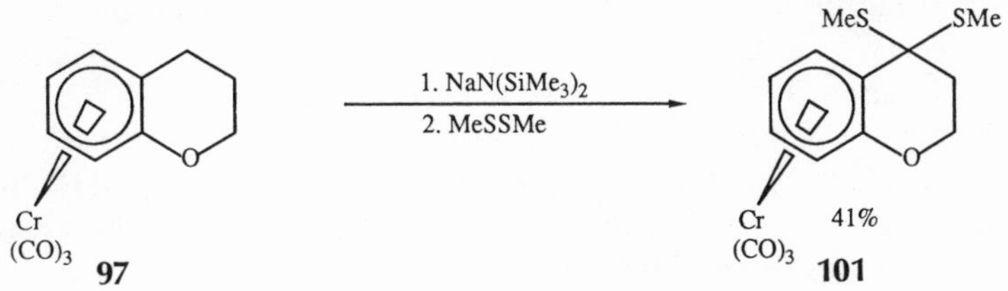

Deprotonation–alkylation of (N-methyltetrahydroquinoline)chromium tricarbonyl **102** yielded the corresponding 4-*exo* derivatives **103**.[39]

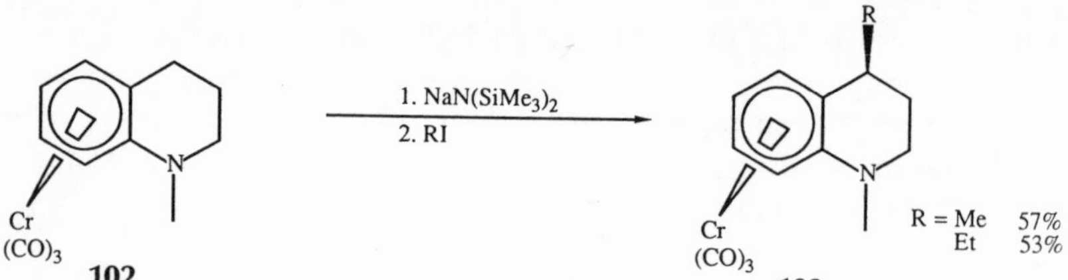

The (N-methyltetrahydrodibenzazepine)chromium tricarbonyl complex **104** undergoes substitution exclusively alpha to the complexed arene ring to yield the 10-*exo* substituted derivatives **105**.[39]

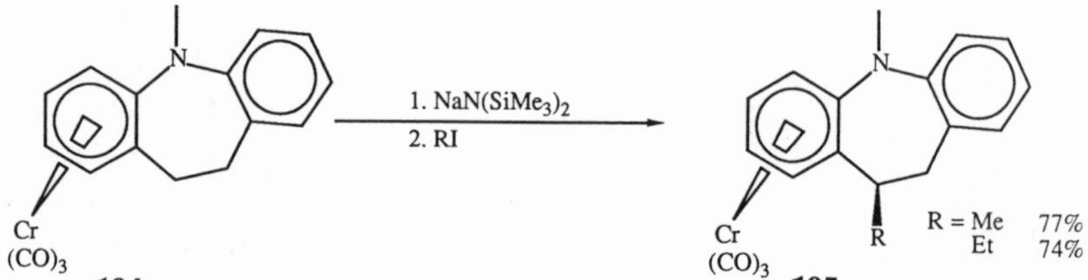

exo(5-Methyl-10-ethyldibenz[b,f]azepine)Cr(CO)₃ (**105**) *(R = Et).*[39] A solution of (5-methyl-10,11-dihydrodibenz[b,f]azepine)Cr(CO)₃ **104** (250 mg, 0.73 mmol) in DMF (2.5 mL) was added dropwise via a cannula to a suspension of sodium hexamethyldisilazide (400 mg, 2.19 mmol) in DMF (2.5 mL) at 0°C. The resulting red solution was stirred (2 h, 0°C) and treated with ethyl iodide (dried over 4 Å molecular sieves, 0.3 mL, 3.75 mmol). After stirring, (4 h, 20°C) water (20 mL) was added and the solution extracted with Et_2O (3 × 40 mL). The combined ether extracts were concentrated and chromatographed (Al_2O_3, Grade V, Et_2O) to give, after recrystallization, the title compound **105** (R = Et) as orange needles (200 mg, 74%).

The (indanone)chromium tricarbonyl derivative (−)-**106** undergoes base catalysed cyclization and dehydration to give via **107**, (+)-**108**. Complete stereocontrol over the α-centre in **107** and hence (+)-**108** is observed as expected from formation of the new C–C bond *exo* to the chromium tricarbonyl. However, no control over the β-hydroxy centre in the intermediate **107**, which is destroyed on going to complex (+)-**108**, is observed. Complete stereocontrol over both the α- and β-centres is, however, observed for the cyclization of the analogous tetralone derivative (−)-**109** to (−)-**110**. In both of these cyclizations only small quantities (< 10%) of Robinson annulated products were observed.[32,48,49]

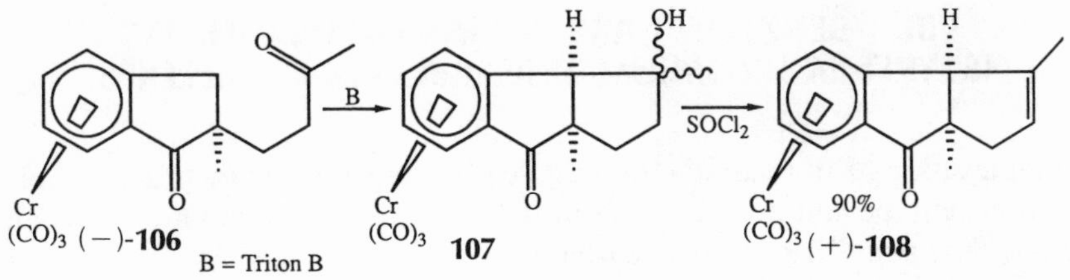

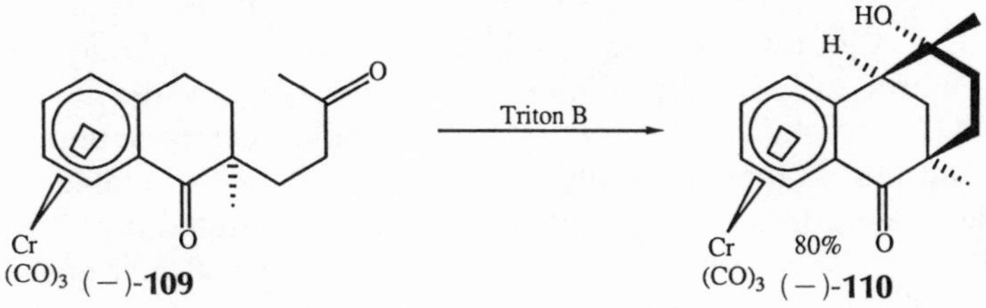

(Fluorene)chromium tricarbonyl **111** gives **112** on treatment with butyllithium at $-20°C$, which, as with other chromium tricarbonyl stabilized benzylic carbanions, undergoes methylation at the α-position to give the *exo*-methyl derivative **113**. At higher temperatures, however, anion **112** undergoes $\eta^6-\eta^5$ ring slippage to generate the chromium anion **114**, which on addition of methyl iodide generates the *endo*-methyl product **115** via rearrangement of a chromium–methyl intermediate.[50]

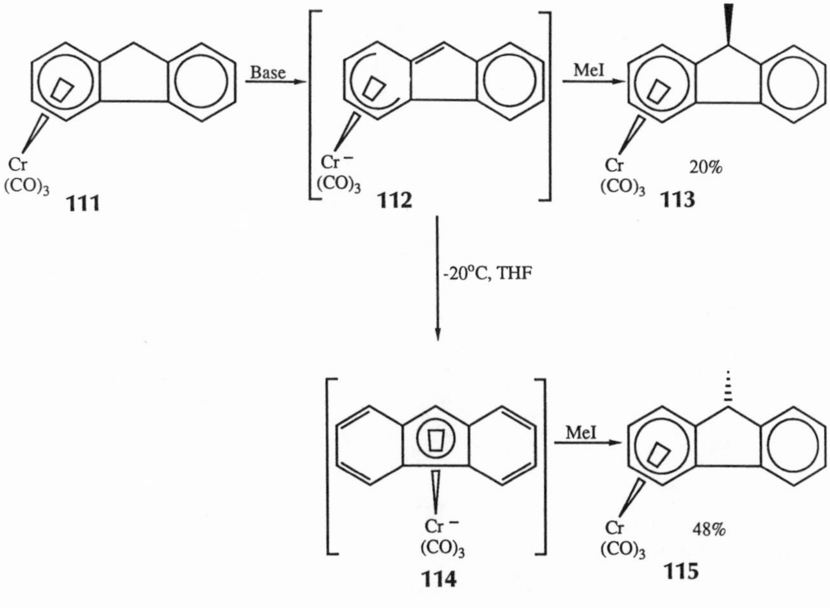

VIII.   BENZYLIC CARBANIONS DERIVED FROM (STYRENE)CHROMIUM TRICARBONYL COMPLEXES

Complexation of styrenes to chromium tricarbonyl renders them susceptible to Michael-type addition reactions at the β-centre. The resultant chromium tricarbonyl stabilized benzylic carbanions may be trapped by a variety of electrophiles thus introducing substituents into both the alpha and beta positions. Addition of the stabilized carbanion 2-lithio-2-methyl-propanonitrile to (styrene)chromium tricarbonyl **116** gave the intermediate anion **117**, which after proton quench and decomplexation yielded compound **118**. Trapping the intermediate **117** with methyl iodide or acetyl chloride produced, after decomplexation, the α-substituted derivatives **119** and **120**, respectively. Other stabilized carbanions have also been shown to react in a similar fashion with complex **116**.[51]

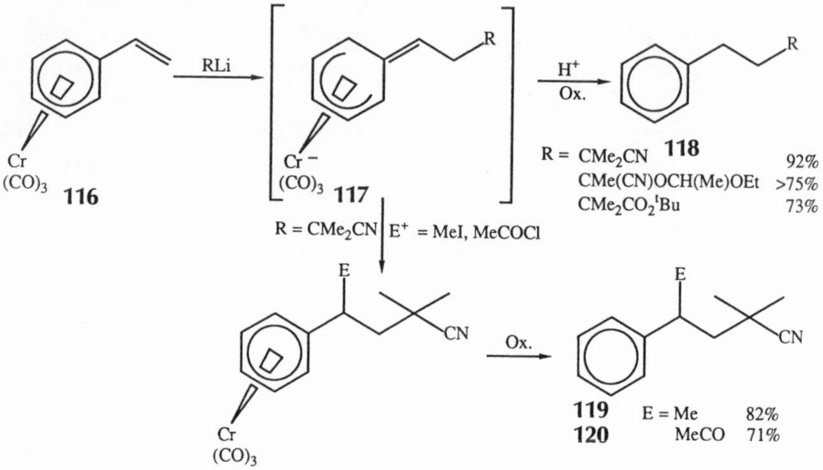

(α-Methylstyrene)chromium tricarbonyl **121** is also susceptible to carbanion addition reactions thus providing an alternative synthesis of compound **119** (E = Me).[51]

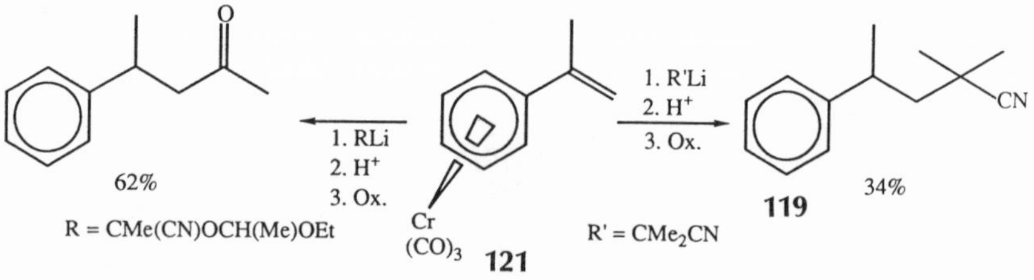

Nucleophilic addition occurred even more readily to [α-(ethylthio)sty-rene]chromium tricarbonyl **122**, the resultant benzylic carbanion being stabilized both by the chromium tricarbonyl and the ethylthio groups. Both simple alkyllithiums and stabilized carbanions react with complex **122**.[51]

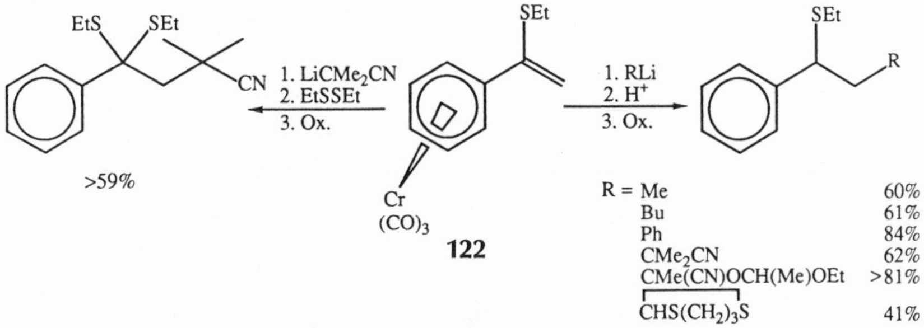

Both the nucleophilic and subsequent electrophilic additions to (dihydronaphthalene)chromium tricarbonyl **123** occur stereoselectively onto the *exo* face to yield, for example, after decomplexation the corresponding *cis*-1,2-disubstituted tetralins **124**.[51] Complex **125** behaves in a similar fashion.[52]

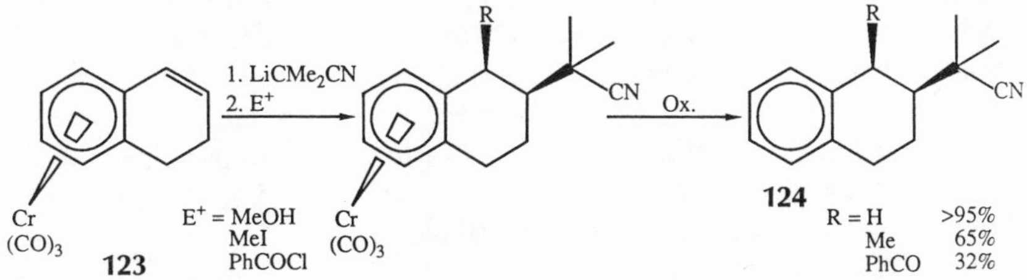

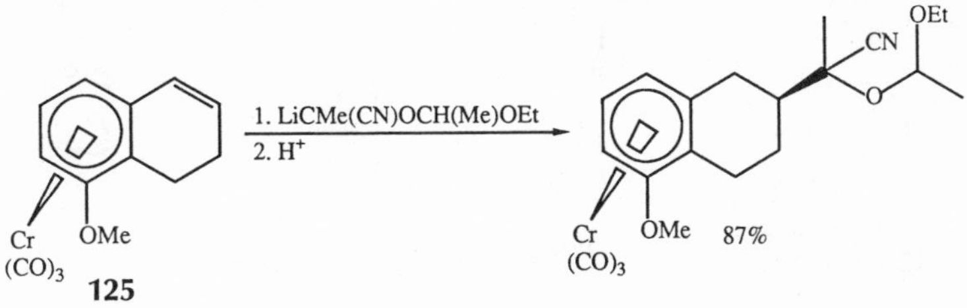

IX. BENZYLIC CARBANIONS DERIVED FROM (β-HETEROSUBSTITUTED ARENE)CHROMIUM TRICARBONYL COMPLEXES

Treatment of uncomplexed arenes possessing β-hetero substituents such as alkoxy or dialkylamino generally results in benzylic deprotonation with concomitant or subsequent elimination of the β-leaving group to form the corresponding styrene. A similar elimination was observed when the (βmethoxyarene)chromium tricarbonyl complex **126** was subject to butyl-lithium and methyl iodide. The actual product isolated was complex **127** arising from nucleophilic addition to the intermediate (styrene)chromium tricarbonyl **116** and subsequent methylation.[53]

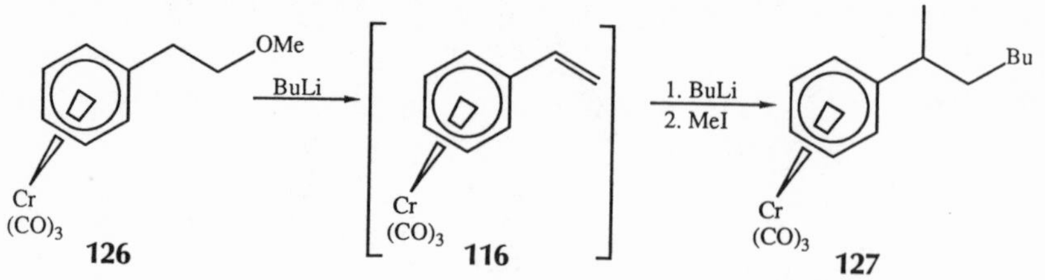

The (dihydrobenzofuran)chromium tricarbonyl **128** and (*N*-pivaloylin-doline)chromium tricarbonyl **129** complexes are particularly prone to base-induced eliminations.[43,54] This is surprising since, for stereoelectronic reasons, elimination from a chromium tricarbonyl stabilized benzylic carbanion requires the β-leaving group to adopt a conformation periplanar with the chromium to arene centroid axis. Such a periplanar conformation cannot be achieved by the anions derived from **128** and **129**. The mechanism probably involves, therefore, a concerted E2 elimination with complexes **128** and **129** being deprotonated in conformations which place an *endo* benzylic C–H bond antiperiplanar to the leaving group rather than antiperiplanar to the chromium. The chromium tricarbonyl moiety would facilitate such eliminations by enhancing the leaving group abilities.

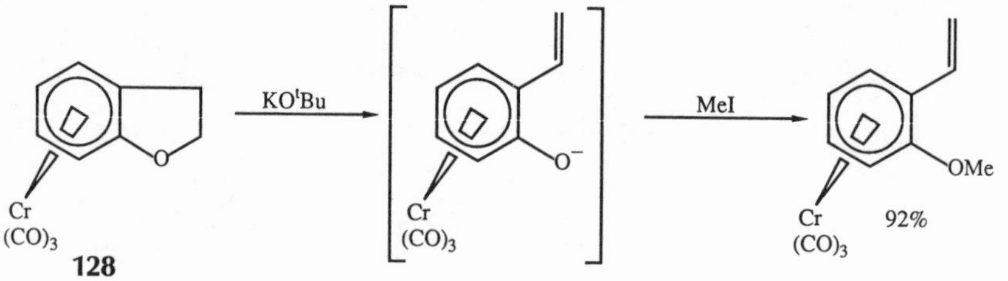

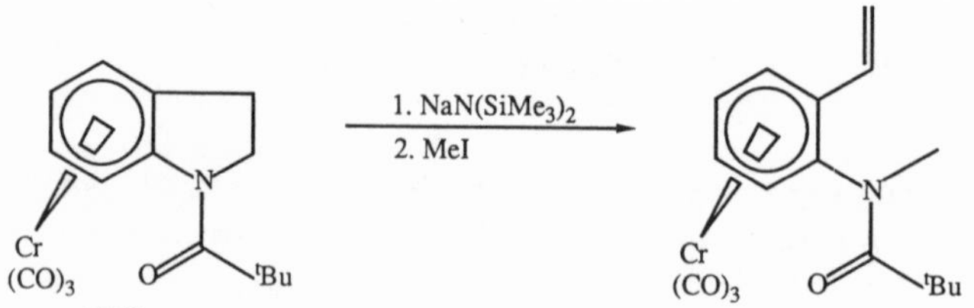

At ambient temperatures benzylic anions derived from (β-dialkylaminoarene)chromium tricarbonyl complexes readily undergo β-elimination.[39,53,55] Below $-40°C$, however, these benzylic carbanions are stable towards elimination and may be trapped by a variety of electrophiles. Thus, treatment of (*N*-phenethylpiperidine)chromium tricarbonyl **130** at $-40°C$ with butyllithium followed by methyl iodide generated the α-methyl derivative **131**.[53] No products were observed corresponding to methylation of the arene ring suggesting that the β-nitrogen is chelating to the butyllithium and delivering it to a proximate benzylic proton.

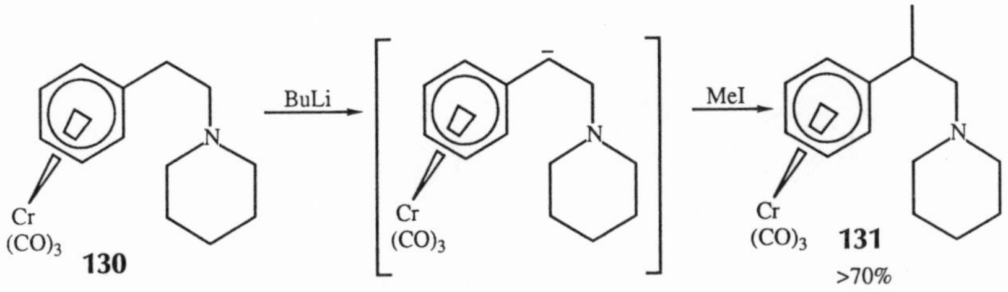

Treatment of $(-)$-(*S*)-(*N*,*N*-dimethylamphetamine)chromium tricarbonyl **132** at $-78°C$ generated the anion **133**. On warming above $-40°C$ anion **133** eliminated dimethylamide to produce (*trans-β*-methylstyrene)chromium tricarbonyl **134**. Quenching anion **133** at $-78°C$ with methanol regenerated $(-)$-**132**. Recovered complex $(-)$-**132** was enantiomerically pure indicating that reversible elimination of dimethylamide was not occurring.[39,55]

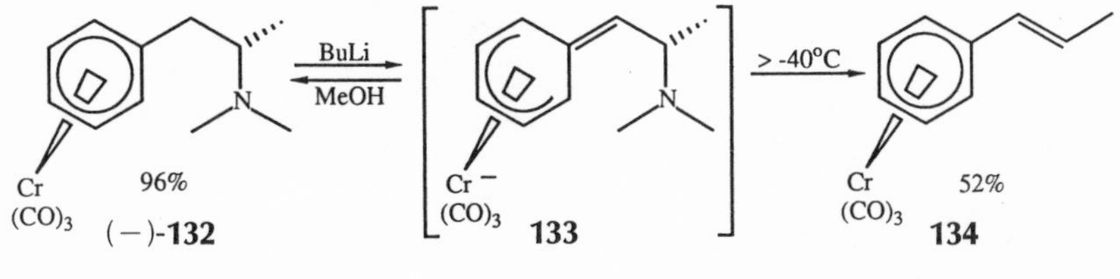

Monodeuteriation of $(-)$-**132** was completely stereoselective producing $(-)$-**135** as a single diastereoisomer. All the deuterium in $(-)$-**135** was removed on treatment with butyllithium followed by proton quench. These results demonstrate that both the deprotonation and the electrophilic addition are highly stereoselective. The *pro*-R benzylic hydrogen is removed from $(-)$-**132** and replaced with retention of configuration. The observed stereoselectivities are consistent with initial coordination of butyllithium to nitrogen, with subsequent delivery of the base to the *pro*-R hydrogen through **136** being easier than to the *pro*-S hydrogen through **137**.[55]

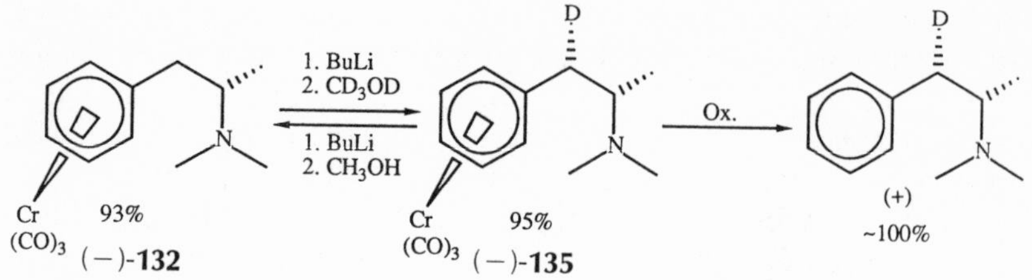

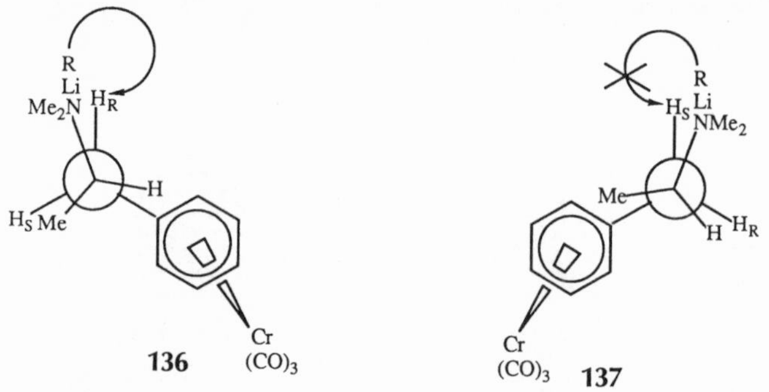

The anion derived from (−)-**132** could be trapped by a variety of electrophiles to give, after decomplexation, the corresponding α-substituted
dimethylamphetamine derivatives. The reactions were, in all cases, completely stereoselective. In this way (+)-(*S*)-*N*,*N*-dimethylamphetamine could be
converted stereoselectively via (−)-**138** into (+)-(1*S*,2*S*)-*N*-methylpseu-
doephedrine **139**.[55]

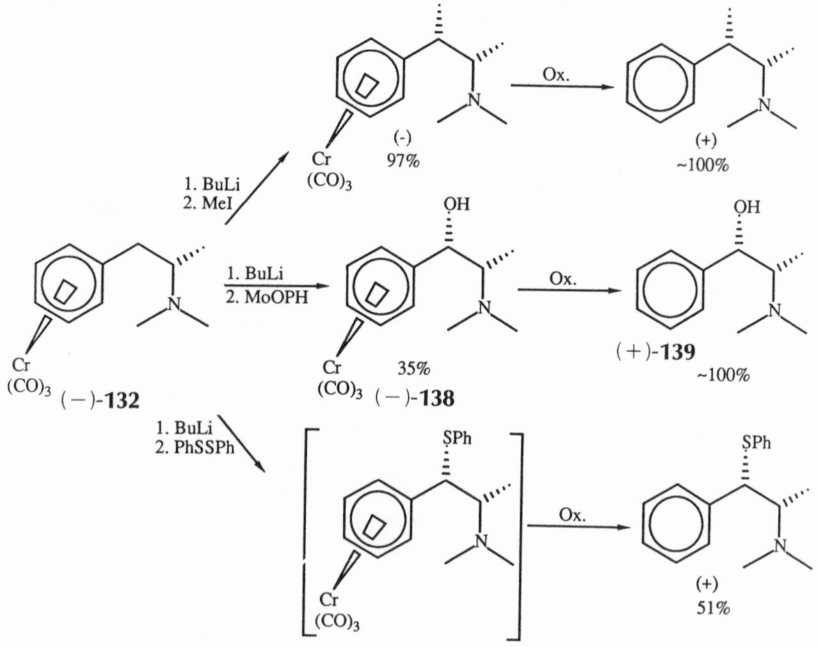

(−)-(1R,2S)-(N-Methylpseudoephedrine)Cr(CO)₃ (**138**).[55] BuLi (1.6 M solution in hexanes, 0.80 mL, 1.28 mmol) was added to a stirred solution of (−)-(*S*)-(*N,N*-dimethylamphetamine)Cr(CO)₃ **132** (300 mg, 1.00 mmol) in THF (20 mL) at − 78°C. After stirring (2 h, − 78°C) MoOPH (700 mg, 1.61 mmol) was added to the crimson solution and the resulting suspension stirred until all of the reagent had dissolved (20 min, − 78°C). Saturated aqueous Na₂SO₃ (10 mL) was added, the mixture warmed to 20°C and treated with water (40 mL). After extraction with Et₂O (3 × 40 mL) and concentration of the combined extracts, column chromatography (Al₂O₃, Grade V, Et₂O) and recrystallization from CH₂Cl₂/hexane gave the title compound (−)-**138** as yellow needles (110 mg, 35%).

Deprotonation of (*N*-methyltetrahydrobenzazepine)chromium tricarbonyl **140** with butyllithium followed by addition of an electrophile resulted in the formation of the corresponding 1-*exo* derivatives **141** (Table 3).[39]

exo-(1-Thiomethyl-3-methyltetrahydro-3-benzazepine)Cr(CO)₃ (**141**) *(E = SMe)*.[39] BuLi (1.6 M solution in hexanes, 0.80 mL, 1.28 mmol) was added to a stirred solution of (3-methyltetrahydro-3-benzazepine)Cr(CO)₃ **140** (300 mg, 1.01 mmol) in THF (25 mL) at − 78°C. After stirring the red solution (2 h, − 78°C) dimethyldisulphide (180 mg, 1.91 mmol) was added and stirring continued (2 h, − 40°C). Methanol was added, the solution warmed to room temperature and evaporated. Column chromatography (Al₂O₃, Grade V, Et₂O) gave the title compound **141** (E = MeS) as a yellow solid (315 mg, 91%).

Complexations of codeine and its O-TBDMS derivative to chromium tricarbonyl are completely stereoselective, the chromium tricarbonyl fragment coordinating to the more accessible face of the arene to give the complexes **142** and (−)-**143**, respectively. Figure 3 illustrates the X-ray crystal structure

Table 3. α-Substitution reactions of (*N*-methyltetrahydrobenzazepine)-Cr(CO)₃ **140**

Electrophile (E⁺)	**141** (E)	Yield (%)
MeI	Me	76
EtI	Et	81
PhCH₂Br	PhCH₂	71
MeSSMe	MeS	91
PhSSPH	PhS	81
MoOPH	OH	33

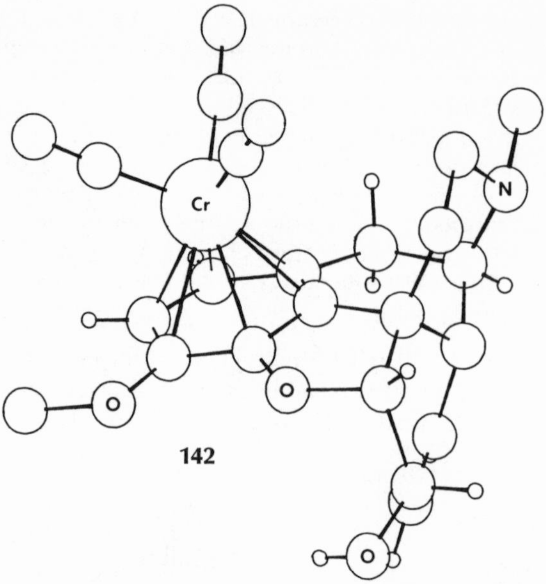
Cr
N
O
O
O
O

142

Figure 3. X-ray crystal structure of (codeine)chromium tricarbonyl **142**.

of (codeine)chromium tricarbonyl **142**. Deprotonation–alkylation of complex (−)-**143** gave, stereoselectively, the (10*S*)-alkyl complexes **144** which after decomplexation yielded the corresponding (10*S*)-alkyl codeine derivatives.[56,57]

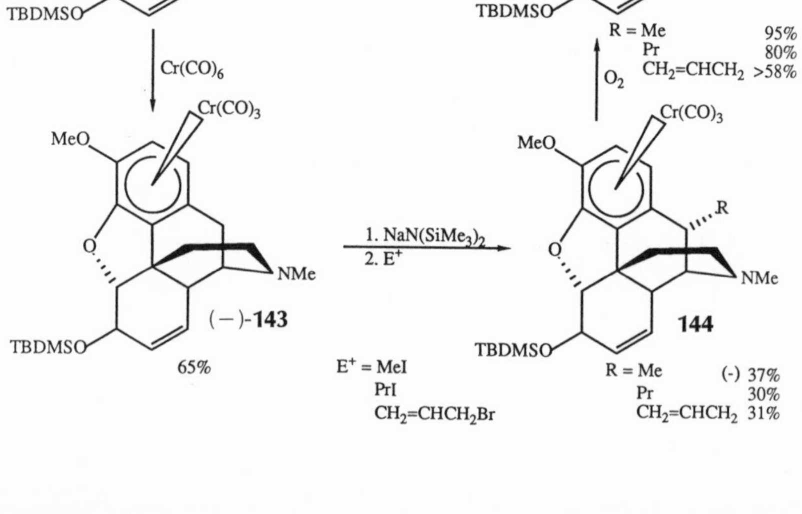
MeO
O
NMe
TBDMSO
Cr(CO)6
Cr(CO)3
(−)-143
65%
1. NaN(SiMe3)2
2. E+
E+ = MeI
PrI
CH2=CHCH2Br
R
144
R = Me (-) 37%
Pr 30%
CH2=CHCH2 31%
O2
R = Me 95%
Pr 80%
CH2=CHCH2 >58%

In an analogous manner morphine was converted stereoselectively via methylation of (−)-**145** to (−)-(10*S*)-methylmorphine **146**.[56]

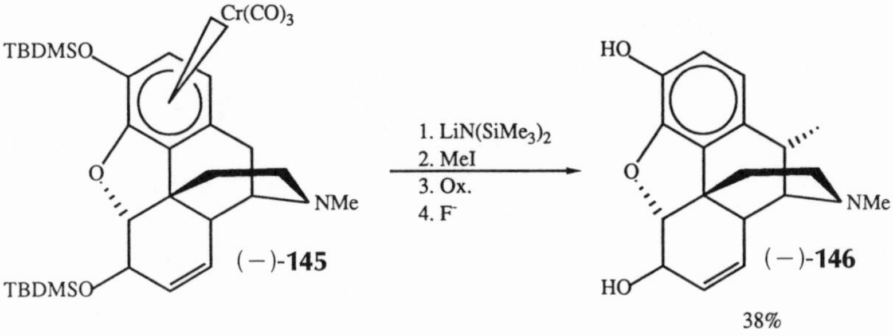

(−)-(10S)-Methylmorphine (**146**).[56] A solution of lithium hexamethyldi-silazide (750 mg, 4.49 mmol) in THF (20 mL) was added to (−)-[*O,O*-bis(t-butyldimethylsilyl)morphine]Cr(CO)₃ **155** (300 mg, 0.46 mmol). After stirring (1 h) at room temperature the mixture was cooled (0°C) and methyl iodide (600 mg, 4.22 mmol) added. The solution was stirred (15 min) at room temperature. Water was added, the mixture extracted with CH₂Cl₂ and the combined extracts evaporated. Column chromatography (SiO₂, CH₂Cl₂) gave (10*S*)-[methyl-*O,O*-bis(t-butyldimethylsilyl)morphine]Cr(CO)₃ as the least polar fraction., The product (180 mg, 0.27 mmol) was dissolved in pyridine (2 mL) and the solution heated at reflux (30 min). Column chromatography (SiO₂, 19:1 CH₂Cl₂/MeOH) gave (10*S*)-methyl-*O,O*-bis(t-butyldimethylsilyl)-morphine (136 mg, 95%). The product was dissolved in THF and treated with excess Bu₄NF·H₂O and the solution stirred (45 min). Evaporation followed by column chromatography afforded (−)-(10*S*)-methylmorphine **146** as a crystalline solid (84%).

X. BENZYLIC CARBANIONS DERIVED FROM (α-HETEROSUBSTITUTED ARENE)CHROMIUM TRICARBONYL COMPLEXES

Treatment of uncomplexed benzylalkyl ethers with strong bases results in benzylic deprotonation followed by rapid migration of the alkyl group from oxygen to carbon, the Wittig rearrangement.[58]

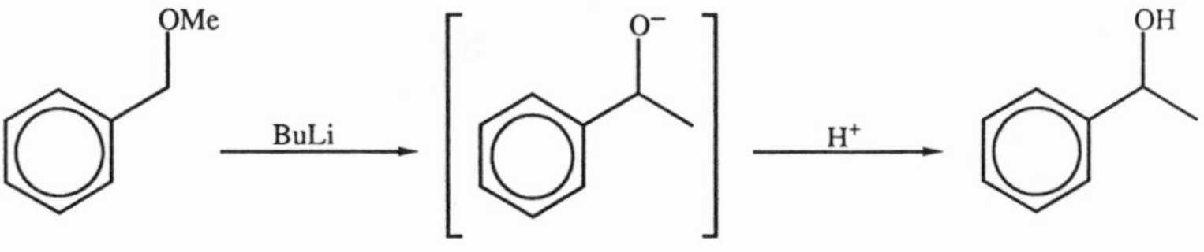

The Wittig rearrangement is, however, completely suppressed on stabilization of the benzylic anion by coordination of the arene to chromium tricar-

Table 4. α-Substitution reactions of (benzylmethyl ether)Cr(CO)$_3$ **147**

Electrophile (E$^+$)	**149** (E)	Yield (%)	Ref.
MeOH	H	92	59
MeI	Me	79	59
EtBr	Et	75	59
^{i}PrI	^{i}Pr	65	59
PhCH$_2$Br	PhCH$_2$	—	59
MeCO$_2$Et	MeCO	67	4
PhCO$_2$Me	PhCO	64	59
MeCHO	MeCHOH	51	59
PhCHO	PhCHOH + PhCO	60	27
Me$_3$SiCl	Me$_3$Si	95	4

bonyl.[59,60] Thus lithiation of (benzylmethyl ether)chromium tricarbonyl **147** with butyllithium at −78°C gave the stabilized anion **148** which was quenched by a variety of electrophiles to produce the α-substituted derivatives **149** (Table 4).[4,59] The regioselective benzylic deprotonation of complex **147** with butyllithium is indicative of the increased acidity of the benzylic protons over the arene protons imparted by the electronegative oxygen subsituent.

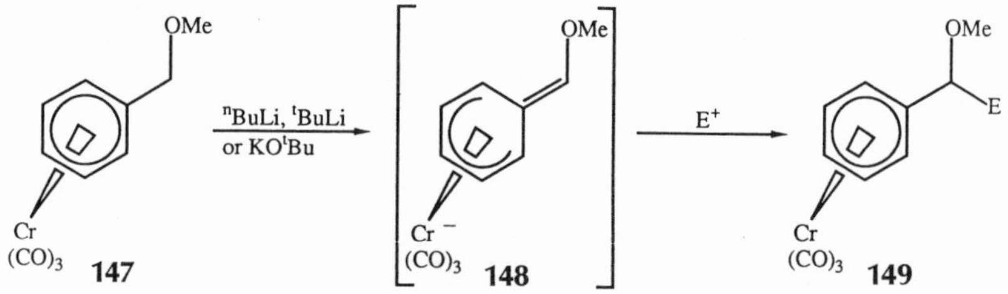

(α-Trimethylsilylbenzylmethyl ether)Cr(CO)$_3$ (**149**) *(E = Me$_3$Si)*.[4] t-BuLi (2.36 M solution in pentane, 1.81 mL, 4.27 mmol) was added to (benzylmethyl ether)Cr(CO)$_3$ **147** (1.00 g, 3.88 mmol) in THF (20 mL) at −78°C and the resultant red solution stirred (−78°C, 2 h). Chlorotrimethylsilane (distilled from CaH$_2$ under N$_2$, 1.48 mL, 11.6 mmol) was added and stirring continued (−78°C, 2 h). Quenching with methanol (2 mL) followed by evaporation of the solvent and column chromatography (Al$_2$O$_3$, Grade V, 1:8 Et$_2$O/petrol) gave, after crystallization from pentane, the title compound **149** (E = Me$_3$Si) as yellow granules (1.22 g, 95%).

Benzylic deprotonation of complex **147** by potassium t-butoxide in the presence of t-butylnitrite generated complex **150** in good yield.[30]

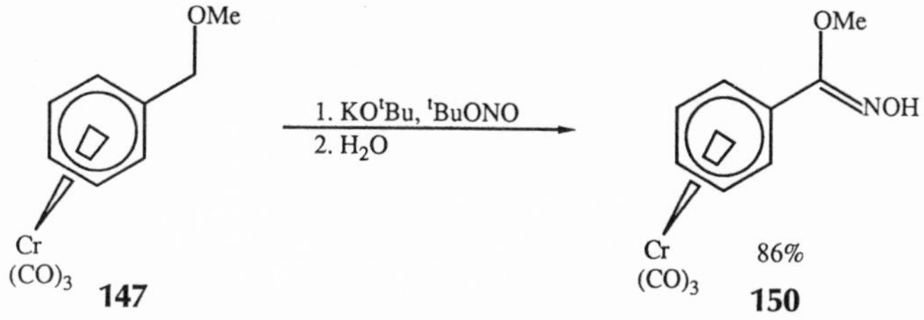

The suppression of the Wittig rearrangement is general for all complexed benzylalkyl ethers, as demonstrated by the deprotonation–methylation of a variety of such complexes **151**. Not even rearrangement of (benzylallyl ether)chromium tricarbonyl is observed. For the complexes **151**, where the alkyl group is chiral (2-butyl or cholesteryl), little stereocontrol is observed during the α-methylations (20% and 52% d.e., respectively).[59]

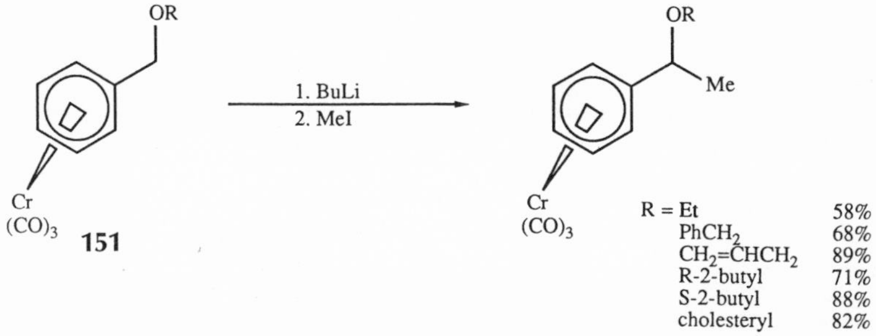

In (benzylalkyl ether)chromium tricarbonyl complexes, where the alkyl substituent bears a remote leaving group, cyclization may occur on exposure to base. For example, (benzyl-3-chloropropyl ether)chromium tricarbonyl **152** smoothly underwent cyclization in the presence of butyllithium at −40°C to give (2-phenyltetrahydrofuran)chromium tricarbonyl **153**. Similarly the 3-chlorobutyl ether complex **154** underwent an analogous cyclization to give **155** as a 3:2 mixture of diastereoisomers.[59]

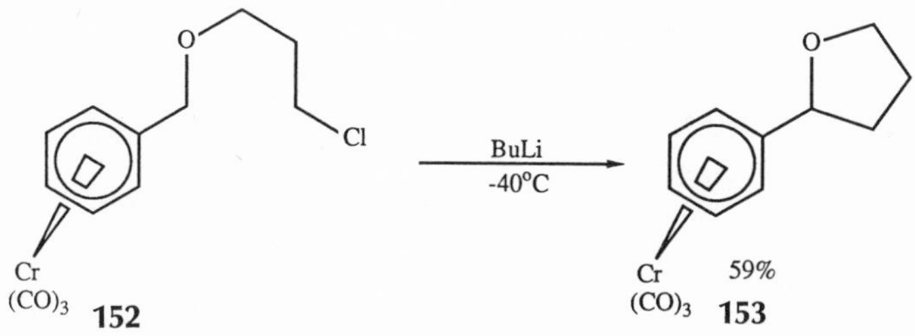

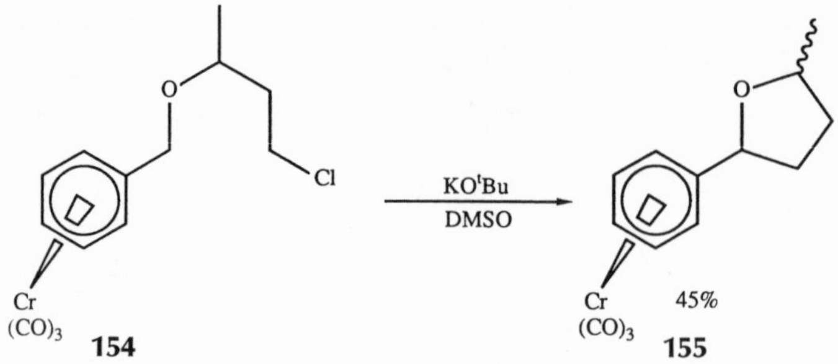

(2-Phenyltetrahydrofuran)Cr(CO)₃ (**153**).[59] To a solution of (3-chloropropyl-benzyl ether)Cr(CO)₃ **152** (290 mg, 0.91 mmol) in THF (25 mL) at −40°C was added BuLi (1.6 M solution in hexanes, 0.90 mL, 1.44 mmol). After stirring (1 h, −40°C), methanol (2 mL) was added, the solution warmed (20°C) and the solvent removed. Column chromatography (Al₂O₃, Grade II, CH₂Cl₂) gave, after evaporation and crystallization from Et₂O/hexane, the title compound **153** as yellow crystals (152 mg, 59%).

(1-Phenethylmethyl ether)chromium tricarbonyl **156** undergoes a clean kinetic deprotonation in the benzylic position by t-butyllithium to give, after electrophilic quench, α,α-disubstituted benzylmethyl ethers.[61] Clean benzylic substitution is also observed using potassium t-butoxide as base.[27]

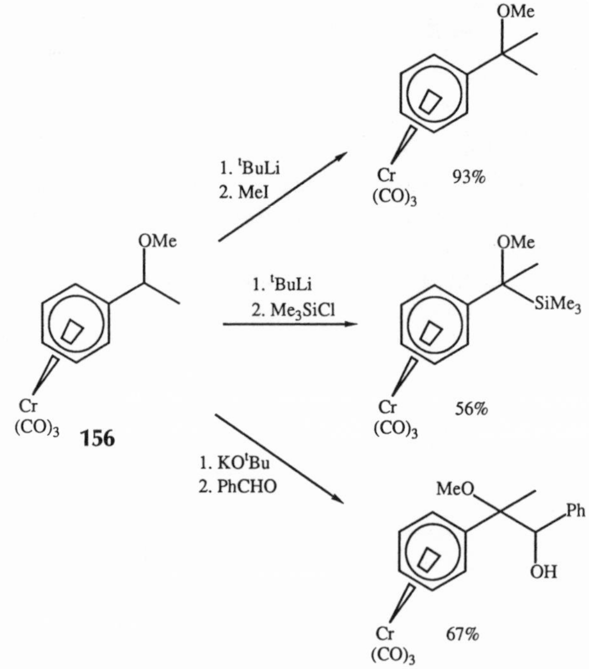

However, alkylation of complex **156** with *n*-butyllithium as base gave predominantly *ortho* methylated products **157** with only a small amount of benzylic methylation. The regioselective *ortho* lithiation of **156** with *n*-butyllithium must be ascribed to chelation of the base to oxygen and subsequent delivery to the proximate *ortho* protons. Presumably t-butyllithium is so reactive that deprotonation occurs before any chelation and with potassium t-butoxide the deprotonation is under thermodynamic control. In the *ortho* lithiation reaction little selectivity between the diastereotopic *ortho* protons is observed.[61]

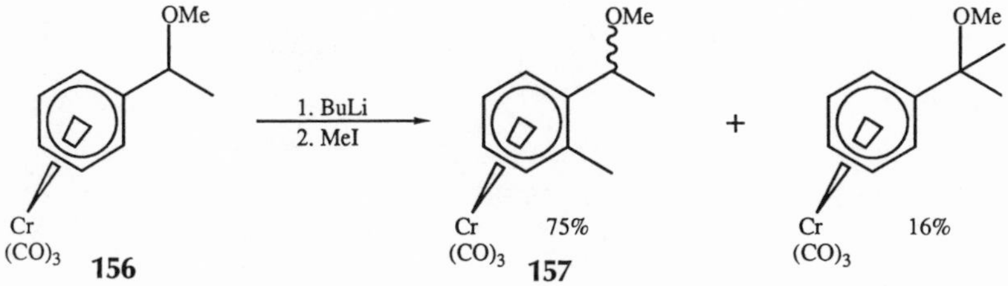

(α-Trimethylsilylbenzylmethyl ether)chromium tricarbonyl **158**, generated by silylation of complex **147**, underwent the Peterson olefination reaction with butyllithium and benzaldehyde to generate, stereoselectively, the enol ether complex (*Z*-α-methoxy-β-phenylstyrene)chromium tricarbonyl **159**.[4]

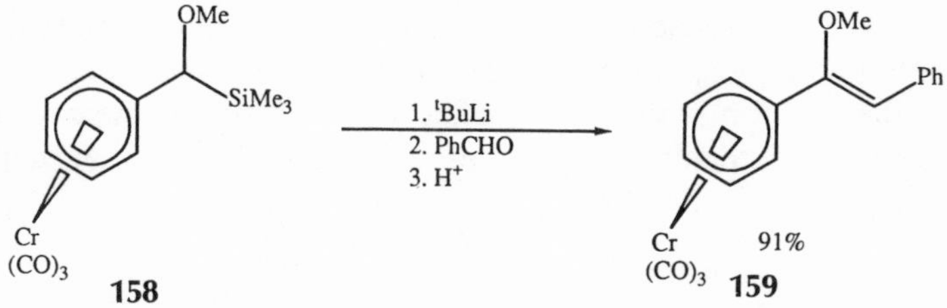

(Z-α-Methoxy-β-phenylstyrene)Cr(CO)$_3$ (159).[4] t-BuLi (2.0 M solution in pentane, 0.30 mL, 0.60 mmol) was added to (α-trimethylsilylbenzylmethyl ether)-Cr(CO)$_3$ **158** (200 mg, 0.61 mmol) in THF (10 mL) at $-78°C$ and the solution stirred (2 h, $-78°C$). Benzaldehyde (distilled *in vacuo* onto 4 Å molecular sieves, 159 mg, 1.50 mmol) was added and the solution stirred (2.5 h, $-78°C$) and then warmed to $-40°C$ (0.5 h). Addition of water (1 mL) followed by warming to 20°C and addition of HCl (1.5 M, 3 mL) with stirring (0.5 h) gave, on evaporation, a red oil. Column chromatography (Al$_2$O$_3$, Grade V, 1:1 Et$_2$O/petrol) and crystallization from pentane gave the title compound **159** as orange crystals (189 mg, 91%).

The (oxazolidine)chromium tricarbonyl complex (+)-**160** derived from (−)-(1R,2R)-pseudoephedrine underwent benzylic deprotonation and methylation to yield an α-methyl derivative as a single diastereoisomer. The stereoselectivity presumably arises as the result of a stereoselective deprotonation to give the configurationally stable anionic intermediate **161**, which methylates with overall retention of configuration to give (+)-**162**.[7]

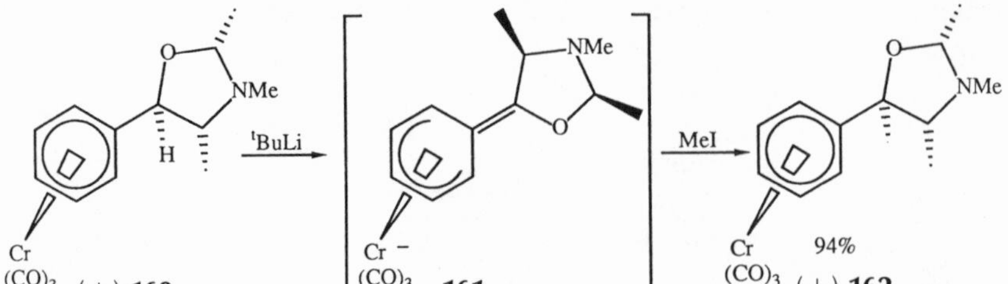

(+)-(2R,4R,5R)-(Trimethyl-5-phenyl-N-methyloxazolidine)Cr(CO)₃ (**162**).[7] t-BuLi (2.62 M in pentane, 0.09 mL, 0.24 mmol) was added to a stirred solution of (+)-(2R,4R)-dimethyl-5R-phenyl-N-methyloxazolidine)Cr(CO)₃ **160** (56.9 mg, 0.17 mmol) in THF (20 mL) at −78°C. After stirring the solution (2 h, −78°C) methyl iodide (0.5 mL, excess) was added and stirring continued (2 h, −78°C). Methanol (1 mL) was added, the solution warmed to room temperature and evaporated. Column chromatography (Al₂O₃, Grade V, CH₂Cl₂) gave the title compound (+)-**162** as a yellow solid (56 mg, 94%).

Treatment of *endo*-(1-methoxytetralin)chromium tricarbonyl **163** with butyllithium followed by methyl iodide gave only the 1-*exo*-methylated derivative **164**; the electron withdrawing influence of the methoxyl substituent making 1-*exo* deprotonation more favourable than 4-*exo* or arene deprotonation.[59]

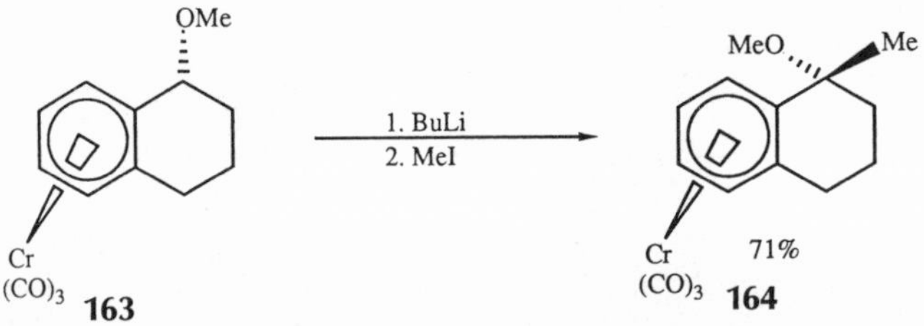

(Phthalan)chromium tricarbonyl **165** undergoes 1-*exo* substitution to give **166** followed by 3-*exo* substitution to produce **167** on repetition of the deprotonation–alkylation reaction. Decomplexation of complexes **167** gave *cis*-1,3-disubstituted phthalan derivatives.[7]

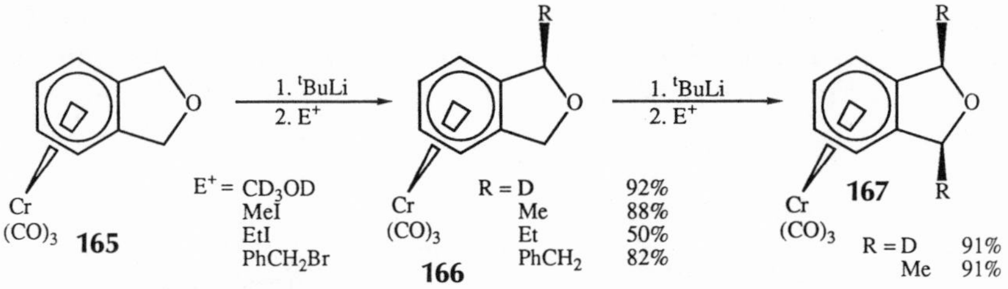

exo-(cis-1,3-Dimethyldihydrobenzo[c]furan)Cr(CO)₃ (**167**) *(R = Me).*[7] t-BuLi (2.62 M in pentane, 0.25 mL, 0.66 mmol) was added to a stirred solution of (1,3-dihydrobenzo[c]furan)Cr(CO)₃ **165** (163 mg, 0.64 mmol) in THF (20 mL) at −78°C. After stirring the solution (2 h, −78°C) methyl iodide (1 mL, excess) was added and stirring continued (2 h, −78°C). Methanol was added, the solution warmed to room temperature and evaporated. Column chromatography (Al₂O₃, Grade V, Et₂O) gave *exo*-(1-methyldihydrobenzo[c]-furan)Cr(CO)₃ **166** (R = Me) as a yellow oil (151 mg, 88%). A solution of complex **166** (R = Me) (124 mg, 0.46 mmol) in THF (20 mL) at −78°C was treated with t-BuLi (2.62 M in pentane, 0.2 mL, 0.52 mmol) and stirred (1 h, −78°C). Methyl iodide (1 mL, excess) was added and stirring continued (2 h, −78°C). Methanol was added, the solution warmed and evaporated. Column chromatography (Al₂O₃, Grade V, Et₂O) gave the title compound **167** (R = Me) as a yellow solid (118 mg, 91%).

Deprotonation–methylation of complex **168** gave the monomethylated derivative **169** as a single diastereoisomer. This diastereoisomer is assumed to arise via *exo* deprotonation of **168** in the sterically most favourable conformation with the two methoxyl groups *anti* to each other, followed by *exo*-methylation. Further deprotonation–methylation of **169** gave the *meso* complex **170** as a single diastereoisomer via an analogous mechanism.[59]

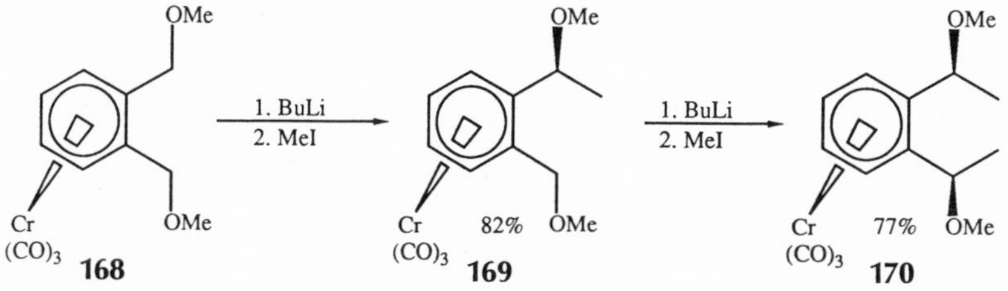

Deprotonation of *endo*-(4-methoxychroman)chromium tricarbonyl **171** with butyllithium and addition of methyl iodide gave exclusively the 4-*exo*-methylated derivative **172**. This was in stark contrast to the epimeric complex *exo*-(4-methoxychroman)chromium tricarbonyl **173**, which underwent regioselective methylation at C-8 under the same conditions as above to give **174**. This complete change of regioselectivity is presumably due to the inaccessability of the 4-*endo*-proton in **173** coupled with the acidity of the C-8 proton being augmented relative to the other arene protons by the inductive effect of the proximate oxygen atom.[62]

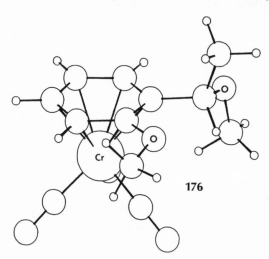

Deprotonation–alkylation reactions of (*o*-methoxybenzylmethyl ether)chromium tricarbonyl **175** are completely stereoselective. Figure 4 illustrates the X-ray crystal structure of complex **176** obtained by deprotonation–methylation of **175**. The observed stereoselectivities are consistent with *exo* deprotonation of **175** occurring in the sterically most favourable conformation with the benzylic methoxyl *anti* to the arene methoxyl followed by

Figure 4. X-ray crystal structure of (*RS,SR*)-(*o*-methoxy-α-methylbenzyl-methyl ether)chromium tricarbonyl **176**.

exo-methylation of the thus formed, configurationally stable, anionic intermediate.[63]

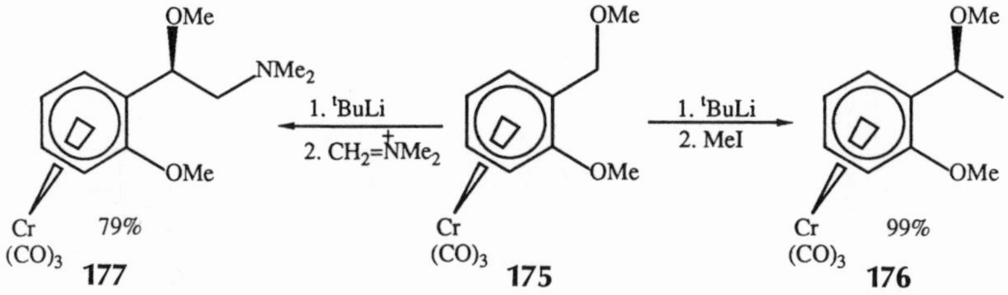

(RS,SR)-[o-Methoxy-α-(N,N-dimethylaminomethyl)benzylmethyl ether]Cr-(CO)$_3$ (177).[63] t-BuLi (2.36 M solution in pentane, 0.29 mL, 0.76 mmol) was added to (o-methoxybenzylmethyl ether)Cr(CO)$_3$ 175 (200 mg, 0.69 mmol) in THF (15 mL) at −78°C and the solution stirred (2 h, −78°C). Eschenmoser's salt (recrystallized from CH$_2$Cl$_2$/hexane, 271 mg, 1.47 mmol) was added and stirring continued (1 h, −78°C–0°C). The solution evaporated to an orange oil. Column chromatography (Al$_2$O$_3$, Grade V, Et$_2$O) gave, after crystallization from CH$_2$Cl$_2$/petrol, the title compound 177 as yellow needles (189 mg, 79%).

In contrast, for (*m*-methoxybenzylmethyl ether)chromium tricarbonyl 178 only ring metallation was observed with butyllithium in the presence of tetramethylethylenediamine to give, following methylchloroformate quench and decomplexation, the ester 179.[60]

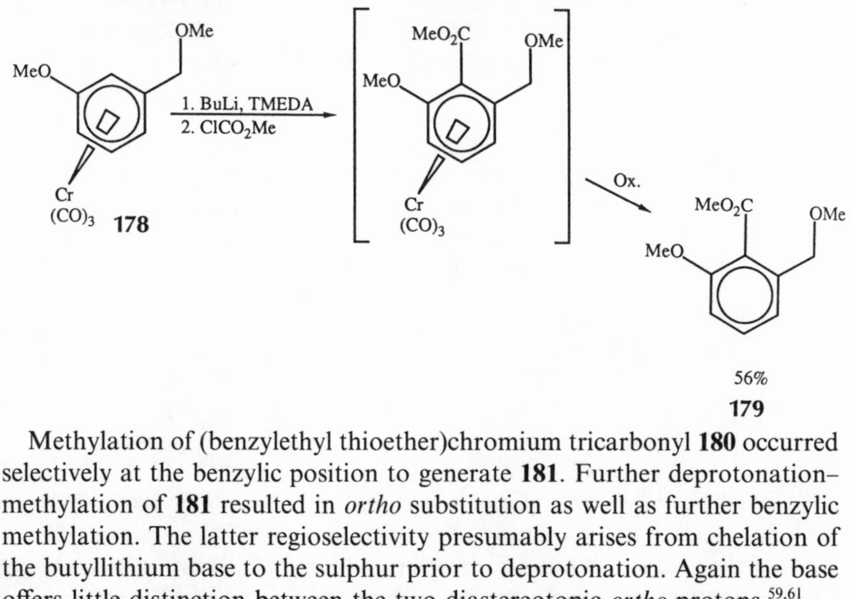

56%
179

Methylation of (benzylethyl thioether)chromium tricarbonyl 180 occurred selectively at the benzylic position to generate 181. Further deprotonation–methylation of 181 resulted in *ortho* substitution as well as further benzylic methylation. The latter regioselectivity presumably arises from chelation of the butyllithium base to the sulphur prior to deprotonation. Again the base offers little distinction between the two diastereotopic *ortho* protons.[59,61]

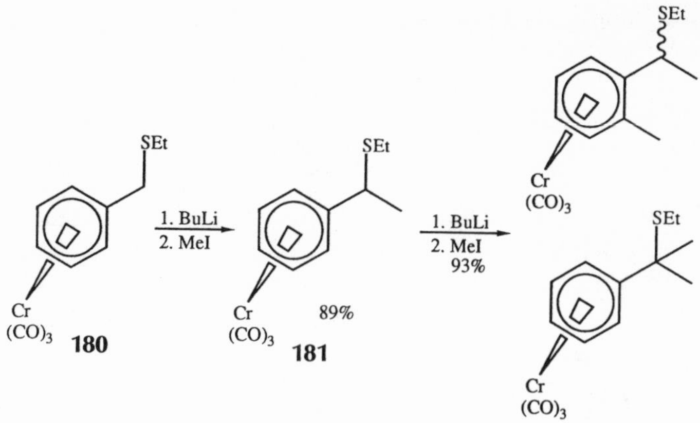

In contrast to the ether and thioether analogues (benzyldimethylamine)chromium tricarbonyl **182** underwent exclusive *ortho* deprotonation with t-butyllithium to generate, after addition of methyl iodide, complex **183**.[4,53]

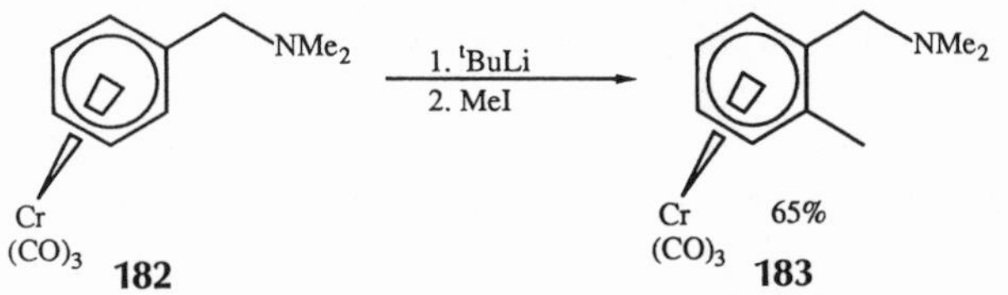

XI.   BENZYLIC CARBANIONS DERIVED FROM (α,β-DIHETEROSUBSTITUTED ARENE)CHROMIUM TRICARBONYL COMPLEXES

(Isochroman)chromium tricarbonyl **184** underwent deprotonation at the 1-position with butyllithium. The 1-*exo* proton being activated by both the oxygen and the chromium tricarbonyl is rendered more acidic than the 4-*exo* or the arene protons. Methylation of the thus formed anion gave *exo*-(1-methylisochroman)chromium tricarbonyl **185**.[62]

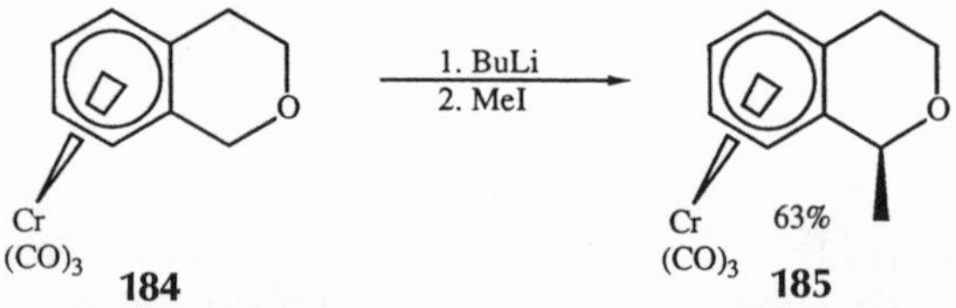

(*N*-Methyltetrahydroisoquinoline)chromium tricarbonyl **186** could potentially undergo deprotonation of an arene, 1-*exo*- or 4-*exo*-proton. Treatment of **186** with butyllithium followed by addition of deuteriomethanol gave regio- and stereoselectively the 4-*exo*-deuterio derivative **187**. Deprotonation–protonation of **187** removed the deuterium and regenerated **186**. Both the deprotonation and electrophilic addition steps had, therefore, occurred completely stereoselectively with the 4-*exo*-hydrogen being removed to generate anion **188** and being replaced with retention of configuration. The regio- and stereoselectivity of these reactions may be understood in terms of an axial lone pair on the nitrogen chelating to the butyllithium and thereby delivering the base to the proximate pseudoaxial 4-*exo*-hydrogen.[64] Chelation of the large butyllithium would also disfavour the 1-*exo*-hydrogen from adopting the required conformation for deprotonation. Consistent with chelation control, the regioselectivity is lost on addition of tetramethylethylenediamine.

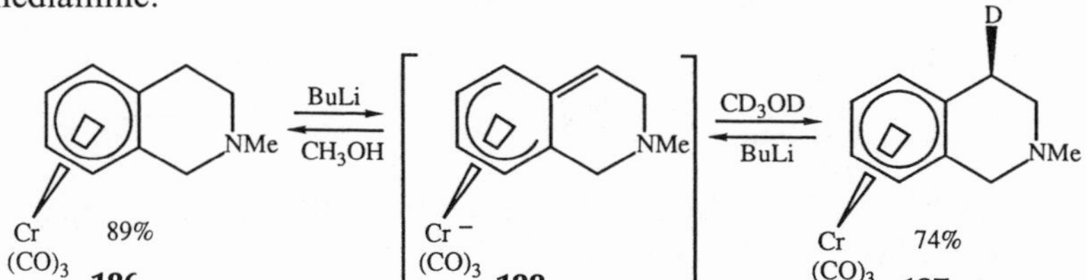

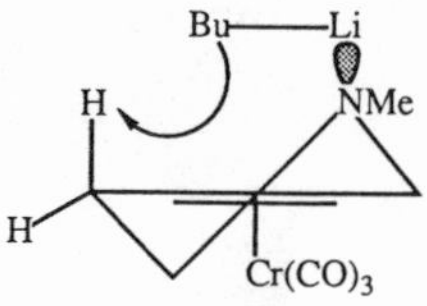

The intermediate anion **188** could be quenched by a variety of electrophiles to give, completely stereoselectively, the corresponding 4-*exo*-derivatives. Decomplexation released the 4-substituted-*N*-methyltetrahydroisoquinoline derivatives **189** (Table 5).[64,65]

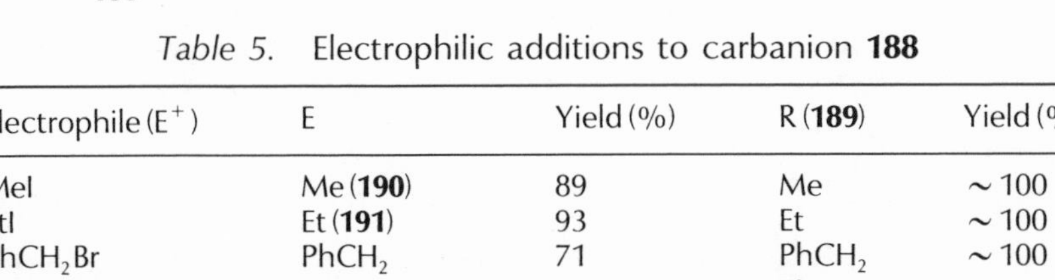

Table 5. Electrophilic additions to carbanion **188**

Electrophile (E^+)	E	Yield (%)	R (**189**)	Yield (%)
MeI	Me (**190**)	89	Me	~100
EtI	Et (**191**)	93	Et	~100
PhCH$_2$Br	PhCH$_2$	71	PhCH$_2$	~100
(C$_6$H$_5$F)Cr(CO)$_3$	[PhCr(CO)$_3$]	—	Ph	30
MoOPH	OH	35	OH	97
Me$_3$SiCl	Me$_3$Si (**192**)	42		
Me$_2$CO	Me$_2$COH	36		

The 4-*exo*-methyl derivative **190** no longer has a 4-*exo*-proton available and deprotonation with t-butyllithium followed by addition of an alkyl halide introduced completely regio- and stereoselectively a further substituent into the 1-*exo*-position. For example, methylation of **190** gave the *cis-exo*-1,4-dimethyl derivative **193**.[65]

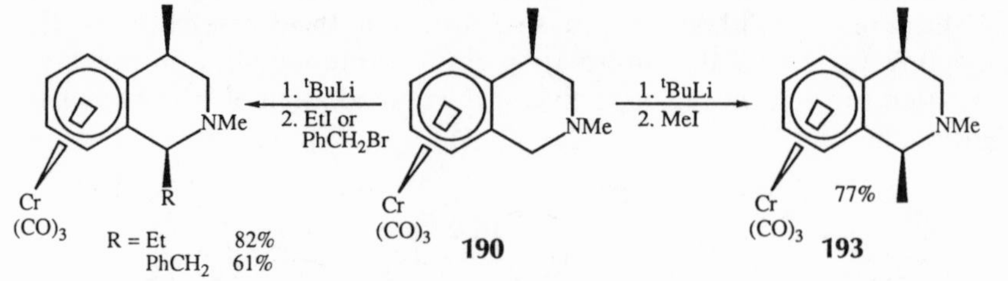

In contrast the 4-*endo*-methyl derivative **194**, prepared as a separable mixture with **190** by direct complexation of 4-methyl-*N*-methyltetrahydroisoquinoline, has a 4-*exo*-hydrogen available and deprotonation–methylation gave the *gem*-dimethyl derivative **195**.[64]

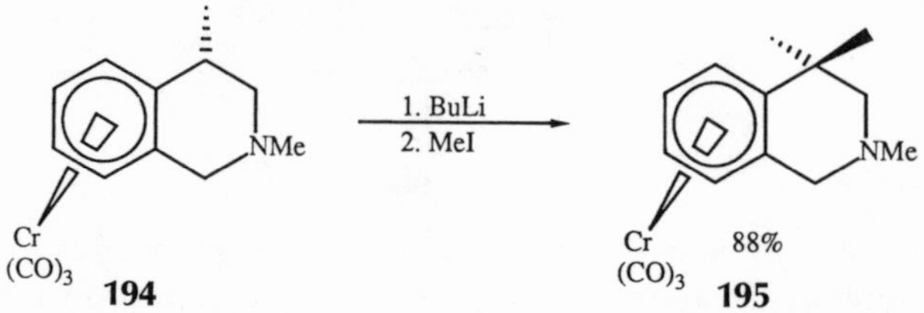

Overall, this methodology allows the sequential introduction of substituents into the 4-*exo* and then 1-*exo* positions of complex **186**, both substituents being introduced from the uncomplexed face. Thus diethylation of **186** gave via **191**, the *cis-exo*-1,4-diethyl derivative **196**.[65]

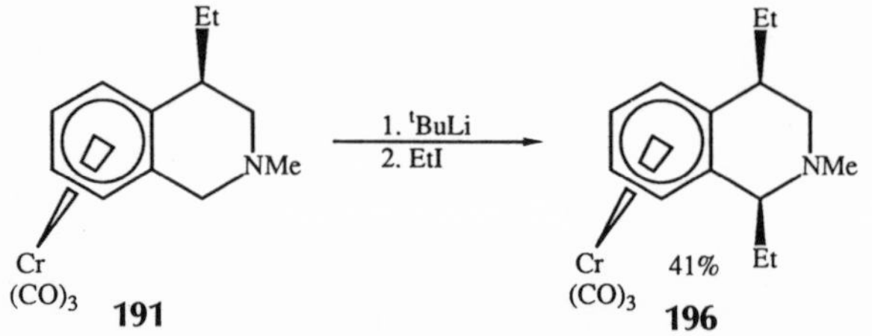

The trimethylsilyl group can serve as temporary protection for the 4-*exo*-position thus allowing the stereoselective synthesis of 1-*exo* derivatives. Methylation of **192** with t-butyllithium–methyl iodide gave **197** as a single diastereoisomer. Fluoride-mediated desilylation of **197** gave the 1-*exo*-

methyl derivative **198**, the desilylation of **197** being favoured by the stabilizing influence of the (arene)chromium tricarbonyl on the intermediate anion.[65]

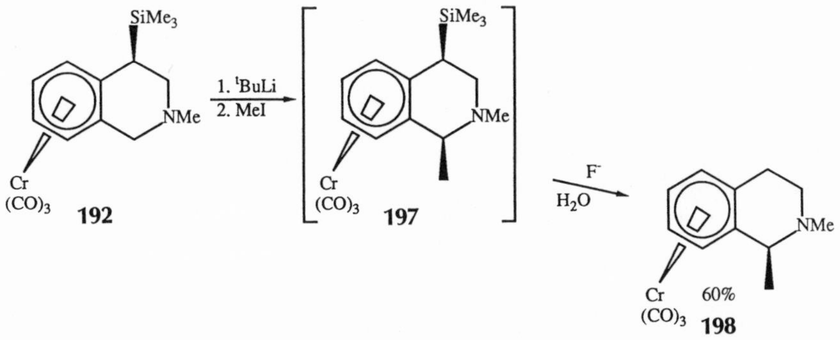

Complexation of 1-methyl-*N*-methyltetrahydroisoquinoline **199** gave a separable mixture of the *exo* and *endo* complexes **198** and **200**. Methylation of **198** provided an alternative route to **193**, while methylation of **200** gave the *trans*-1-*endo*-4-*exo*-dimethyl derivative **201**. Both **193** and **201** were produced as single diastereoisomers.[65]

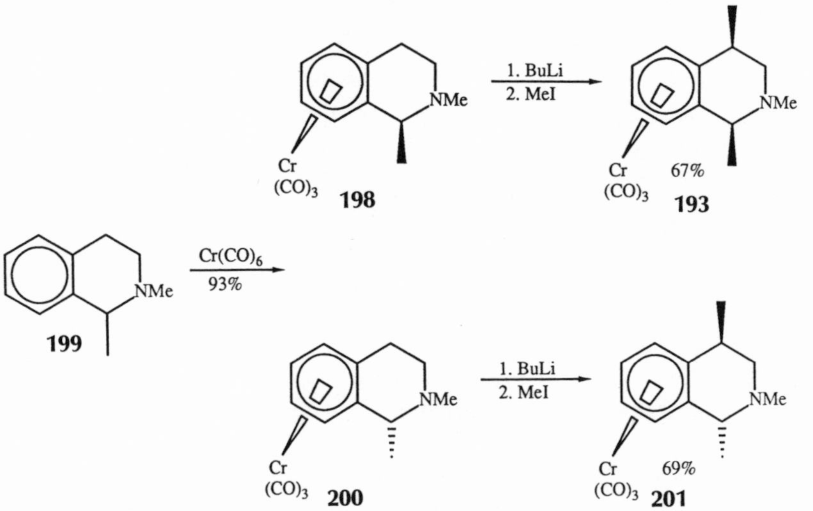

Complexation of the (+)-(*S*)-amphetamine-derived (+)-(3*S*)-methyl-*N*-methyltetrahydroisoquinoline **202** gave a mixture of the *exo* and *endo* complexes **203** and **204**. Stereospecific double *exo*-1,4-dimethylation of **203** and **204** gave (+)-**205** and (+)-**206**, respectively, each of which was isolated diastereoisomerically and enantiomerically pure. The X-ray crystal structure of the all *cis*, all *exo* complex (+)-**205** is illustrated in Figure 5.[66] Decomplexation liberated the corresponding free arenes.

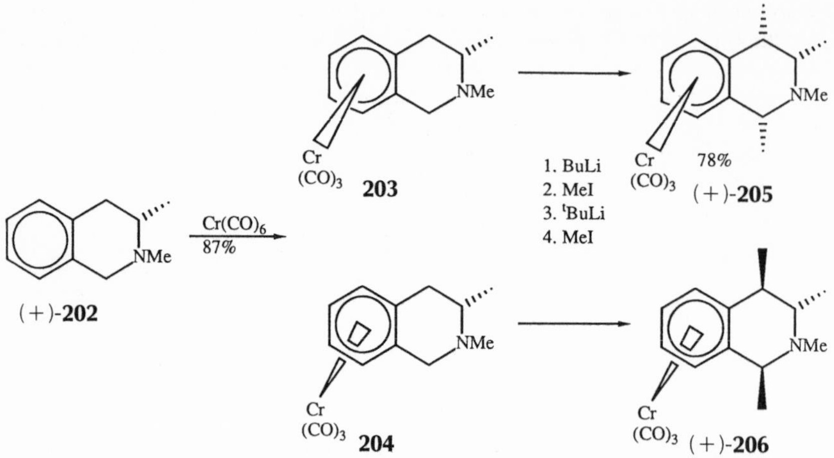

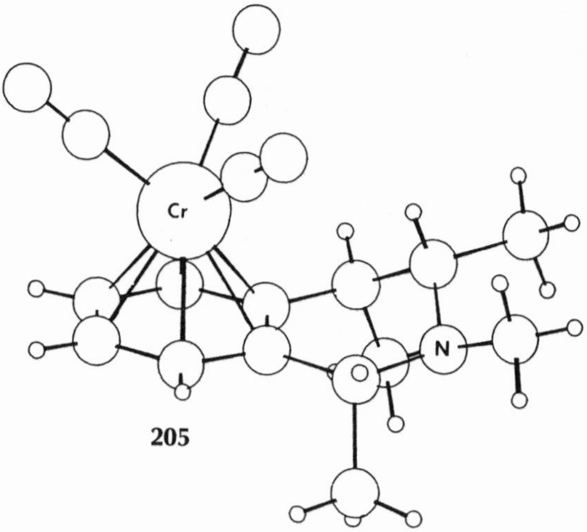

Figure 5. X-ray crystal structure of (+)-exo-(1R,3S,4R)-(trimethyl-*N*-methyltetrahydroisoquinoline)chromium tricarbonyl **205**.

(+)-exo-(1R,3S,4R,N)-(Tetramethyltetrahydroisoquinoline)Cr(CO)₃ **(205)**.[66]
A mixture of *exo* and *endo*-(3*S,N*)-(dimethyltetrahydroisoquinoline)Cr(CO)₃ **203** and **204** [ratio 1.5:1, as derived from complexation of (+)-(3*S,N*)-dimethyltetrahydroisoquinoline **202**, 2.74 g, 9.22 mmol] in THF (60 mL) at −68°C was treated with BuLi (1.65 M solution in hexanes, 6.15 mL, 10.1 mmol) and the red solution stirred (2 h, −68°C). Methyl iodide (1.7 mL, 27.3 mmol) was added and stirring continued (1 h, −68°C). After addition of methanol the solution was warmed to room temperature and evaporated. Column chromatography (Al₂O₃, Grade V, CH₂Cl₂) gave a mixture of the 4-*exo*-alkylated complexes as a yellow oil (ratio 1.5:1, 2.53 g, 88%), which was dissolved in THF (60 mL) and cooled to −70°C. t-BuLi (2.62 M solution in pentane, 3.25 mL, 8.5 mmol) was added and the resultant red solution stirred

(2 h, −70°C). Methyl iodide (1.6 mL, 25.7 mmol) was added and stirring continued (2 h, −70°C). After addition of methanol the solution was warmed and evaporated. Column chromatography (Al$_2$O$_3$, Grade V, CH$_2$Cl$_2$) gave a mixture of (+)-*exo*-(1*R*,3*S*,4*R*,*N*)-(tetramethyltetrahydroisoquinoline)Cr(CO)$_3$ **205** and (+)-*exo*-(1*S*,3*S*,4*S*,*N*)-(tetramethyltetrahydroisoquinoline)Cr(CO)$_3$ **206** as a yellow solid (ratio 1.5:1, 2.36 g, 89%), which were separable by flash chromatography (SiO$_2$, Et$_2$O).

Complex **207** underwent exclusive 4-*exo*-methylation with butyllithium and methyl iodide to give **208**. The expected arene deprotonation adjacent to oxygen being prevented in this case by the bulk of the triisopropylsilyl protecting groups.[39]

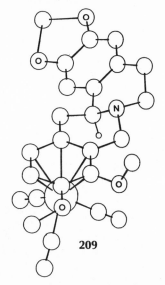

Complexation of (−)-canadine gave a separable mixture of the *exo*- and *endo*-complexes (−)-**209** and (−)-**210**. Figure 6 illustrates the X-ray crystal structure of the *exo*-complex (−)-**209** . Initial deprotonation of (−)-**209** occurred at C-12 promoted by the methoxyl group at C-11. In order to get benzylic deprotonation this position had to be protected by silylation, which

Figure 6. X-ray crystal structure of (−)-*exo*-(canadine)chromium tricarbonyl **209**.

gave (−)-**211**. The trimethylsilyl group not only protects C-11 but also activates the 8-*exo*-proton, and therefore treatment of (−)-**211** with t-butyllithium followed by methyl iodide regio- and stereoselectively gave the 8-*exo*-methyl derivative **212** as a single diastereoisomer. Desilylation and decomplexation released (−)-(8*S*)-methylcanadine **213**. A similar sequence of reactions starting from complex (−)-**210** gave, via (−)-**214** and **215**, the epimer (−)-(8*R*)-methylcanadine **216**.[13]

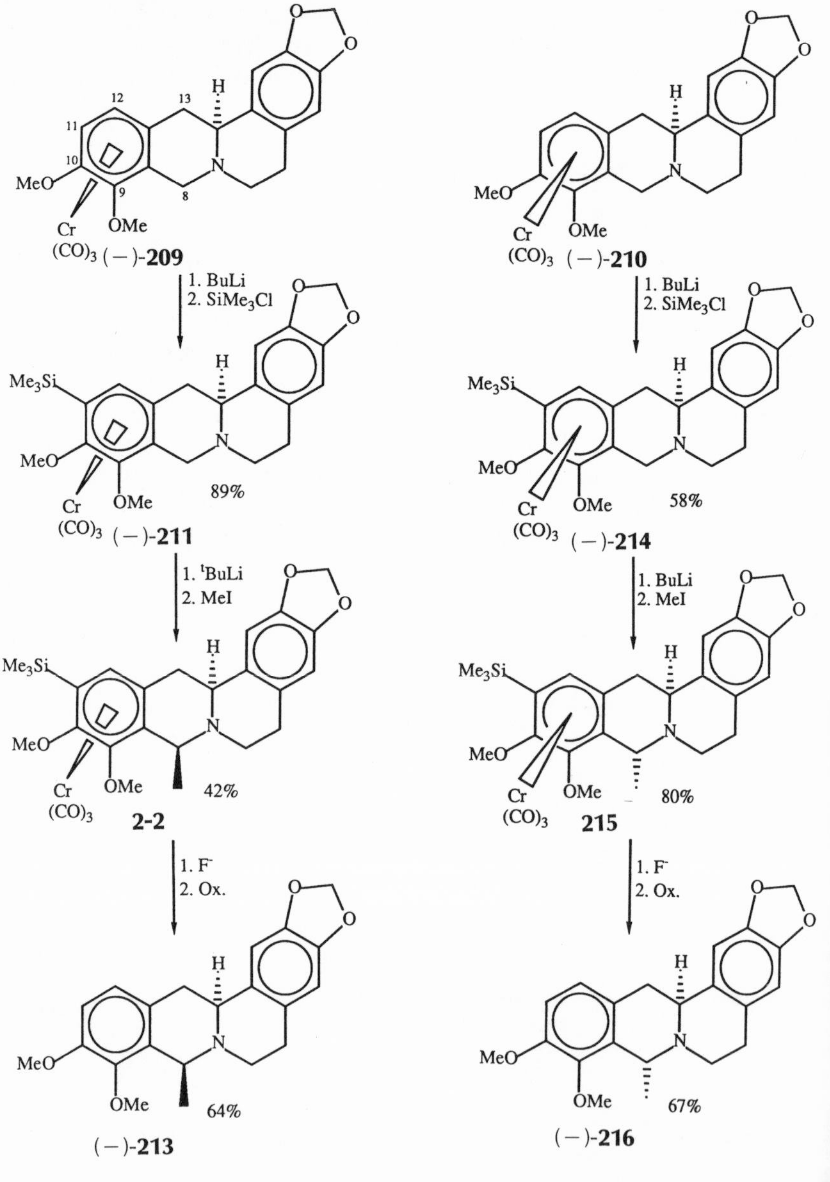

exo-(8R,14S)-(8-Methyl-11-trimetylsilylcanadine)Cr(CO)₃ **(212)**.[13] BuLi (1.6 M solution in hexanes, 0.40 mL, 0.64 mmol) was added to a stirred solution of (−)-*exo*-(11-trimethylsilylcanadine)Cr(CO)₃ **211** (220 mg, 0.40 mmol) in THF (20 mL) at − 78°C. After stirring (2 h, − 78°C) methyl iodide (456 mg, 3.23 mmol) was added and stirring continued (2 h, − 78°C). After addition of methanol the solution was warmed to room temperature and evaporated. Column chromatography (Al₂O₃, Grade V, Et₂O) gave the title compound **212** as a yellow foam (180 mg, 80%).

XII. CONCLUSIONS

The potential for synthesis employing benzylic carbanions stabilized by complexation to chromium tricarbonyl is evident from the results described above. (Arene)chromium tricarbonyl complexes are easily prepared and benzylic carbanions may be generated and trapped by electrophiles completely regio- and stereoselectively. Although many aspects remain unexplored, enough information is currently available on the generation and reactivity of chromium tricarbonyl complexed benzylic carbanions for the reactivity of novel complexes to be reliably predicted. This chemistry is now ripe for exploitation in synthesis, the greatest potential being in the areas of asymmetric and enantiospecific syntheses of homochiral materials.

REFERENCES AND NOTES

1. S. G. Davies, "Organotransition Metal Chemistry: Applications to Organic Synthesis"; Pergamon Press, Oxford, 1982.
2. B. Rees and P. Coppens, *Acta Cryst.*, 1973, *B29*, 2516; M. F. Bailey and L. F. Dahl, *Inorg. Chem.*, 1965, *4*, 1314.
3. J. Y. Saillard, D. Grandjean, P. Le Maux and G. Jaouen, *Nouv. J. Chim.* 1981, *5*, 153; G. A. Sim, *Ann. Rev. Phys. Chem.*, 1967, *18*, 57.
4. S. G. Davies and C. L. Goodfellow, unpublished results.
5. B. Nichols and M. C. Whiting, *J. Chem. Soc.*, 1959, 551.
6. S. Top and G. Jaouen, *J. Org. Chem.*, 1981, *46*, 78.
7. S. G. Davies and S. J. Coote, unpublished results.
8. C. A. L. Mahaffy and P. L. Pauson, *Inorg. Synth.*, 1979, *19*, 154.
9. D. Seyferth, J. S. Merola and C. S. Eschbach, *J. Am. Chem. Soc.*, 1978, *100*, 4124.
10. For an example of the use of MeCN see W. S. Trahanovsky and D. K. *J. Am. Chem. Soc.*, 1969, *91*, 5870. For an example of the use of EtCN see M. Stobbe, O. Reiser, T. Thiemann, R. G. Daniels and A. de Meijere, *Tetrahedron Lett.*, 1986, *27*, 2353.
11. D. E. F. Gracey, W. R. Jackson, W. B. Jennings and T. R. B. Mitchell, *J. Chem. Soc. (B)*, 1969, 1204.
12. E. P. Kundig, C. Perret, S. Spichiger and G. Bernardinelli, *J. Organomet. Chem.*, 1985, *286*, 183.
13. P. D. Baird, J. Blagg, S. G. Davies and K. H. Sutton, *Tetrahedron*, 1988, *44*, 171.
14. S. G. Davies and C. L. Goodfellow, *J. Organometal. Chem.*, 1988, *340*, 195.
15. M. F. Semmelhack, H. T. Hall, M. Yoshifuji and G. Clark, *J. Am. Chem. Soc.*, 1975, *97*, 1247.

16. K. M. Doxsee, R. H. Grubbs and F. C. Anson, *J. Am. Chem. Soc.*, 1984, *106*, 7819.

17. G. Jaouen and A. Meyer, *J. Am. Chem. Soc.*, 1975, *97*, 4667.

18. A. Ceccon, A. Gambaro and A. Venzo, *J. Organometal. Chem.*, 1984, *275*, 209.

19. J. Lebibi, J. Brocard and D. Couturier, *Bull. Soc. Fr.*, 1982, *Part II*, 357.

20. M. C. Senechal-Tocquer, D. Senechal, J.-Y. Le Bihan, D. Gentric and B. Caro, *J. Organometal. Chem.*, 1985, *291*, C5.

21. R. J. Card and W. S. Trahanovsky, *J. Org. Chem.*, 1980, *45*, 2560.

22. A. Creccon, F. Piccini and A. Venzo, *Gazz. Chim. Ital.*, 1978, *108*, 705.

23. M. F. Semmelhack, G. R. Clark, J. L. Garcia, J. J. Harrison, Y. Thebtaranonth, W. Wulff and A. Yamashita, *Tetrahedron*, 1981, *37*, 3957.

24. C. D. Broaddus, *J. Org. Chem.*, 1970, *35*, 10.

25. G. Simonneaux and G. Jaouen, *Tetrahedron*, 1979, *35*, 2249.

26. J. Brocard, J. Lebibi and D. Couturier, *J. Chem. Soc., Chem. Commun.*, 1981, 1264.

27. M.-C. Senechal-Tocquer, D. Senechal, J.-Y. Le Bihan, D. Gentric and B. Caro, *J. Organometal. Chem.*, 1987, *321*, 353.

28. M. F. Semmelhack, J. Bisha and M. Czarny, *J. Am. Chem. Soc.*, 1979, *101*, 769.

29. B. Caro, J.-Y. Le Bihan, J.-P. Guillot, S. Top and G. Jaouen, *J. Chem. Soc., Chem. Commun.*, 1984, 602.

30. D. Senechal, M.-C. Senechal-Tocquer, D. Gentric, J.-Y. Le Bihan, B. Caro, M. Gruselle and G. Jaouen, *J. Chem. Soc., Chem. Commun.*, 1987, 632.

31. J.-F. Halet, J.-Y. Saillard, B. Caro, J.-Y. Le Bihan, S. Top and G. Jaouen, *J. Organometal. Chem.*, 1984, *267*, C37.

32. G. Jaouen, *Ann. N. Y. Acad. Sci.*, 1977, *295*, 59.

33. W. S. Trahanovsky and R. J. Card, *J. Am. Chem. Soc.*, 1972, *94*, 2897.

34. T. G. Traylor and M. J. Goldberg, *J. Am. Chem. Soc.*, 1987, *109*, 3968.

35. G. Jaouen, M. Meyer and G. Simonneaux, *J. Chem. Soc., Chem. Commun.*, 1975, 813.

36. H. des Abbayes and M.-A. Boudeville, *J. Org. Chem.*, 1977, *42*, 4104.

37. J. Brocard and J. Lebibi, *J. Organometal. Chem.*, 1987, *320*, 295.

38. J. Brocard, L. Pelinski and J. Lebibi, *J. Organometal. Chem.*, 1986, *309*, 299.

39. J. Blagg, D.Phil thesis, Oxford, 1986.

40. J. Brocard, A. Laconi, D. Couturier, S. Top and G. Jaouen, *J. Chem. Soc., Chem. Commun.*, 1984, 475.

41. G. Jaouen, S. Top, A. Laconi, D. Couturier and J. Brocard, *J. Am. Chem. Soc.*, 1984, *106*, 2207.

42. For an example of the *ortho* directing nature of the methoxyl group see P. J. Beswick, S. J. Leach, N. F. Masters and D. A. Widdowson, *J. Chem. Soc., Chem. Commun.*, 1984, *46*, and references therein.

43. J. Brocard and J. Lebibi, *J. Organometal. Chem.*, 1986, *310*, C63.

44. S. Top, G. Jaouen, A. Vessieres, J.-P. Abjean, D. Davoust, C. A. Rodger, B. G. Sayer and M. J. McGlinchey, *Organometallics*, 1985, *4*, 2143.

45. S. Top, A. Vessieres, J.-P. Abjean and G. Jaouen, *J. Chem. Soc., Chem. Commun.*, 1984, 428.

46. H. Kunzer and M. Thiel, *Tetrahedron Lett.*, 1988, *29*, 1135; J. C. Gill, B. A. Marples and J. R. Traynor, *ibid*, 1987, *28*, 2643.

47. B. E. Mobbs, D.Phil thesis, Oxford, 1985.

48. A. Meyer and O. Hofer, *J. Am. Chem. Soc.*, 1980, *102*, 4410.

49. G. Jaouen and A. Meyer, *Tetrahedron Lett.*, 1976, 3547.

50. A. N. Nesmeyanov, N. A. Ustynyuk, L. N. Novikova, T. N. Rybina, Y. A. Ustynyuk, Y. F. Oprunenko and O. I. Trifonova, *J. Organometal. Chem.*, 1980, *184*, 63.

51. M. F. Semmelhack, W. Seufert and L. Keller, *J. Am. Chem. Soc.*, 1980, *102*, 6584.

52. M. Uemura, T. Minami and Y. Hayashi, *J. Chem. Soc., Chem. Commun.*, 1984, 1193.

53. S. G. Davies and B. E. Mobbs, unpublished results.

54. P. J. Dickens, A. M. Z. Slawin, D. A. Widdowson and D. J. Williams, *Tetrahedron Lett.*, 1988, *29*, 103.
55. J. Blagg and S. G. Davies, *Tetrahedron*, 1987, *43*, 4463.
56. H. B. Arzeno, D. H. R. Barton, S. G. Davies, X. Lusinchi, B. Meunier and C. Pascard, *Nouv. J. Chim.*, 1980, *4*, 369.
57. N. Mathews and M. Sainsbury, *J. Chem. Res. (S)*, 1988, 82.
58. U. S. Schollkopf, *Angew. Chem., Int. Ed. Engl.*, 1970, *9*, 763 and references therein.
59. J. Blagg, S. G. Davies, N. J. Holman, C. A. Laughton and B. E. Mobbs, *J. Chem. Soc., Perkin Trans.*, 1986, 1581.
60. M. Uemura, N. Nishikawa, K. Take, M. Ohnishi, K. Hirotsu, T. Higuichi and Y. Hayashi, *J. Org. Chem.*, 1983, *48*, 2349.
61. J. Blagg, S. G. Davies, C. L. Goodfellow and K. H. Sutton, *J. Chem. Soc., Perkin Trans. 1*, 1987, 1805.
62. N. J. Holman, D.Phil thesis, Oxford 1984.
63. S. G. Davies, C. L. Goodfellow and K. H. Sutton, unpublished results.
64. J. Blagg, S. J. Coote and S. G. Davies, *J. Chem. Soc., Perkin Trans. 1*, 1986, 2257.
65. J. Blagg, S. J. Coote, S. G. Davies, D. Middlemiss and A. Naylor, *J. Chem. Soc., Perkin Trans. 1*, 1987, 689.
66. S. J. Coote, S. G. Davies and K. H. Sutton, *J. Chem. Soc., Perkin Trans. 1*, 198, 1481.

PALLADIUM-MEDIATED ARYLATION OF ENOL ETHERS

G. Doyle Daves, Jr.

I. INTRODUCTION

The detailed study of palladium-mediated reactions of enol ethers which has been underway in our laboratory for more than a decade began innocently and without any appropriate background in organometallic chemistry. We

Advances in Metal-Organic Chemistry, Volume 2, pages 59–99
Copyright © 1991 JAI Press Ltd
ISBN: 0-89232-948-3

had been interested in the development of a general, efficient synthetic route to C-glycosides[1,2] (C-nucleosides). In this connection we noted a report, in 1976 by Bergstrom and Ruth,[3] that reaction of a pyrimidinylmercuric salt with olefins in the presence of Pd(II) achieved pyrimidine C-5 alkylation in good yields.

This report,[3] the first involving a palladium-mediated reaction at C-5 of pyrimidine, suggested to us the possibility of developing a conceptually new and remarkably direct route for synthesis of C-nucleosides and C-glycosides.[1,2] We envisaged that similar palladium-mediated reaction of a carbohydrate-derived enol ether (i.e., a 1,2-unsaturated carbohydrate or glycal) with a heterocyclic or aromatic mercuric salt, or other appropriate metallo derivative, might establish the glycosidic bond and yield a C-glycoside directly. We began experimental work soon after and this initial expectation was realized after 18 months of diligent research and we reported[4] in 1978 a facile C-glycoside synthesis by a palladiium-mediated, regio- and stereospecific coupling of carbohydrates with aglycones.

During the course of this initial study, we developed a fascination for the directness and strategic simplicity of synthetic applications of organometallic chemistry and gained a rudimentary understanding of basic organometallic reaction mechanisms. We have continued to study palladium-mediated arylation reactions of enol ethers and related systems within the confines of our interest in synthesis of C-glycoside antitumor and antiviral agents. Because the organometallic chemistry involved is complex and strict regio- and stereochemical control of C–C bond formation is essential in synthesis, it has been necessary for us to carry out a detailed study of underlying reaction mechanisms. Happily, enol ethers, and particularly chiral, cyclic enol ethers derived from carbohydrates, have provided an excellent system for study of fundamental mechanisms of organometallic reactions. In this chapter, we (a) discuss the understanding of the underlying chemistry involved in enol ether arylation which has emerged during the past decade, (b) indicate critical problems which remain to be solved and (c) illustrate with examples, the rich potential of these palladium-mediated coupling reactions for organic synthesis.

II. BACKGROUND: HECK ARYLATION

Palladium-mediated reaction of an aryl or heteroaryl metallo derivative with an olefin is a version of a generic organometallic reaction type known as 'Heck arylation' or more broadly as the 'Heck reaction.' This new organometallic reaction type was announced by Heck in 1968 and described in impressive detail in seven simultaneous reports.[5-11] From this birth, a medley of C–C bond forming reactions of broad scope and great synthetic utility has developed.[12,13]

The basic chemistry involved in the Heck reaction, coupling of an organopalladium species (R–Pd; R = aryl, vinyl, heterocycle, benzyl, methyl) with an olefin, is generally well understood.

$$R-Pd-X + CH_2=CH_2 \rightarrow RCH=CH_2 + HX + Pd(0)$$

It is convenient to consider the overall process as involving four discrete organometallic reaction steps:[14] (1) organopalladium reagent formation, (2) π-complex formation between the organopalladium reagent and the olefin, (3) collapse of the π-complex by insertion of the olefin into the R–Pd bond to form a σ-adduct and (4) decomposition of the σ-adduct with product formation (Scheme 1).

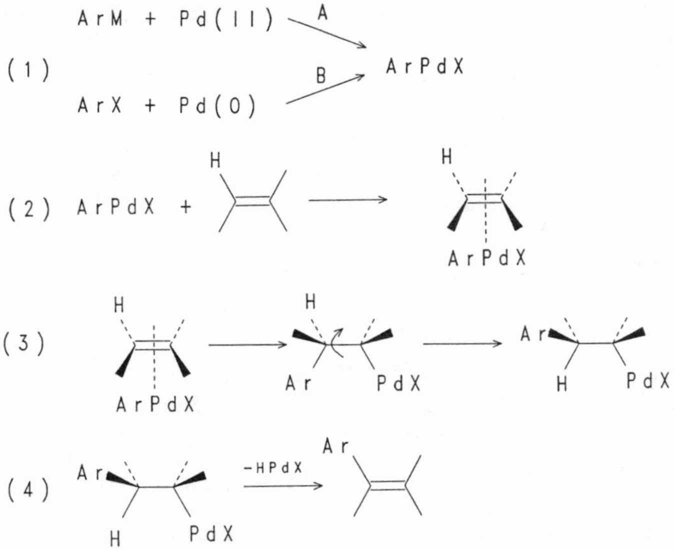

Scheme 1

III. PALLADIUM-MEDIATED ARYLATION OF ENOL ETHERS

Successful use of Heck arylation, particularly in synthesis of complex molecules like C-glycosides, requires that each of the four component steps of the overall reaction be controlled effectively. Study of these processes has led to a level of understanding which permits palladium-mediated enol ether arylation reactions to be carried out under conditions with regio- and stereochemical control of C–C bond formation and preselection of σ-adduct decomposition mode.

A. Formation of 'Ar-Pd' Reagent

Evidence is strong that the initial step of a palladium-mediated coupling reaction is the formation of an organopalladium(II) reagent (Scheme 1, step 1).[12-14] These organopalladium reagents are almost always generated *in situ* and evidence for their structures is largely circumstantial. In several cases, we have been successful in obtaining direct spectroscopic evidence for aryl and heteroaryl palladium reagents in solution by analysis of reaction mixtures using fast atom bombardment mass spectrometry.[15] In this way, mass spectra were recorded which exhibit ions characteristic of the aryl-bound palladium species present in a reaction mixture as a result of a transmetallation process.[15] Several arylpalladium reagents, bearing stabilizing ligands (e.g., triphenylphosphine) have been isolated and characterized.[16-18] These organometallic reagents are square planar complexes of tetracoordinate Pd(II);[16] the reagent used in Heck-type coupling reactions is formulated as Ar–Pd(L$_2$)–X where X is an anion, e.g. Cl$^-$ or AcO$^-$, and L is an added neutral ligand like triphenylphosphine or, if such ligands are not provided, solvent (e.g. CH$_3$CN).

Two principal methods are used for preparation of the reactive organopalladium reagent:[12,13] (a) transmetallation by reaction of an organometallic compound (e.g. an organomercurial or stannane)[5,13,14,19-20] with a palladium(II) salt (Scheme 1, step 1a) or (b) oxidative addition of Pd(0) into an Ar–X bond (Scheme 1, step 1b).

1. Transmetallation

In transmetallation, formation of the reactive organopalladium reagent is accomplished by transfer of an aryl ligand to palladium(II) from an arylmercurial, arylstannane or other arylmetallo derivative.[5,13,14,19-20]

$$Ar–Hg(II) + Pd(II) \rightarrow Ar–Pd(II) + Hg(II)$$

This process typically occurs at room temperature or below and affords the advantage that this method of organopalladium reagent preparation permits coupling reactions with enol ethers or other olefins to be carried out at room temperature.

A disadvantage of preparing the reactive organopalladium reagent by transmetallation is that, because it is based on palladium(II), for coupling reactions to be catalytic in palladium, an oxidant (e.g. $CuCl_2$, O_2, benzoquinone[21]) is required. Decomposition of the σ-adduct formed in the coupling reaction by β-hydride elimination[12,13] yields an unstable Pd(II) hydride species which undergoes reductive elimination forming Pd(0) which, in the absence of an oxidant, stops the reaction.

$$H-Pd(II)-X \rightarrow H-X + Pd(0)$$

It should be noted that, under some conditions, Hg(II) is used to oxidize Pd(0) to Pd(II).[22] However many, indeed most, reactions involving formation of an organopalladium reagent by transmetallation from the corresponding organomercurial require stoichiometric palladium if a second oxidant is not present. The Hg(II) present in the reaction mixture is apparently in a form, e.g. complexed or colloidal, which is unreactive or reacts only slowly. The Pd(0) which forms, quickly aggregates and often forms a metallic mirror; bulk palladium reacts with oxidants too slowly to maintain a catalytic reaction.

The relatively high cost of palladium is a factor in some applications; however, if the use of a palladium-mediated coupling reaction saves several steps when compared with alternative synthetic routes or when costly organic intermediates are involved, as is the case in our C-glycoside syntheses, use of palladium in stoichiometric quantities is easily justified.

2. Oxidative Addition

Oxidative addition by insertion of palladium into an aryl–substituent σ-bond, is the other primary reaction by which the reactive organopalladium reagent is formed.[12,13] Although organometallic reagent formation by insertion of palladium into an aryl–hydrogen bond (direct metallation) is well known, it is usually most convenient to use aryl halides.[12,13]

$$Ar-X + Pd(0) \rightarrow Ar-Pd-X$$

Though this oxidative addition reaction requires Pd(0), Pd(II) can be, and often is, used as Pd(II) is readily reduced to Pd(0) by olefin or other species present in a typical reaction mixture.[13,14]

Typically, elevated temperatures (80–120°C) and aryl iodides and bromides are employed in oxidative addition reactions; bromides often require activation by the presence of electron-withdrawing substituents on the aryl ring or phosphine ligands on palladium.[13] Aryl chlorides are usually unreactive; however, some exceptions have been noted.[23,24] Jeffery[25] has found that use of tetrabutylammonium chloride in dimethylformamide solvent markedly facilitates the reaction permitting organopalladium reagent formation from aryl iodides and bromides at or near room temperature. Arylpalladium reagents are also prepared from aroyl halides,[26-29] by oxidative addition of palladium(0) into the carbonyl carbon–halogen bond followed by decarbonylation. Recently, aryl triflates have been used successfully as precursors of the arylpalladium reagent;[30-33] presumably this reaction involves oxidative addition of Pd(0) into the aryl-oxygen bond.

B.   π-Complex Formation

1.   The Enol Ether Structure

The critical questions concerning the structures of enol ethers are (a) how much interaction (conjugation) exists between the olefinic -electrons and the lone-pair p-electrons on oxygen, (b) what conformation(s) are important in solution and (c) what is the effect of substitution. These questions have been the focus of numerous studies of enol ethers using infrared,[34-36] microwave,[37,38] photoelectron,[39] nuclear magnetic resonance (^{1}H,[40] ^{13}C,[41-43] and ^{17}O[44]) spectroscopies, gas phase protonation[45,46] and other techniques.[47,48] In addition, theoretical calculations have been carried out.[45,49-50] These studies confirm that enol ether bonding involves a four electron π-system with significantly more electron density on the olefinic β-carbon than on the α-carbon as depicted by the simple valence bond picture.[42,44]

$$H_2C = CH - O - R \leftrightarrow H_2\overset{-}{C} - CH = \overset{+}{O} - R$$

The lowest energy conformation of simple acyclic enol ethers in solution is the s-*cis* conformation; usually this conformation is in equilibrium with a second conformation for which the non-planar *gauche* and planar s-*trans* conformations are limiting forms.[42,44]

S-CIS   GAUCHE   S-TRANS

A β-substituent *cis* to the ether oxygen destabilizes the s-*cis* conformation and leads to a mixture of gauche and s-*trans* conformations with a consequent diminution of conjugation of oxygen p-electrons with the olefinic double bond. Enol ethers with both a substituent on the olefinic α-carbon and a β-substituent cis to ether oxygen are denied both the s-*cis* and the s-*trans* conformations for which p–π conjugation is maximal.[42,44]

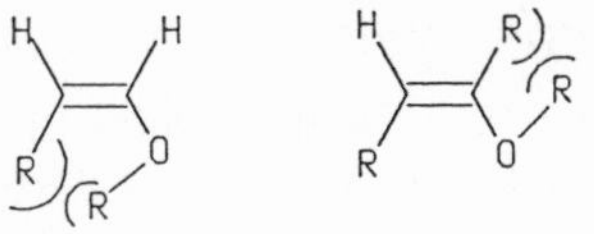

Incorporation of the enol ether into a ring introduces other constraints.[42] Maximal p–π conjugation in a cyclic enol ether requires the ether oxygen, the α and β olefinic carbons and the γ carbon to be coplanar. In six-membered ring systems this is achieved by a half-chair conformation[42,43] and in five-membered rings by a puckered or envelope conformation.[37,42] The C-5 carbon of 2,3-dihydrofuran has been estimated to be 19° out of plane in the gas phase.[37] p–π Conjugation is somewhat better in five-membered ring enol ethers than in six.[42]

Pyranoid glycals (1,5-anhydro-2-deoxyhex-1-enitols), enol ethers derived from six-membered ring carbohydrates by introduction of a double bond between carbons 1 and 2, favor either a $^4H_5$ or $_4H^5$ conformation depending on substitution pattern.[43] Comparison of derivatives of D-glucal and D-allal, both of which favor a $^4H_5$ conformation,[43,51] permits the effect of a substituent at the γ (allylic) carbon to be examined. In the glucal series, the allylic substituent adopts a *pseudo*-equatorial orientation whereas in the allal series this substituent occupies a *pseudo*-axial position. It was found that the electron density at C-1 of glucals was greater than that at C-1 of the C-3 epimeric allals; conversely, the electron density at C-2 was greater for the allals.[43] That is, the carbon–carbon π-bond was found to be more stabilized (less polarized) when the allylic C–O bond was coplanar with the π-system and more polarized when the C–O bond was perpendicular to the π-system.[43,52] Such stereoelectronic effects,[53,54] which include the anomeric effect[55] and the vinylogous anomeric effect,[56,57] have received increasing attention in recent years.

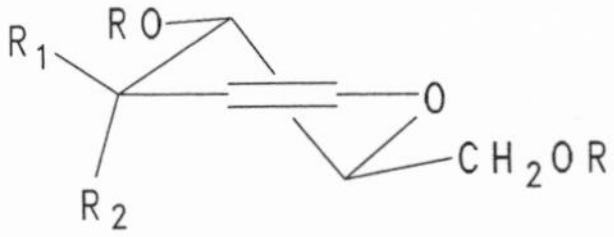

GLUCAL SERIES R_1 = OR, R_2 = H

ALLAL SERIES R_1 = H, R_2 = OR

2.   Complexation of Pd(II) with an Enol Ether

When an enol ether and a Pd(II) salt are mixed in solution complexation occurs to produce stable crystalline complexes which have been characterized.[58]

$$2\ Cl_2Pd(PhCN)_2\ +\ 2\ RHC{=}CH{-}O{-}R'$$

$$\rightarrow\ [Cl_2Pd(RHC{=}CH{-}O{-}R')]_2\ +\ 4\ PhCN$$

The composition of these complexes has importance for understanding the bonding between palladium and an enol ether. In these complexes, which involve bridging chlorines between two tetracoordinate square planar Pd(II) centers,[22,59] the enol ether occupies a single ligand site. In what appears to be the only reported crystal structure of an enol ether–Pd(II) complex, the enol ether is also a dihapta ligand.[59]

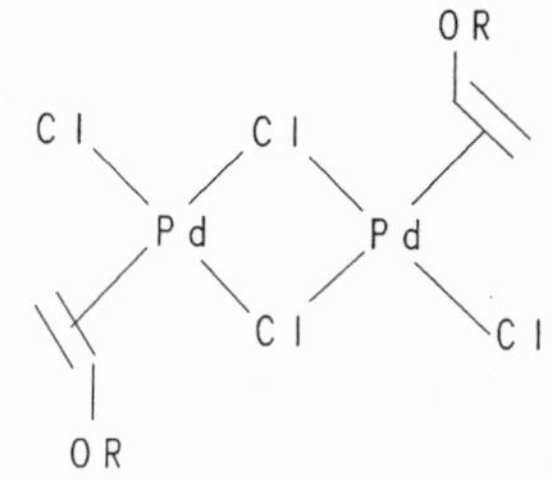

[Cl₂Pd(MeOCH=CHOMe)₂.[58] To 200 mg of $Cl_2Pd(PhCN)_2$ in 20 mL of toluene was added 0.05 mL of 1,2-dimethoxyethene. Stirring was continued for 1 h and the precipitated product which accumulated was collected and washed with toluene.

For an enol ether in which an oxygen p-electron pair interacts conjugatively with the C–C double bond π–electrons, the bonding system available for interaction with Pd(II) is a p–π four electron system with molecular orbitals related to those of allyl anion.[60,61] It is well established that allyl anion forms stable complexes with Pd(II) in which the allyl anion is a trihapta ligand.[22,62] The failure of an enol ether to function as a trihapta ligand for Pd(II), and only as a dihapta ligand, is consistant with the asymmetry of the ψ_1 and ψ_2 bonding orbitals[60] and the well-known weakness of the oxygen → palladium bond.[63]

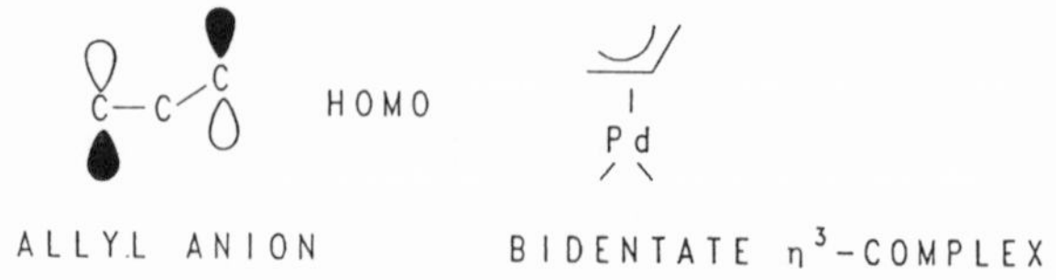

ALLYL ANION   BIDENTATE η^3–COMPLEX

ENOL ETHER   MONODENTATE η^2-COMPLEX

Saegusa and coworkers, in studying reactions of silyl enol ethers with Pd(II) salts, isolated palladium-containing complexes which were initially formulated as oxo-π-allylpalladium(II) species[64] in which the enolate is a trihapta ligand. Subsequent analyses and spectroscopic studies showed the complexes to be oligomeric Pd(II) enolates.[65]

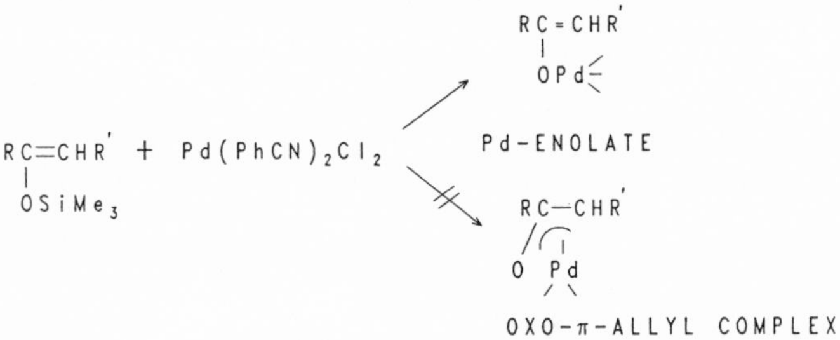

OXO-π-ALLYL COMPLEX

Additional insight into the bonding of enol ether–Pd(II) complexes is provided by the observation that isomerization of complexed (E) enol ethers to complexes of (Z) enol ethers, which are thermodynamically more stable,[66] is unusually facile.[58,59] Further, in the crystal structure of the enol ether–Pd(II) complex,[59] the olefinic β-C–Pd distance (2.14 Å) is significantly shorter than the α-C–Pd distance (2.26) and the midpoint of the olefin bond is considerably displaced from the Pd-ligand square plane. All these facts suggest a significant contribution of the σ-bonding canonical form in enol ether–Pd(II) complexes;[59] the bonding of enol ethers to platinum[67] and iron[68] appears to be similar.

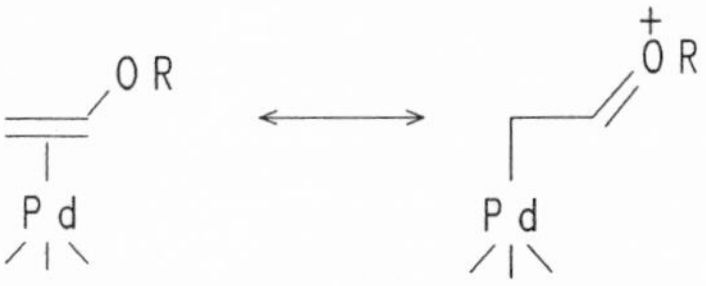

3. Stereochemistry of π-Complexation

Complexation of palladium(II) with a chiral cyclic enol ether involves discrimination between the two faces of the ring. Early in our study of palladium-mediated coupling reactions of carbohydrate derived enol ethers, we discovered[4] that this discrimination often results in completely stereospecific reaction. Thus, reaction of (1,3-dimethyl-2,4-tetrahydropyrimidine-dione-5-yl)mercuric acetate[69] (1) with 3,4,6-tri-O-acetyl-D-glucal[70] (2) in the presence of Pd(II)[4,71–74] yielded a single σ-adduct (4), derived from a single π-complex (3). Addition of triphenylphosphine to the reaction mixture after the adduct had formed, permitted 4 to be isolated and characterized spectroscopically.[72] The structure of adduct 4 indicates that the precursor π-complex (3) was formed by attack of the the organopalladium reagent, derived from 1 by transmetallation, on enol ether 2 from the face of the ring opposite the allylic acetate substituent.

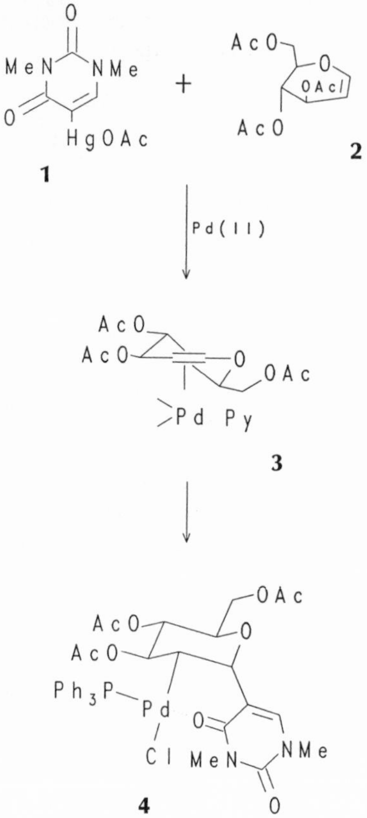

(1,3-Dimethyl-2,4-pyrimidinedion-5-yl)mercuric acetate (1).[69] A mixture of 14 g (0.1 mol) of 1,3-dimethyluracil and 31.8 g (0.1 mol) of mercury(II) acetate in 200 mL of methanol containing several drops of 70% perchloric acid was stirred at room temperature for 12 h. The product which had precipitated was collected and washed with cold methanol (36.4 g, 93%) and used without further purification.

Chloro *[1,3-dimethyl-5-(3,4,6,tri-O-acetyl-2-deoxy-α-D-arabino-hexopyran-osyl)- 2,4-pyrimidinedonnato] (triphenylphosphine)palladium* (**4**).[72] (1,3-Di-methyl-2,4-pyrimidinedion-5-yl)mercuric acetate (**1**) (800 mg, 2.01 mmol) was added to a preformed mixture of Pd(II) acetate (450 mg, 2.00 mmol), anhydrous lithium chloride (170 mg, 4.01 mmol) and acetonitrile (50 mL). After 5 min, 3,4,6-tri-O-acetyl-D-glucal (**2**)[70] (600 mg, 2.21 mmol) was added with vigorous stirring. The reaction mixture was stirred for three days at room temperature and then filtered through Celite. Triphenylphosphine (788 mg, 3.00 mmol) was added to the filtrate and the resulting yellowish solution was allowed to stand under nitrogen for 2 h. The precipitate which formed was removed and the resulting filtrate was evaporated to yield an oil which was separated by flash chromatography using ethyl acetate. Rechromatography followed by precipitation of the product from benzene using ether gave 607 mg (38%) of a slightly yellowish powder, m.p. 138 C.

Almost all conclusions concerning the structures of Pd(II)–enol ether π-complexes have been derived by examination of reaction products formed in specific reactions or, rarely as in the case of **3**, from structures of derived σ-adducts. Reversible formation of isomeric π-complexes by attack of Pd(II) on both faces of the enol ether π-system seems reasonable and cannot be ruled out. However, in most reactions studied, the isolated products have been derived from a single π-complex.

Study of other cyclic six-membered ring carbohydrate-derived enol ethers (pyranoid glycals) related to **2** has shown that stereospecific π-complexation by organo-palladium reagents is general. When palladium-mediated coupling reactions of these chiral enol ethers were carried out at room temperature, in every case studied,[4,71-74] π-complexation had occurred by attack of the organopalladium reagent on the enol ether from the side opposite the allylic substituent. Dunkerton has reported[75] a series of reactions of pyranoid glycals with methanol in the presence of palladium chloride which may involve π-complexation by attack of palladium on the cyclic enol ether from the same face occupied by an allylic acetate group. However, the mechanism of these methoxypalladation reactions is not clear and firm conclusions concerning the stereochemistry of π-complexation cannot be reached.

Even at temperatures significantly above room temperature, stereospecificity of π-complexation is preserved. Czernecki and coworkers[76,77] have carried out a number of palladium-mediated arylation reactions of carbohydrate-derived enol ethers using the aryl component (benzene, anisole or 1,3-dimethoxybenzene) as solvent in the presence of palladium acetate and acetic acid at temperatures of 80–120°C. Under these conditions, the isolated products (e.g. **5** + **6**; **8**) were derived from a single π-complex formed by arylpalladium reagent attack on one face of the enol ether.

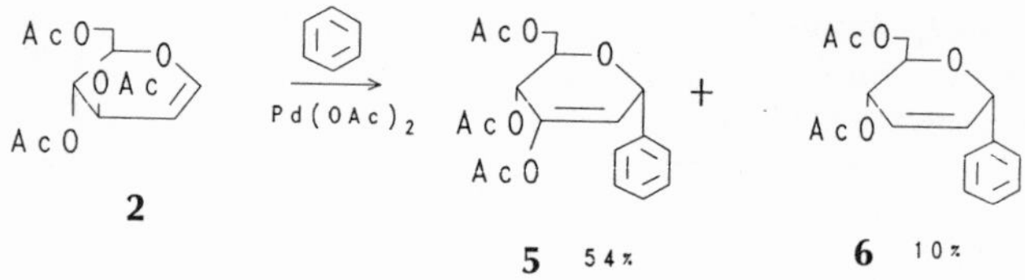

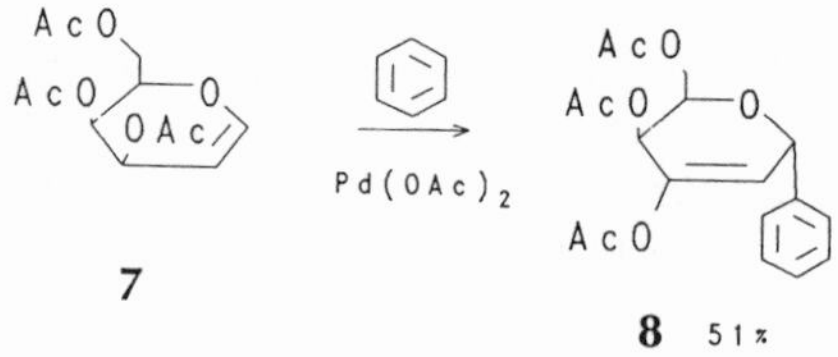

Arylation of 3,4,6-tri-O-acetyl-D-glucal (2) with benzene.[76] A solution of 3 mmol of glycal **2** and 3 mmol of Pd(II) acetate in 24 mL of acetic acid and 500 mmol of benzene was heated for 8 h under reflux. The reaction mixture was filtered to recover 240 mg (75%) of Pd and the residue obtained upon evaporation of solvents was separated by chromatography to yield 87 mg (10%) of 4,6-di-O-acetyl-2,3-dideoxy-α-D-erythro-hex-2-enopyranosylbenzene **(6)** and 565 mg (54%) of 3,4,6-tri-O-acetyl-2-deoxy-α-D-erythro-hex-2-enopyranosyl-benzene **(5)**.

In one example of a six-membered ring enol ether with the allylic carbon unsubstituted **(9)**, palladium-mediated coupling with pyrimidinylmercurial **1** yielded two products **(10 and 11)** which are derived from isomeric π-complexes.[78]

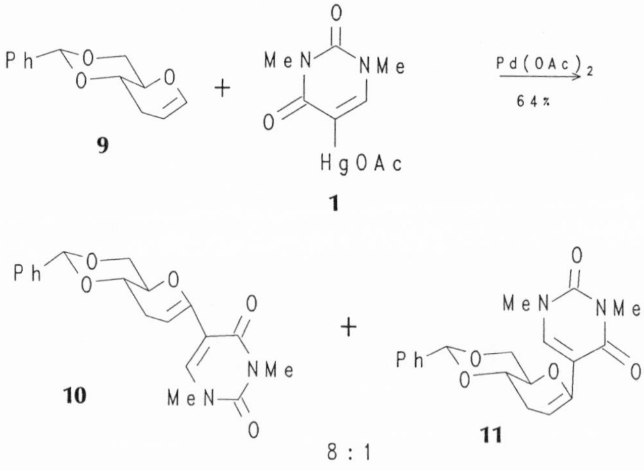

Table 1. Stereochemistry of π-Complex Formation in Palladium-Mediated Reactions of (1,3-Dimethyl-2,4-Tetrahydropyrimidinedione-5-yl)Mercuric Acetate (**1**) with Substituted Furanoid Glycals[a]

12–17

Compound	Substituents		Complex formed	Yield of coupled product (%)
	R_1	R_2		
12	H	H	α + β	28 (α), 45 (β)
13	H	CH_2OCH_3	β	65
14	CH_2OCH_3	H	α	78
15	CH_2OCH_3	CH_2OCH_3	β	71
16	CH_2OCH_3	$Si(iPr)_3$	β	92
17	$Si(iPr)_3$	$Si(iPr)_3$	β	51

[a]Data taken from Refs 79 and 80.

It is not clear why such effective stereochemical discrimination occurs during π-complexation of these cyclic enol ethers. The half-chair conformation[42,43] of these enol ethers leaves the olefinic bond relatively unencumbered sterically so that both faces of the ring are accessible to bulky reagents. It appears that π-complex formation is sensitive to the allylic oxy substituent and involves a stereoelectronic effect.[43,52-57] We have observed[71] that yields of palladium-mediated coupling reactions are higher when the allylic oxy substituent is pseudo-equatorial than when it is pseudo-axial. This suggests that π-complexation might be more favorable with pseudo-equatorial allylic oxygen in which the enol ether C–C double bond is somewhat less polarized.[43]

In contrast to six-membered ring cyclic enol ethers, stereoselectivity of π-complex formation between a five-membered ring enol ether and Pd(II) is readily rationalized on the basis of steric factors alone. All the available experimental results are consistent with π-complexation by attack of the Pd(II) reagent on the olefinic bond from the least sterically hindered face of the cyclic system. Data in Table 1 are illustrative and permit several conclusions to be drawn concerning the stereochemistry of complexation of five-membered ring enol ethers with an organopalladium reagent. First, unless the ring substituents are quite small, complete stereoselectivity is observed. Second, the substituent on the allylic carbon is more important than the more remote C-5 substituent in determining which face of the enol ether will experience attack by the organopalladium reagent. Thus, when the two substituents are comparable in size, complexation occurs from the face opposite the allylic (C-3) substituent (see **15**, **16** and **17**). Third, selective complexation of either face of the cyclic enol ether is readily achievable by manipulation of the relative steric bulks of the substituents on the two faces of the ring (compare **13** and **14**). Fourth, an unbranched substituent on the face of the ring experiencing attack provides little interference and permits high yield coupling (**15**, **16**) while a bulky, branched substituent diminishes the yield (compare **17** with **15** and **16**). Finally, hydroxyl substituents do not direct complex formation and their presence on the enol ether does not adversely affect the subsequent coupling reaction.

(2′R)-cis-5-[2′,5′-Dihydro-5′-[(methoxymethoxy)-methyl]-4′-(methoxy-me-thoxy)-2′-furanyl]-1,3-dimethyl-2,4-pyrimidinedione.[79] A mixture of Pd(OAc)$_2$ (90 mg, 0.4 mmol), 1,3-dimethyl-2,4-pyrimidindione-5-ylmercuric acetate **1**[169] (160 mg, 0.4 mmol) and acetonitrile (15 mL) was stirred 5 min. Then glycal **14** (102 mg, 0.4 mmol) was added; after an additional 10 min sodium bicarbonate (170 mg, 2 mmol) was added and the mixture was stirred for 24 h at room temperature. The reaction mixture was filtered through glass wool and the volatiles were removed *in vacuo*. Preparative thin layer chromatography (silica) using ether gave 97 mg (71%) of C-nucleoside, m.p. 116–117°C.

C. σ-Adduct Formation

1. Two Routes: Syn Migration and Anti Attack

Two mechanistically and stereochemically different routes have been established for the process π-complex → σ-adduct. Collapse of an enol ether-arylpalladium π-complex by insertion of the C–C double bond into a C–Pd bond (addition of arylpalladium across the double bond) gives rise to an adduct with Pd(II) σ-bonded to carbon (see, e.g. **3** → **4**). Abundant evidence has established that this route to a σ-adduct is a *syn* (four-center) process in which Pd and the transfered ligand form σ-bonds to the two respective olefinic carbons (reaction a).[12,13,22,62] In instances where an uncomplexed nucleophile is present in the reaction mixture, *anti* attack by the nucleophile on the palladium-activated double bond (reaction b) can occur, providing a second path from a π-complex to a σ-adduct.[81,82]

Selection between these two processes (reactions a and b) is dependent on the nature of the highest occupied molecular orbital (HOMO) of the nucleophilic species. Nucleophiles with low energy (HO^-, R_2NH) or intermediate energy ($Na^{+-}CH(COOEt)_2$, Cl^-) HOMOs add to a Pd(II)-complexed olefin by *anti* attack (path a) whereas those with high energy HOMOs (Ph^-, CH_3^-, H^-) bond to the Pd(II) center and add *cis* to the olefin (path b).[82-83]

Many examples of reactions involving *anti* attack by alcohols or carboxylic acids on Pd(II)-complexed enol ethers or enol esters are known.[22] Typical are reactions of ethyl vinyl ether with *n*-butanol in the presence of $(PhCN)_2PdCl_2$,[84,85] in which ether exchange occurs at $-40°C$; at room temperature a mixture of acetals is obtained.

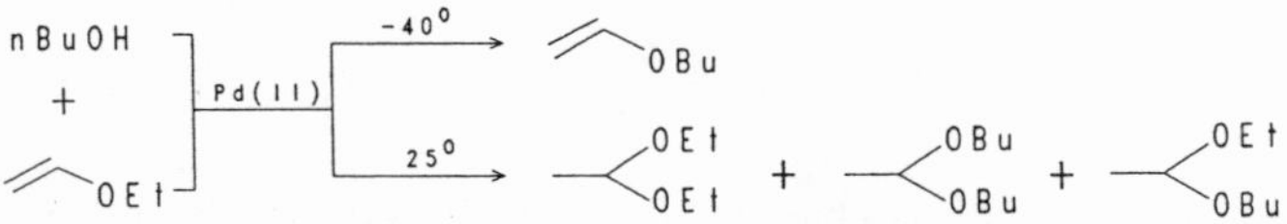

Most studies involving palladium-mediated reactions of enol ethers with carbon nucleophiles have utilized aryl, i.e. high energy (soft) carbanions, which undergo *syn* addition characteristic of the Heck reaction.[12,13] However, a few examples involving reactions of enol ethers with stabilized carbanions have been reported.[86] Reaction of 2,3-dihydrofuran with sodio diethylmalonate in the presence of Pd(II) gave a good yield of the C-1 substituted product **18**. The exocyclic double bond in **18** is strong evidence that the precursor σ-adduct results from attack of the malonate *anti* to palladium since adduct decomposition occurs by *syn* β-hydride elimination.[12,13,22,62] The corresponding product **19**[86] probably also involves *anti* attack of the carbanion on the π-complex of the enol ether but, in this case, the selectivity for β-hydride elimination is different.

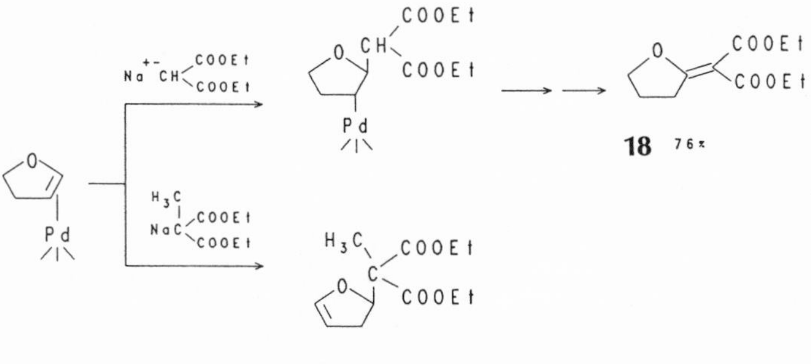

2. Regiochemistry of σ-Adduct Formation

Most significant about σ-adduct formation from a π-complex is that it is this step which determines the regiochemistry of the over-all olefin substitution reaction. An Ar–Pd–enol ether π-complex can collapse to give either of two σ-adducts (path a or path b) or a mixture containing both regioisomers. It is obvious that the utility of palladium-mediated coupling reactions of enol ethers for synthetic applications is dependent on attaining high regioselectivity. As a result, we and others have studied factors which influence regioselectivity in these reactions.

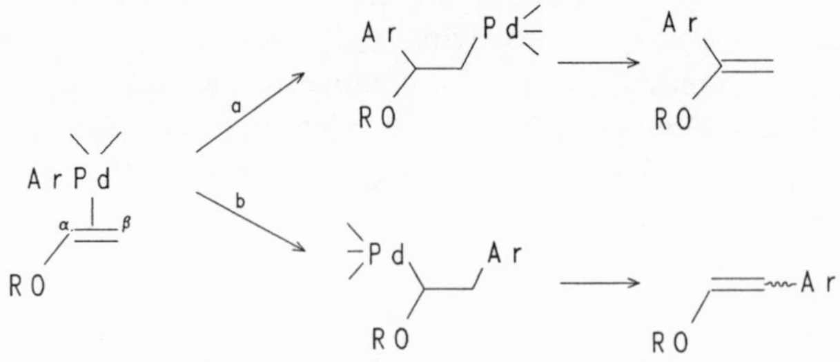

In our earliest studies[4,69] which involved cyclic enol ethers, we were pleased to discover that palladium-mediated coupling reactions occurred with completely regiospecific coupling at the α-carbon (path a). Attempts to extend this reaction to acyclic enol ethers gave very different results.[87] Whereas palladium-catalyzed reaction of iodobenzene with 3,4-dihydro(2-H)pyran gave a single coupled product (20) by regiospecific arylation of the enol ether α-carbon, reaction with either ethyl vinyl ether or vinyl acetate produced mixtures of products (21–23 and 24–27 respectively) resulting from arylation at both the α and β carbons. Similar results were obtained by others[8,88–90] in earlier, related studies.

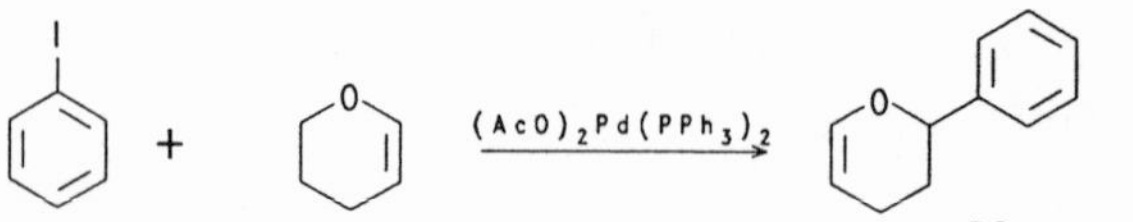

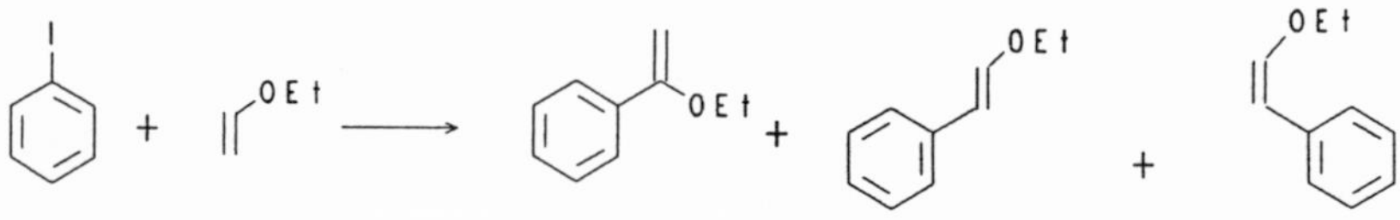

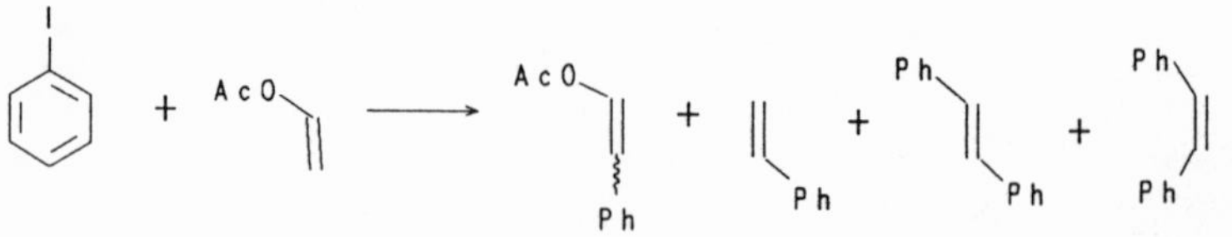

2-Phenyl-3,4-dihydro-2H-pyran (**20**).[87] A mixture of 2 g (10 mmol) of iodobenzene, 10 mL of 3,4-dihydro-2*H*-pyran, 1.5 g (15 mmol) of triethylamine and 0.07 g (0.01 mmol) of diacetatobis-(triphenylphosphine)-Pd(II) in a sealed tube was heated at 100°C for 3 h. The cooled reaction mixture was partitioned between water and chloroform. The chloroform solution was evaporated and the residue was dissolved in hexane and chromatographed on silica gel to yield 0.96 g (63%) of **20** as an oil.

Surprisingly, and unaccountably, palladium-catalyzed reactions of vinyl acetate with the iodopyrimidines **28** and **29** did not yield mixtures of products derived from regioisomeric σ-adducts, but rather, single products (**30** and **31** respectively) resulting from coupling of the pyrimidinyl moiety at the vinyl acetate α-carbon followed by loss of palladium acetate.[91]

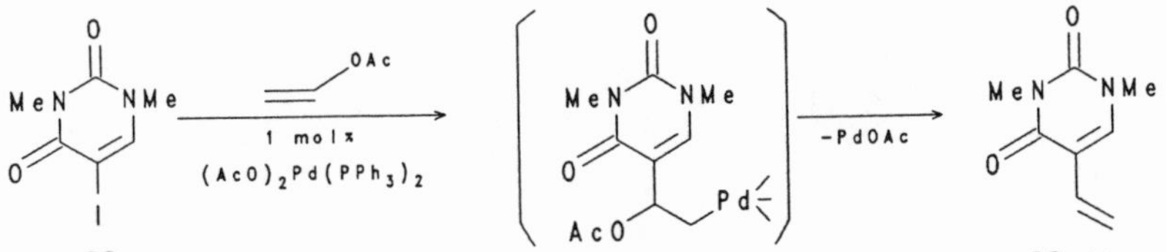

Study of factors which affect the regiochemistry of palladium-mediated reactions of enol ethers began with a report by Hallberg[92] involving arylation reactions of methyl vinyl ether in the presence of palladium acetate. In this report, it was noted that (a) an electron withdrawing group (NO_2) on aryl favored β-arylation, while an electron donating group (OMe) resulted in exclusive α-arylation and (b) the presence of triphenylphosphine in the reaction mixture reduced the yield of β-aryl product significantly. These results are in accord with studies by Heck[93-97] in which an effect of electron density in the aryl ring on regiochemistry of olefin arylation was noted. Following further demonstrations of the utility of the nitroaryl system for regioselective β-arylation of acyclic enol ethers[98,99] a more extensive and systemmatic study[14] was undertaken to determine which reaction variables affected regiochemistry of palladium-mediated enol ether arylations. Results from this and related studies are summarized in Tables 2–8 and in the accompanying reaction schemes.

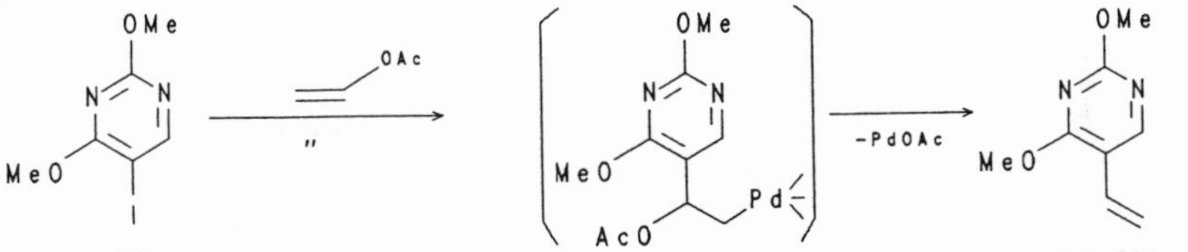

Table 2. Regio-chemistry of Enol Ether
Arylation: Effect of Arylating Agent Structure.[a]

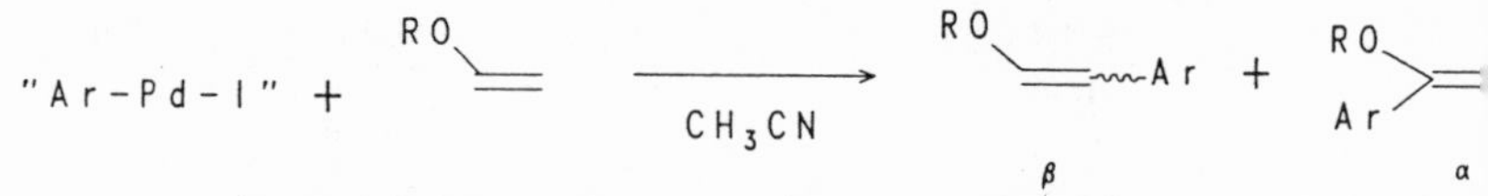

Ar	Ratio β/α
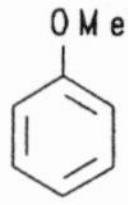 OMe (phenyl ring)	only α
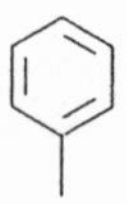 (methyl-substituted phenyl ring)	0.5
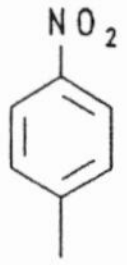 NO$_2$ (nitro-substituted phenyl ring)	5

[a]Data taken from Refs 14, 87 and 92; see also Refs
98–100.

As noted, the electron density of the aromatic ring in the arylating agent
'Ar–Pd–X' strongly affects the regiochemistry of enol ether arylation (Table
2). When the aryl group is phenyl,[87] little regioselectivity is observed in
arylation of acyclic enol ethers. However, addition of an electron-donating
group (OMe) on the phenyl ring permits regiospecific α-arylation to be
achieved whereas addition of an electron-withdrawing para nitro group
results in a five to one selectivity for β-arylation (Table 2).

Significant effects on enol ether arylation regiochemistry result from variation of the reaction medium-catalyst system (Table 3). The data establish that
catalytic arylation of n-butyl vinyl ether using either iodobenzene or 4-nitro-
iodobenzene occurs readily in the presence of Pd(II) acetate or chloride or
using soluble (Pd(PPh$_3$)$_4$) or insoluble (palladium on carbon) Pd(0) catalysts.
Arylation regiochemistry is insensitive to which form of catalyst is used,
particularly if reactions are carried out in a solvent (acetonitrile) which
coordinates palladium. When a poorly coordinating solvent (toluene) is used,
the ratio of β to α arylation increases; similarly, addition of triphenylphos-

Table 3. Regiochemistry of Enol Ether Arylation: Effect of Reaction Medium–Catalyst System[a]

Arl	Catalyst	Solvent	Yield (%) β	Yield (%) α	Ratio β/α
Phl	Pd(OAc)$_2$	CH$_3$CN	31	56	0.5
	PdCl$_2$	CH$_3$CN	32	61	0.5
	Pd/C	CH$_3$CN	34	52	0.6
	Pd(PPh$_3$)$_4$	CH$_3$CN	42	53	0.8
	Pd(OAc)$_2$/2PPh$_3$	CH$_3$CN	38	60	0.6
	Pd/C	toluene	37	24	1.5
p–O$_2$N–Phl	Pd(OAc)$_2$	CH$_3$CN	83	17	5.0
	Pd/C	CH$_3$CN	84	16	5.0
	Pd/C/2PPh$_3$	CH$_3$CN	70	28	2.5
	Pd/C	toluene	78	10	7.8

[a]Data taken from Ref 14; 1 mol% catalyst, 16 h at 100°C; yields determined by gas chromatography.

phine to the reaction mixture as a coordinating ligand decreases β-arylation. The pattern which emerges (Table 3) is that β-arylation is favored by poorly coordinated palladium while α-arylation is enhanced when a coordinating solvent or added ligand is present. Three to five-fold changes in the β/α arylation ratio are achievable by adjustment of these variables. When the effects of an electron deficient aryl group (Table 2) are combined with use of a non-coordinating reaction medium-catalyst system selectivities for β-arylation as high as eight to one are achieved (Table 3).

The nature of the anion bound to palladium in the arylpalladium reagent 'Ar–Pd–X' also has an important effect on enol ether arylation regiochemistry (Table 4). When *n*-butyl vinyl ether was arylated using preformed *p*-nitrophenylpalladium reagents in which the anion (X) was I, Br and Cl,[14] the observed ratios for β/α arylation were 1.0, 4.1 and 10.2, respectively. The regiochemical product ratios (Table 4) correlate with the relative strengths of Pd(II)–X bonds which are in the order I > Br > Cl ≫ OAc[22] and further support the view that β-arylation is favored by electron deficient palladium. Yields for reactions involving *p*-nitrophenylpalladium reagents (Table 4) were diminished because a competing side reaction occurred involving migration of phenyl from a triphenylphosphine ligand to palladium followed by reaction of this newly formed 'Ph–Pd–X' reagent with enol ether.[14] This

Table 4. Regiochemistry of Enol Ether Arylation: Effect of Anion (X) in Reactions of "Ar–Pd–X" with n-Butyl Vinyl Ether[a]

"Ar–Pd–X"	Solvent	Yield (%)		ratio
		β	α	β/α
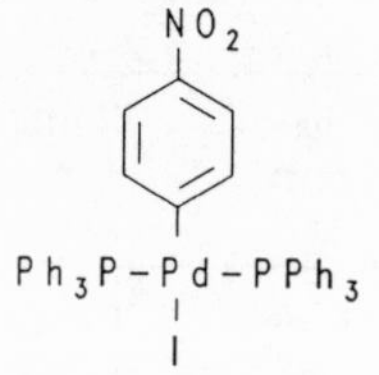 NO$_2$ (ring) Ph$_3$P–Pd–PPh$_3$ I	toluene	20	21	1.0
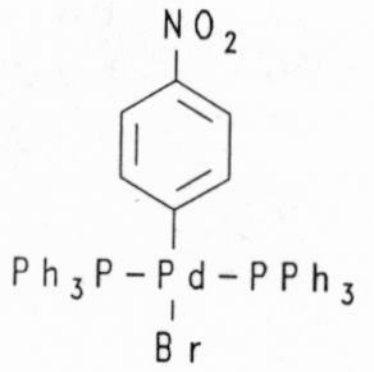 NO$_2$ (ring) Ph$_3$P–Pd–PPh$_3$ Br	toluene	37	9	4.1
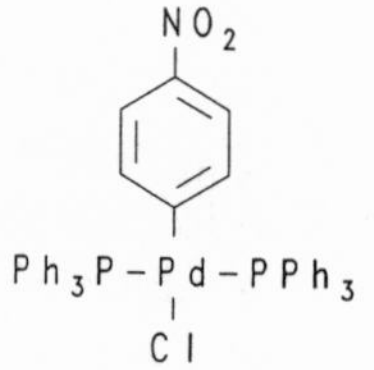 NO$_2$ (ring) Ph$_3$P–Pd–PPh$_3$ Cl	toluene	41	4	10.2
 (ring) L–Pd–L Cl	CH$_3$CN	13	53	0.2
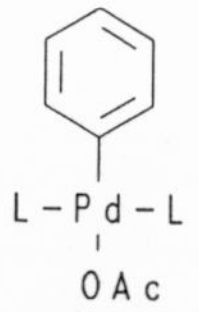 (ring) L–Pd–L OAc	CH$_3$CN	44	50	0.9

[a]Data taken from Ref. 14; stoichiometric Pd; reactions in toluene run 4h at 100°C; reactions in acetonitrile run 4h at 25°C.

Table 5.   Regiochemistry of Enol Ether Arylation: Effect of Added Salt on Arylations with Aryl Triflates[a]

Catalyst system	Solvent	Aryl product yield (%)	Ratio β	α
1% Pd(OAc)$_2$	CH$_3$CN	4	0.5	1
1% Pd(OAc)$_2$	DMF	82	1.5	1
1% Pd(OAc)$_2$ + 100% Bu$_4$NCl	CH$_3$CN	57	8	1
1% Pd(OAc)$_2$ + 200% LiCl	DMF	68	13	1

[a]Data taken from Ref. 33.

reaction, which involves the replacement of p-nitrophenyl by the more electron-rich phenyl moiety in the arylpalladium reagent, emphasizes the electron deficient nature of palladium in these reagents.

Andersson and Hallberg[33] used the strategy of electron depletion at palladium to enhance β-arylation regioselectivity in selection of aryl triflates as precursors for arylpalladium reagent formation[30-33] (Table 5). This study[33] produced the highest ratios of selectivity for β-arylation of enol ethers ($\beta/\alpha = 13$, 14) reported to date. Achievement of good yields of coupled product required use of dimethylformamide (DMF) as solvent rather than acetonitrile and addition of a halide salt. In the absence of halide ion yields were unacceptably low.[3] Arylation regioselectivity was also low; this was unexpected since, as a very poor ligand for palladium, triflate was expected to produce high β-regioselectivity. Echavarren and Stille[101] suggested that triflate is so poor as a ligand that it fails to support palladium-mediated reactions unless a more effective ligand is provided. Only when the reaction of aryl triflates and alkyl vinyl ethers was carried out in dimethylformamide with chloride or bromide ion present were impressive regioselectivities attained (Table 5).

Another effective strategy for β-arylation[28] involves preparation of the 'Ar–Pd–Cl' reagent from the corresponding aroyl chloride[26] since only a few successful examples of arylpalladium reagent formation from aryl chlorides have been reported.[23,24] This indirect technique for arylpalladium chloride reagent formation permitted Andersson and Hallberg[28] to achieve regioselective β-arylation of enol ethers in good yield (Table 6).

Table 6. Regiochemistry of Enol Ether Arylation: Formation of "Ar-Pd-Cl" from Aroyl Chorides[a]

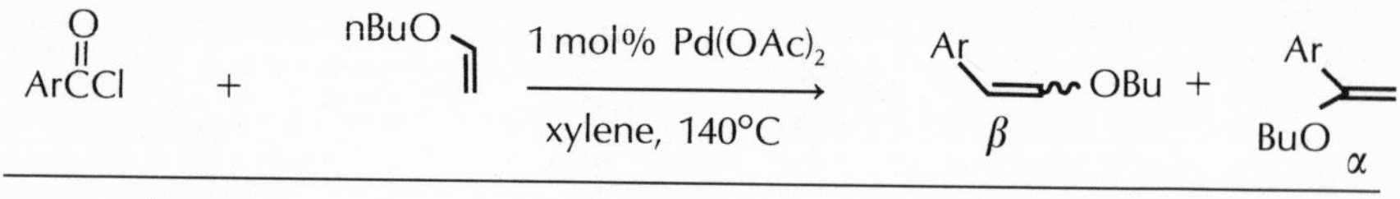

Ar	Ratio β/α	β-Aryl product yield (%)
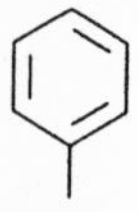 (phenyl)	3	53
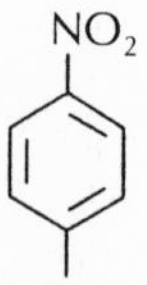 (4-NO$_2$-phenyl)	10	60
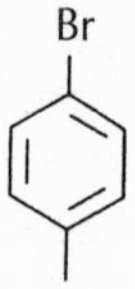 (4-Br-phenyl)	2.5	55
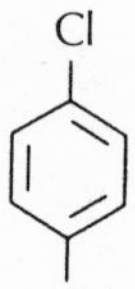 (4-Cl-phenyl)	3.0	60
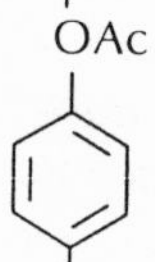 (4-OAc-phenyl)	2.7	40
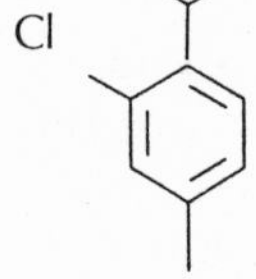 (2-Cl-biphenyl)	3.7	43
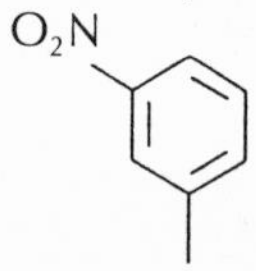 (3-O$_2$N-phenyl)	4.1	44

Table 6   (Continued)

$$\underset{ArCCl}{\overset{O}{\|}} \quad + \quad \underset{\|}{\overset{nBuO}{\diagdown}} \quad \xrightarrow[\text{xylene, } 140°C]{1 \text{ mol\% Pd(OAc)}_2} \quad \underset{\beta}{Ar\diagdown\!\!=\!\!\sim\!\!OBu} \quad + \quad \underset{BuO}{\overset{Ar}{\diagdown}}\!\!\underset{\alpha}{=}$$

Ar	Ratio β/α	β-Aryl product yield (%)
AcO—⟨benzene ring⟩		trace
O₂N—⟨benzene ring⟩		trace

^a Data taken from Ref. 28.

β-Arylation of n-butyl vinyl ether using aroyl chlorides.[28] The reactions were carried out on a 20 mmol scale in a 50-mL flask equipped with a condenser with a drying tube. A mixture of palladium acetate (0.045, 0.2 mmol) and *N*-ethyl-morpholine (2.8 g, 24 mmol) in 20 mL of xylene was stirred until solution was attained. Then *n*-butyl vinyl ether (4 g, 40 mmol) and 20 mmol of the appropriate benzoyl chloride was added and the mixture was heated under reflux. After 3 h, the reaction mixture was cooled and 50 mL of ether was added to precipitate the amine hydrochloride. The resulting crude product was separated by flash chromatography. Product ratios and yields are contained in Table 6.

This work has been extended to synthesis of a series of aroylated vinyl ethers.[29] Reaction of an aroyl chloride (**32**) with enol ethers and palladium in refluxing xylene (140°C) produces β and α arylated products (Table 6). Lowering the reaction temperature to 50–70°C permits formation of aroyl-palladium reagent **33** by oxidative insertion of Pd(0) into the carbonyl-chlorine bond of the aroyl chloride; however, at these temperatures decarbonylation to the derived arylpalladium reagent **34** does not occur.

At the lower reaction temperature, the initially formed aroylpalladium reagent (**33**) undergoes efficient and regiospecific β-coupling with formation of β-aroylated vinyl ethers (masked aryl 1,3-dicarbonyl compounds, **35**). This aroylation process is the first completely β-selective palladium-mediated coupling reaction of enol ethers to be developed; no product of α-coupling is observed.[29] Several (*E*)-3-*n*-butoxy-1-arylpropene-1-ones (**36–40**) have been prepared in this way.[29]

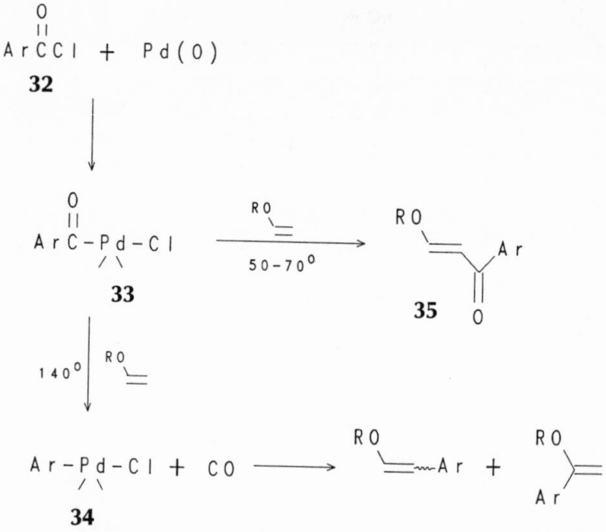

β-Aroylation of alkyl vinyl ethers.[29] Palladium acetate (0.1 mmol), an alkyl vinyl ether (5–10 mL) and triethylamine (12 mmol) were placed in a 50-mL thick-walled tube fitted with a teflon-lined screw cap and stirred until a clear orange solution was attained (10 min). Then the appropriate aroyl chloride (10 mmol) was added, the tube was closed and heated at 60–70°C for 24 h. The resulting black slurry was diluted with 100 mL of ether and filtered to remove triethyl-amine hydrochloride; the resulting solution was concentrated and the product isolated by flash chromatography on silica using ether/pentane.

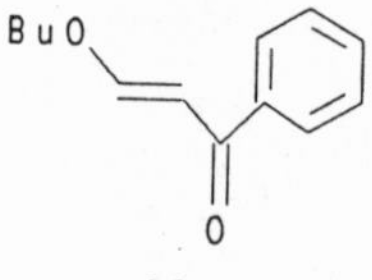

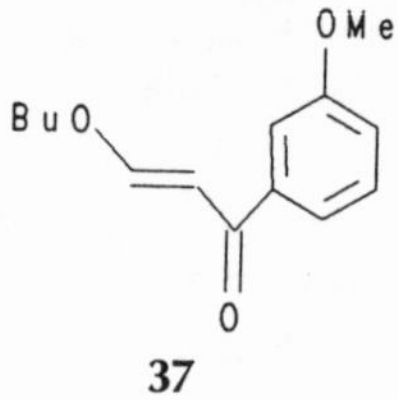

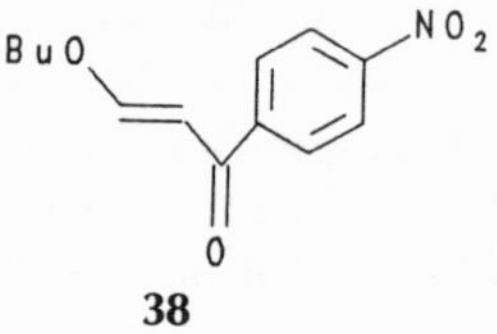

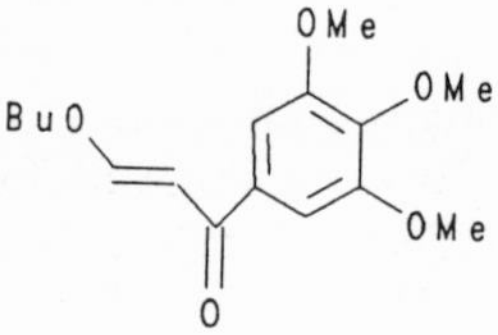

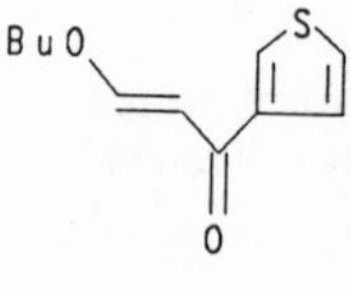

Table 7. Regiochemistry of Enol Ether Arylation: Effect of Enol Ether Structure[a]

		Yield %		Ratio
Enol ether	Arl	β	α	β/α
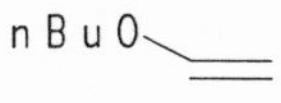 n B u O— (vinyl)	PhI	34	52	0.6
	p–O$_2$N–PhI	84	16	5.0
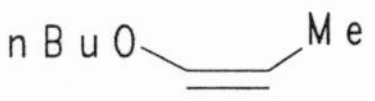 n B u O—Me	PhI	17	83	0.2
	p–O$_2$N–PhI	37	32	1.2
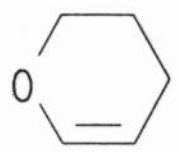 O (dihydropyran ring)	PhI	0	76	–
	p–O$_2$N–PhI	0	16	–
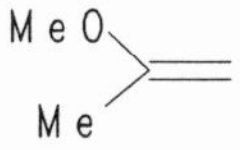 Me O / Me (isopropenyl)	PhI	0	low	–
	p–O$_2$N–PhI	0	low	–

[a]Data taken from Ref. 14; 1 mol% Pd/C in CH$_3$CN at 100°C, 16–24 h.

Perhaps the most dramatic effects on regiochemistry of palladium-mediated enol ether arylation results from changes in the structure of the enol ether (Table 7). Regioisomeric mixtures result when simple, acyclic enol ethers (methyl vinyl ether,[92] ethyl vinyl ether[87] and *n*-butyl vinyl ether[14]) are arylated. Substitution of an alkyl group at the β olefinic carbon increases α-arylation and incorporation of a β-substituent into a ring completely eliminates β-arylation and results in regios(pecific α-arylation (Table 7).[69,87,100] Surprisingly, an alkyl group on the enol ether α-carbon also supresses β-arylation.[14,100] This latter result is completely inconsistent with the concept that steric factors govern arylation regioselectivity.[12,13,22,62]

Palladium-mediated arylations of exo-methylene carbohydrates **41**[102] and **45**[78] have produced low yields of β-aryl products (**43** and **46**, respectively). The fact that the major product formed in arylation of **41** was the α-aryl isomer **44** in which the new C–C bond is formed at the sterically conjested disubstituted olefinic carbon of **41** is further evidence for the dominance of electronic factors in determining the regiochemistry of these arylation reactions.

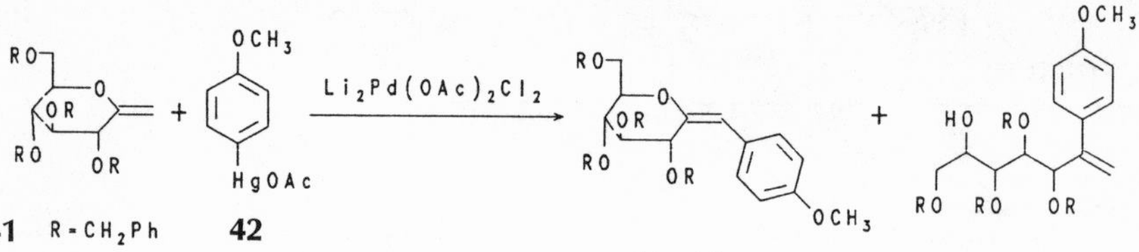

RO—⟨OR, RO, OR⟩ + [OCH₃ aryl HgOAc] — Li₂Pd(OAc)₂Cl₂ → RO—⟨OR, RO, OR⟩=CH–aryl–OCH₃ + HO RO RO RO OR (aryl OCH₃)

41 R=CH₂Ph   **42**

43 13%   **44** 49%

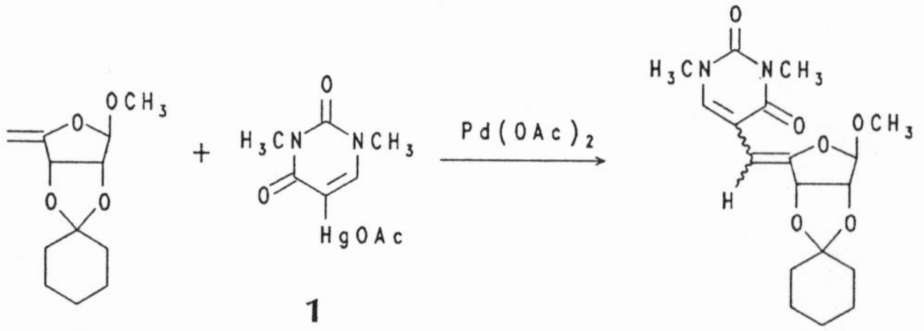

[45 structure] OCH₃ + H₃CN–NCH₃ (1) HgOAc — Pd(OAc)₂ → **46** 9%T

45   **1**

Reaction of 3,4-dihydro(2-*H*)pyran with a vinyl triflate (**47**) in the presence of catalytic Pd(PPh₃)₄ yielded the corresponding α-coupled product **48**.[103] Thus, even preparation of a highly electron-deficient organopalladium reagent by use of a triflate precursor (Table 5) did not achieve β-coupling to a cyclic enol ether.

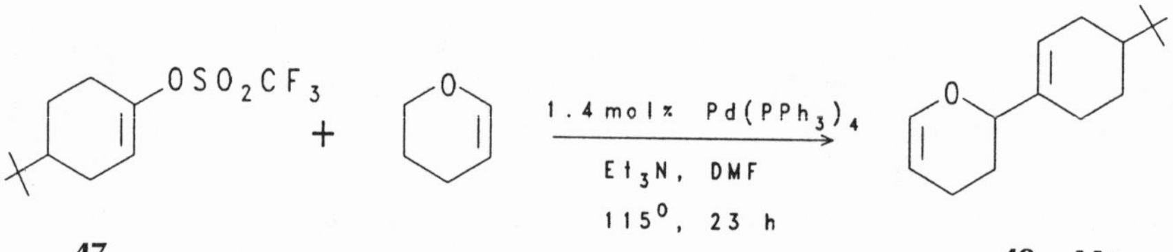

[47 structure] OSO₂CF₃ + [O pyran] — 1.4 mol% Pd(PPh₃)₄, Et₃N, DMF, 115°, 23 h → **48** 53%

47   **48** 53%

The failure of steric factors to account for the regiochemistry observed in palladium-mediated reactions of enol ethers is further made evident by the data contained in Table 8.[100] These data show that steric crowding of the site of the coupling reaction by α-substitution of a cyclic enol ether or by placing a substituent on the arylating agent adjacent to the reactive *ipso* carbon does not lead to β-arylation. Indeed, even when both these steric effects are involved in a single reaction, only α-arylation is observed. Although the yields of coupled products diminish as a result of steric interference at the reaction sites, regiospecific α-arylation occurs.

Table 8.   Effect of Steric Hinderance at the Reaction Site[a]

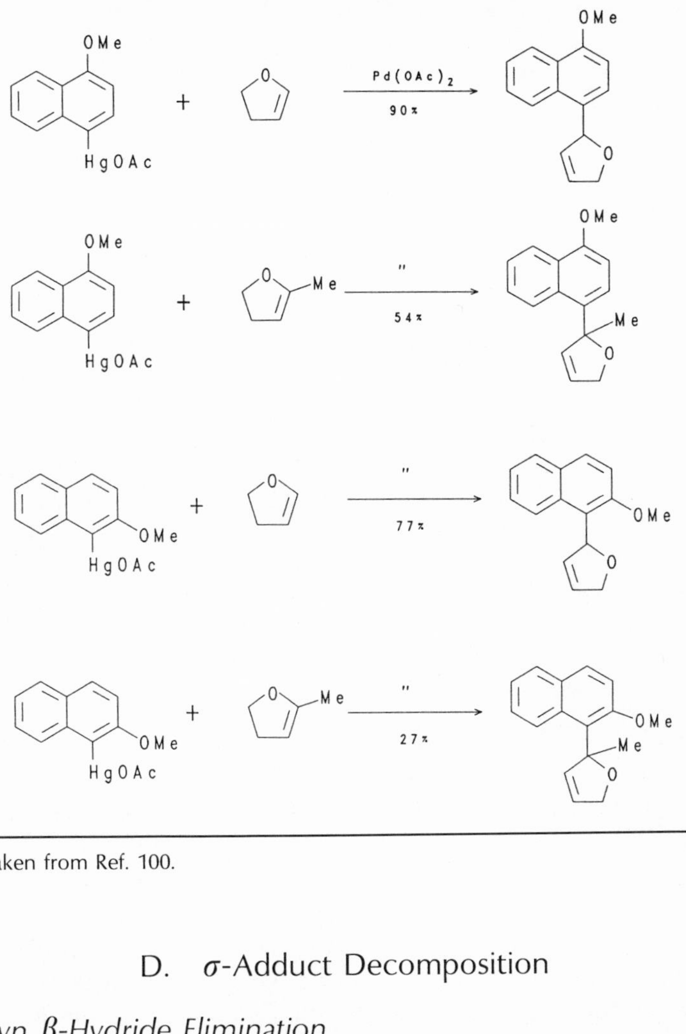

[a]Data taken from Ref. 100.

D.   σ-Adduct Decomposition

1.   Syn β-Hydride Elimination

Formation of enol ether arylation products occurs by σ-adduct decomposition with elimination or replacement of palladium. In Heck reactions involving simple olefins[12,13] decomposition involves loss of palladium and β-hydrogen.[12,13,22,62] This 'β-hydride' elimination is a *syn* process which occurs via a well studied[82,83,104–109] four-center transition state. *Syn* β-hydride elimination occurs via an intermediate π-complex which can reform the same or an isomeric σ-adduct in a series of equilibria.[93] In this way double bond migration can occur[69] as shown in Scheme 2.[100]

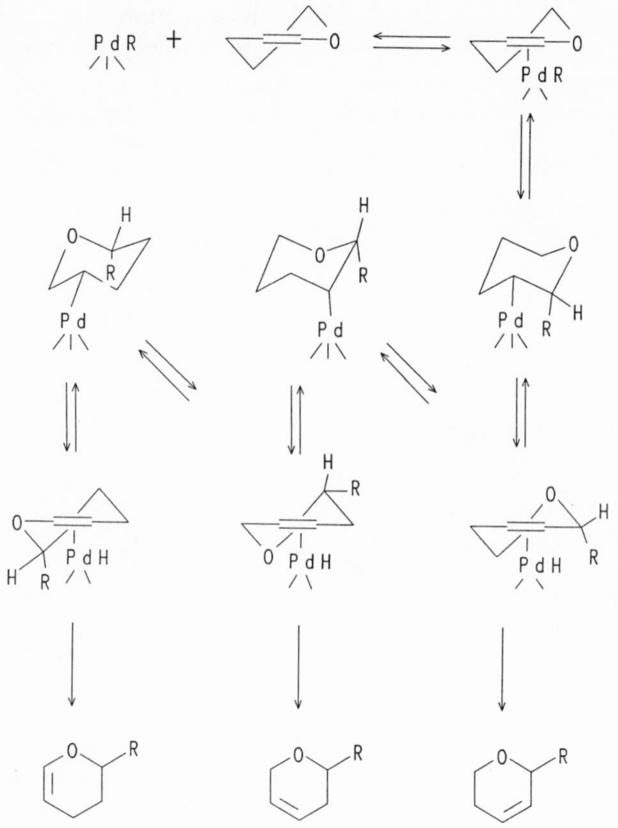

Scheme 2. Double bond migration in σ-adduct decomposition by *syn* β-hydride elimination.

Decomposition of the π-complex formed upon β elimination of 'Pd–H' from the σ-adduct to yield uncomplexed olefin is irreversible since reductive elimination of the free 'Pd–H' thus formed occurs producing Pd(0) (required in the catalytic cycle for oxidative addition to Ar–X to form the arylpalladium reagent).

$$\text{H–Pd(II)–X} \rightarrow \text{Pd(0)} + \text{HX}$$

As a result, elimination of 'Pd–H' followed by readdition to the opposite face of the olefinic double bond does not occur. Thus, in cyclic enol ether arylation reactions (Scheme 2), β-hydride elimination does not form a product with a double bond in conjugation with the aryl substituent.[14,100]

In a few instances, palladium-mediated reactions of enol ethers have yielded products which cannot be rationalized by assuming σ-adduct formation and subsequent β-hydride elimination. Reaction of benzo[*b*]furan (**49**) with an arylmercuric salt in the presence of palladium(II) results in regiospecific α-arylation with formation of **50**.[110]

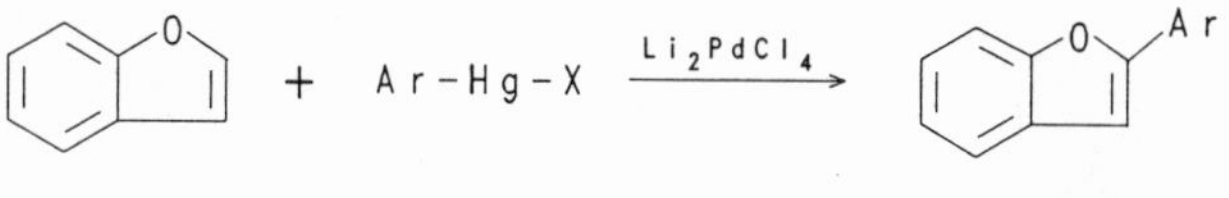

49   **50**

In this reaction, formation of **50** by decomposition of a precursor σ-adduct formed by *syn* addition of Ar–Pd to the enol ether double bond would require *anti* β-hydrogen elimination. It is doubtful that this occurs; it seems more likely that the reaction proceeds by an alternative mechanism like that established for coupling of two aryls to form biphenyls.[12,13,22]

$$\text{Ar–Pd–X} + \text{Ar'–H} \rightarrow \text{Ar–Ph–Ar'} + \text{HX} \rightarrow \text{Ar–Ar'} + \text{Pd(0)}$$

Arylation of 4-chromanone enol acetates (**51**) in the presence of palladium acetate[111] yielded the corresponding isoflavonone product **52**. To rationalize this result the authors[111] have postulated elimination of an acyl palladium species.

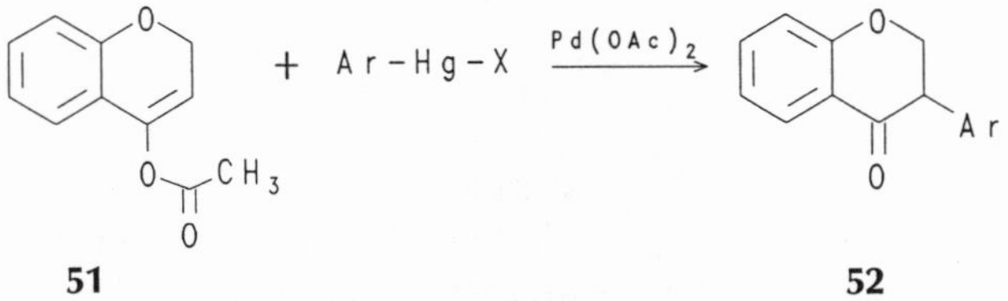

51   **52**

σ-Adducts derived by β-arylation of acyclic enol ethers decompose with formation of mixtures of ($E$$E$) and ($Z$) olefinic products. For decomposition to occur by β-hydride elimination, the initially formed σ-adduct must undergo a bond rotation of about 120° (in either direction) to bring a β-hydrogen into the syn periplanar relationship with palladium necessary for elimination.[82,83,104-109]

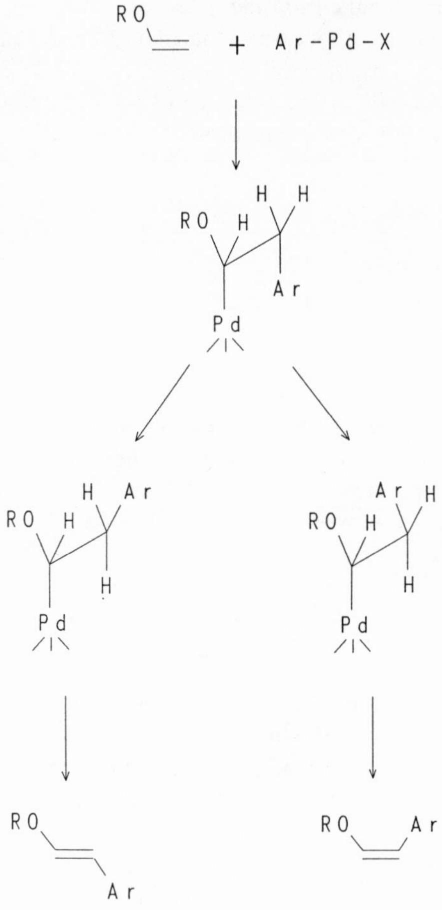

Hallberg[14,28] has shown that the ratios of (E) to (Z) olefins formed in palladium-catalyzed enol ether arylation reactions vary with reaction conditions. (E)/(Z) ratios vary from about 0.5 to 2.5 and are sensitive to (a) the electron density of the aryl group, (b) the presence or absence of added triphenylphosphine ligands for palladium and (c) the nature of the halide ion present in the reaction mixture. Control experiments[14] have shown that the (E)/(Z) ratios observed result from the primary σ-adduct decomposition reaction and not from later isomerization processes, presumably because no soluble Pd(II) is present in the reaction mixtures.[58] At present, the underlying reasons for the different ratios observed are not understood since, *a priori*, bias in the direction of C–C bond rotation which is sensitive to reagent structure and reaction conditions was not expected.

2. Other σ-Adduct Decomposition Modes

When more complex enol ethers, such as carbohydrate-derived glycals, are involved, processes for σ-adduct decomposition other than β-hydride elimination are often observed. The isolation[74] and characterization[72] of the σ-adduct **(4)** resulting from reaction of (1,3-dimethyl-2,4-tetra-hydropyrimidinedione-5-yl)mercuric acetate **(1)** with 3,4,6-tri-*O*-acetyl-D-glucal **(2)** in the presence of palladium acetate and lithium chloride followed by addition of triphenylphosphine provided an excellent opportunity to study adduct decomposition reactions. Four different products were obtained by selection of the reaction conditions for decomposition of σ-adduct **4**.[71,74] In each case, a single product was formed in nearly quantitative yield.[74]

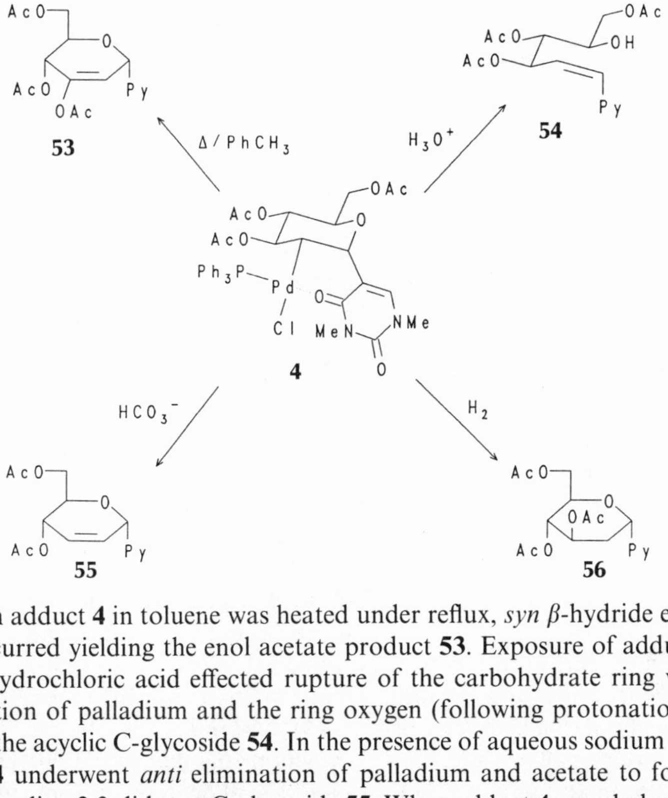

When adduct **4** in toluene was heated under reflux, *syn* β-hydride elimination occurred yielding the enol acetate product **53**. Exposure of adduct **4** to dilute hydrochloric acid effected rupture of the carbohydrate ring via *anti* elimination of palladium and the ring oxygen (following protonation) producing the acyclic C-glycoside **54**. In the presence of aqueous sodium bicarbonate, **4** underwent *anti* elimination of palladium and acetate to form the corresponding 2,3-dideoxy C-glycoside **55**. When adduct **4** was shaken under two atmospheres of hydrogen, palladium was replaced by hydrogen yielding **56**.[74]

The preparation and isolation of σ-adduct **4** depended on the presence of chloride ion to stabile the adduct sufficiently so that it accumulated in the reaction mixture. In related reactions, the σ-adduct formed by coupling of **1** and **2** was decomposed in the reaction mixture by treatment with hydrogen sulfide[4,73] or was permitted to decompose as formed,[71,73] owing to the availability of only poorly coordinating[63] acetate anions. Under these conditions, mixtures of products resulted since several decomposition reactions of the intermediate σ-adduct occurred at similar rates.[71] Included in the mixture of products isolated[71] under these conditions was an enol acetate product **57** not seen when **4** was decomposed under controlled conditions.[74] Formation of enol acetate **57**, isomeric with **53** which was formed by decomposition of **4** under controlled conditions,[74] requires acetate migration to the Pd-bearing carbon;[112] the mechanism of this process is not clear.

57

The specific reaction conditions used in these experiments affected the relative amounts of the several σ-adduct decomposition products which formed. If only acetate anions are present in the reaction mixture the product of ring rupture **54** was not observed.[71,73] Interestingly, if only chloride anions were present the coupling reaction was completely suppressed[73] and pyrimidine dimerization took place.[113]

The σ-adduct, **4**, is stable in the solid state and moderately stable in solution at room temperature although metallic palladium is formed over a period of hours.[72] This stability is impressive in view of the presence of a *cis* β-hydrogen,[12,13,22,62] the weakness of the oxygen-palladium bond[63] in the unusual six-membered ring metallocycle[114,115] and the multiple modes of decomposition available. The stability of **4** is suggestive that σ-adduct decomposition reactions require in the respective transition states, precise geometric alignment of bonds to be broken.

Another process for decomposition of a σ-adduct has been observed in the furanoid glycal (five-membered ring enol ether) series.[83,117] The σ-adduct (**59**) formed from palladium-mediated reaction of the pyrimidinylmercuric acetate **1** with an O-5 protected furanoid glycal **58** derived from ribonolactone[116] was unusually stable and did not decompose spontaneously in the reaction mixture.[117] The origin of this stability is apparent; **59** does not possess a *syn* β-hydrogen or an *anti* β-acetate and the *anti* periplanar arrangement of Pd and ring oxygen is not accessible in a five-membered ring as it is in six-membered rings.[73]

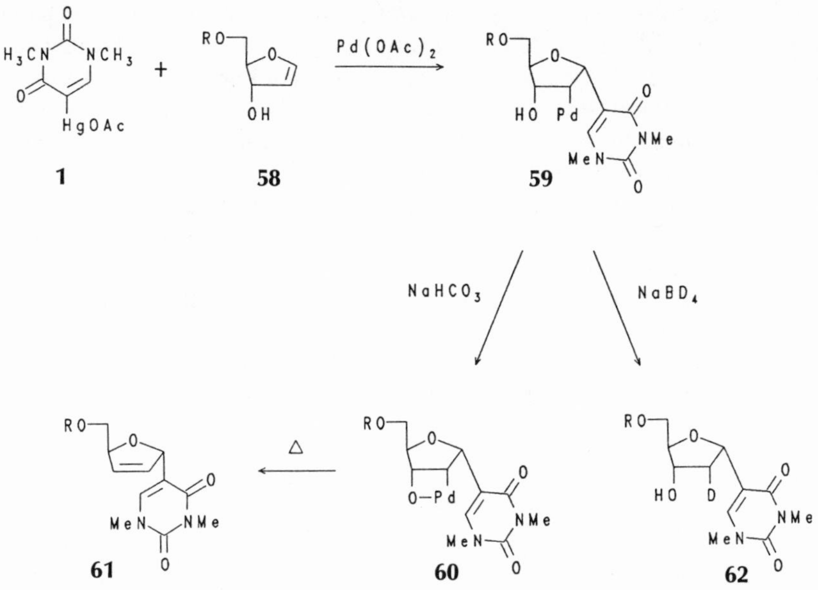

Decomposition of σ-adduct **59** was accomplished by warming or by adding sodium bicarbonate to the reaction mixture.[117] The sole coupled product isolated was **61**, formed by elimination of palladium and the *cis* β-hydroxyl group. The elimination reaction is formulated[117] as occurring via an intermediate palladaoxacyclobutane (**60**) since the reaction is base-assisted. Similar mechanistic suggestions have been made to account for related organometallic reactions.[118-121] Decomposition of σ-adduct **59** was also accomplished by shaking the reaction mixture under a hydrogen atmosphere. In this way, the corresponding 2-deoxy-α-pseudouridine derivative was produced; if deuterium was used instead of hydrogen, the deutero analog **62** was formed, establishing the hydrogen replacement reaction as *syn*.[22]

3. Strategies for Control of σ-Adduct Decomposition Modes

Utilization of enol ether arylation reactions in synthesis requires that formation of product mixtures in σ-adduct decomposition reactions be avoided, or at least minimized. Three strategies for accomplishing this goal in palladium-mediated reactions of structurally complex glycals have been devised. If the glycal lacks an allylic (C-3) oxy-substituent (see enol ether **9**),[78] σ-adduct decomposition must occur solely by β-hydride elimination. Unfortunately, since the allylic oxy-substituent is needed to direct stereochemistry of π-complex formation, use of this strategy for controling σ-adduct decomposition led to mixture of products derived from stereoisomeric π-complex formation.[78a]

σ-Adduct decomposition can be controlled by selection of the leaving group ability of the allylic (C-3) oxy substituent present on the glycal. Thus, palladium-mediated reactions of organomercurial 1 with glycals 15 or 16 which possess poor leaving groups at C-3, yielded, in each case, single products formed by σ-adduct decomposition by β-hydride elimination.[122] However, if the C-3 oxy substituent is a good leaving group, i.e., a carboxylate, 65, the directive effect on stereochemistry of π-complex formation was retained but σ-adduct decomposition now occured solely by *anti* loss of palladium and the carboxylate group forming 66.[122] The yield of 66 (20%) was low owing to competitive decomposition of 65 to the corresponding furan.[123] These preliminary results indicate that, when suitable procedures are developed for preparation and handling of furanoid glycals with good leaving groups at C-3, palladium mediated coupling of properly designed glycals will permit facile preparations of either 2-deoxy- or 2,3-dideoxy-C-glycosides.

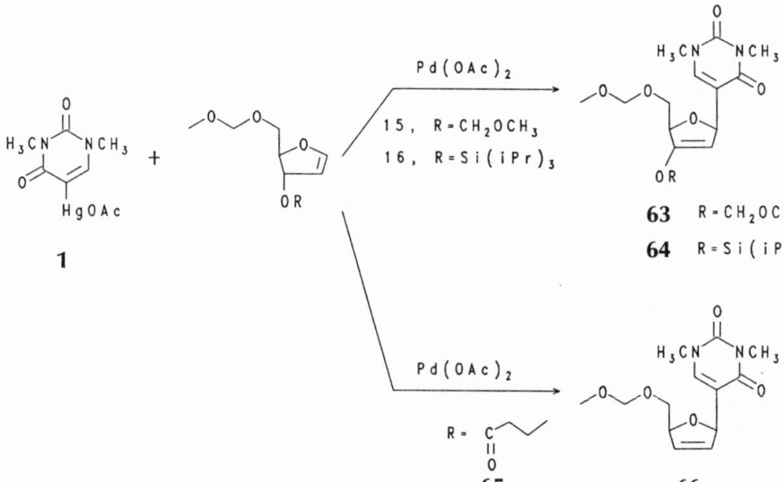

The selective decomposition of the intermediate σ-adduct by anti loss of palladium and carboxylate is particularly gratifying since prediction of preferred conformations in five-membered rings is exceptionally difficult. In five-membered rings, it is typical that multiple, essentially isoenergetic conformations coexist in solution[124] and the nature, number and positions of substituents affect conformational preferences.[125] σ-Adduct decomposition to yield products 63 or 64 and 66 are assumed to involve conformations A and A' respectively. In conformation A, which has the large groups at C-1 and C-4 in favorable pseudoequatorial positions, palladium and β-hydrogen are essentially *syn* periplanar as required for β-hydride elimination. The agostic interaction between palladium and hydrogen is an added factor in stabilizing

conformation A.[105] These favorable factors account for the facility of the
β-hydride eliminations leading to **63** and **64**.[79,80,122] In the σ-adduct formed
from furanoid glycal **65**, the effectiveness of carboxylate as a leaving group
diverts the reaction to palladium carboxylate elimination with formation of
66 which, presumably, requires a conformational change to **A'** in which
palladium and carboxylate are *anti* periplanar.[122]

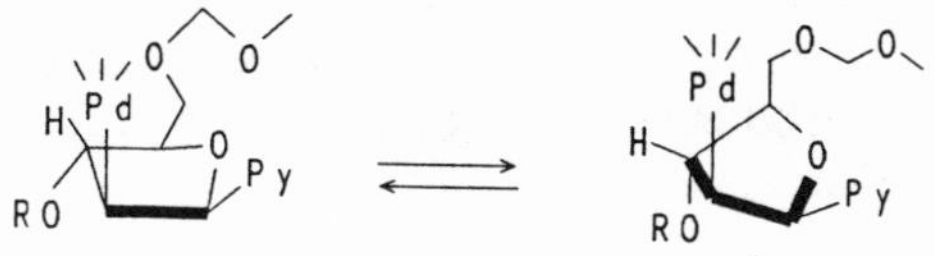

| A | A' |

Similar results were obtained with six-membered ring enol ethers. Palladium-mediated reaction of **1** with pyranoid glycal **67** resulted in formation
of a σ-adduct which decomposed by both *anti* palladium acetate elimination
and *syn* β-hydride elimination to give a mixture of products (**68** and **69**,
respectively).[122] When the C-3 substituent was changed to a poor leaving
group (trialkylsiloxy **70**), only β-hydride elimination occurred yielding **71** as
the sole product of the coupling reaction.[122]

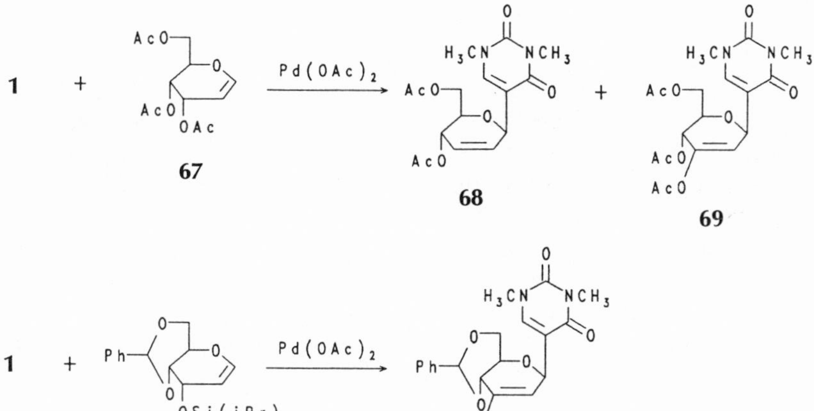

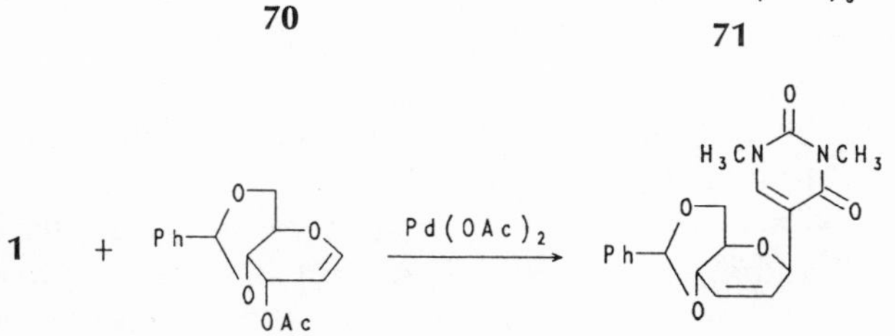

Glycal **70** possesses not only a poor leaving group at C-3 but also conformation-restricting cycloderivatization of the C-4 and C-6 oxygens. Such conformational restriction makes possible selective σ-adduct decomposition by *anti* palladium acetate loss (compare **72**→ **73** with **67** → **68** + **69**).[122] Conformation **B** has palladium and the C-3 oxy substituent in the *anti* periplanar orientation necessary for elimination and, when R = acetyl β-hydride elimination is not seen. However, if R = alkyl or trialkylsilyl the resulting C-3 substituent is such a poor leaving group that σ-adduct decomposition occurs by the alternative, high energy conformation **B'** in which palladium and C-3H are *syn* periplanar.[4,73,122]

IV. CONCLUSIONS

Detailed investigation of the mechanisms of the discrete, sequential steps involved in palladium-mediated reactions of enol ethers, recognition of factors which influence the stereo- and regiochemistry of these component steps and development of strategies to maintain control throughout the process has led to a procedure for C–C bond formation with impressive synthetic utility. The reaction is general; it proceeds as readily with complex carbohydrate-derived enol ethers as with the simplest alkyl vinyl ethers. Structurally complex organopalladium reagents are effective in enol ether arylation. The development of procedures which permit regioselective arylation at either the α- or β-olefinic carbon of acyclic enol ethers is an important advance.

The utility of enol ether arylation reactions in synthesis is evident from two recent examples. The first synthetic benzo[*d*]naphtho[1,2-*b*]-pyran-6-one C-glycosides[1] have been prepared recently by a palladium-mediated enol ether arylation reaction.[20] In this synthesis, the organopalladium reagent prepared from organostannane **74** by transmetallation underwent regio- and stereospecific coupling with glycals **16** and **17** to form, in each case, a single C-glycoside product, **75** and **76**, respectively.[20] This direct and facile C-glycoside synthesis presages the preparation of a wide variety of compounds related to gilvocarcin V and other important anticancer antibiotic anthracycline C-glycosides.[1]

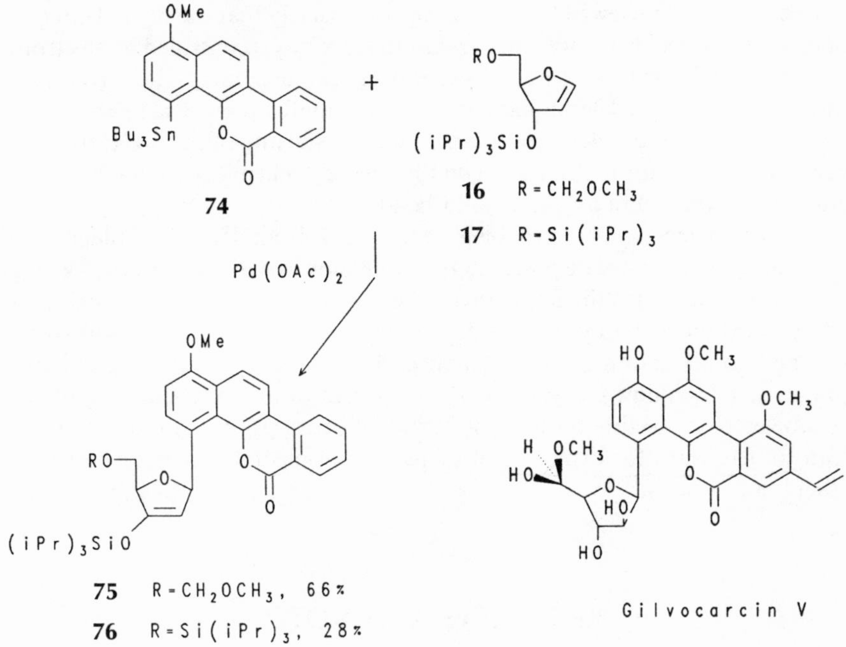

74

16 R = CH$_2$OCH$_3$

17 R = Si(iPr)$_3$

Pd(OAc)$_2$

75 R = CH$_2$OCH$_3$, 66%

76 R = Si(iPr)$_3$, 28%

Gilvocarcin V

In another application of the chemistry reviewed here, palladium-catalyzed, regioselective β-arylation of methyl vinyl ether using π-nitrophenylbromide as arylating agent followed by reduction of the resulting nitrophenyl olefin produced the substituted aniline **78** which served as a key intermediate in an efficient synthesis[99] of metoprolol, a β$_1$-blocker useful for the treatment of hypertension.

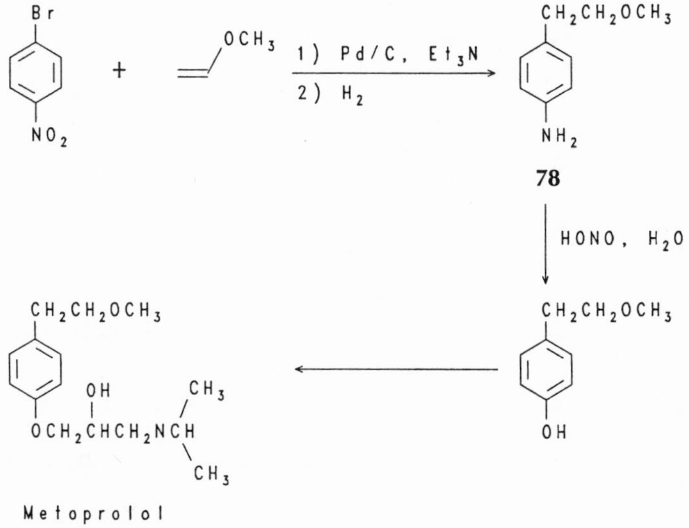

78

HONO, H$_2$O

Metoprolol

Recent results reviewed here establish conclusively that the regiochemistry of palladium-mediated enol ether arylation reactions is controled by electronic factors. An often encountered generalization concerning Heck reactions asserts that the new C–C bond forms preferentially at the least substituted olefinic carbon regardless of polarization by substituents;[13,62] clearly, this generalization is not useful for predicting the regiochemical outcome of enol ether arylation reactions and should be abandoned.

Significant mechanistic problems remain unsolved; several of them have been noted at appropriate places in the discussion. Two of the more puzzling questions concern (a) the impressive differences between arylation regiochemistries exhibited by cyclic and acyclic enol ethers and (b) the surprising finding that alkylation of the α-olefinic carbon of an acyclic enol ether seems to completely inhibit β-arylation. Development of an understanding of the stereoelectronic factors responsible for these observations will undoubtedly lead to new ways for exploitation of palladium-mediated reactions of enol ethers in synthesis.[126,127]

REFERENCES AND NOTES

1. Hacksell, U.; Daves, G. D., Jr. *Prog. Med. Chem.* 1985, *22*, 1.
2. Daves, G. D., Jr.; Cheng, C. C. *Prog. Med. Chem.* 1976, *13*, 303.
3. Bergstrom, D. E.; Ruth, J. L. *J. Am. Chem. Soc.* 1976, *98*, 1587.
4. Arai, I.; Daves, G. D., Jr. *J. Am. Chem. Soc.* 1978, *100*, 287.
5. Heck, R. F. *J. Am. Chem. Soc.* 1968, *90*, 5518.
6. Heck, R. F. *J. Am. Chem. Soc.* 1968, *90*, 5526.
7. Heck, R. F. *J. Am. Chem. Soc.* 1968, *90*, 5531.
8. Heck, R. F. *J. Am. Chem. Soc.* 1968, *90*, 5535.
9. Heck, R. F. *J. Am. Chem. Soc.* 1968, *90*, 5538.
10. Heck, R. F. *J. Am. Chem. Soc.* 1968, *90*, 5542.
11. Heck, R. F. *J. Am. Chem. Soc.* 1968, *90*, 5546.
12. Heck. R. F. "Organo-transitional Metal Chemistry"; Academic Press: New York, 1974.
13. Heck, R. F. "Palladium Reagents in Organic Synthesis"; Academic Press: New York, 1985.
14. Andersson, C.-M.; Hallberg, A.; Daves, G. D., Jr. *J. Org. Chem.* 1987, *52*, 3529.
15. Kalinoski, H. T.; Hacksell, U.; Barofsky, D. F.; Barofsky, E; Daves, G. D., Jr. J. Am. Chem. Soc. 1985, 107, 6476.
16. Wisner, J. M.; Bartczak, T. J.; Ibers, J. A.; J. J. Low; Goddard, W. A., III *J. Am. Chem. Soc.* 1986, *108*, 347.
17. Fitton, P.; Rick, E. A. *J. Organomet. Chem.* 1971, *28*, 287.
18. Nenitzescu, C. D.; Isacescu, D. A.; Gruescu, C. *Bull. Soc. Chim. Romania* 1938, 20, 127; *Chem. Abs.* 1940, *34*, 1977.
19. Stille, J. K. *Angew. Chem. Int. Ed. Engl.* 1986, *25*, 508.
20. Outten, R. A.; Daves, G. D., Jr. *J. Org. Chem.* 1987, *52*, 5064.
21. Bäckvall, J.-E.; Nordberg, R. E.; Wilhelm D. *J. Am. Chem. Soc.* 1985, *107*, 6892.
22. Henry, P. M. "Palladium Catalyzed Oxidation of Hydrocarbons"; D. Reidel: Dordrecht, 1980.

23. Davison, J. B.; Simon, N. M.; Sojka, S. A. *J.* Mol. Catal. 1984, 22, 349.
24. Julia, M.; Duteil, M. *Bull. Soc. Chim. Fr.* 1973, 2590.
25. Jeffery, T. *J. Chem. Soc. Chem. Commun.* 1984, 1287.
26. Blaser, H. U.; Spencer, A. *J. Organomet. Chem.* 1982, *233*, 267.
27. Biavati, A.; Chiusoli, G. P.; Costa, M.; Terenghi, G. *Transition Metal Chem.* 1979, *4*, 398.
28. Andersson, C. M.; Hallberg, A. *J. Org. Chem.* 1988, *53*, 235.
29. Andersson, C. M.; Hallberg, A. *Tetrahedron Lett.* 1987, *28*, 4215.
30. Cacchi, S.; Ciattini, P. G.; Morera, E.; Ortar, G. *Tetrahedron Lett.* 1986, *27*, 3931.
31. Cacchi, S.; Ciattini, P. G.; Morera, E.; Ortar, G. *Tetrahedron Lett.* 1986, *27*, 5541.
32. Chen, Q. Y.; Yang, Z. Y. *Tetrahedron Lett.* 1986, *27*, 1171.
33. Andersson, C. M.; Hallberg, A. *J. Org. Chem.* 1988, *53*, 2112.
34. Crandall, J. K.; Centeno, M. A. *J. Org. Chem.* 1979, *44*, 1183.
35. Sakakibara, M.; Inagaki, F.; Hakada, I.; Shimanouchi, T. *Bull. Chem. Soc. Jpn.* 1976, *49*, 46.
36. Trofimov, B. A.; Shergina, N. I.; Atavin, A.S.; Kositsina, E. I.; Gusarov, A. V.; Gavrilova, G. M. *Izv. Akad. Nauk SSSR, Ser. Khim.* 1972, 116.
37. Durig, J. R.; Li, Y. S.; Tong, C. K. *J. Chem. Phys.* 1972, *56*, 5692.
38. Cahill, P.; Gold, L. P.; Owen, L. *J. Chem. Phys.* 1968, *48*, 1620.
39. Bloch, M.; Brogli, F.; Heilbronner, E.; Jones, T. B.; Prinzbach, H.; Schweikert, O. *Helv. Chim. Acta* 1978, *61*, 1388.
40. Trofimov, B. A., Kalabin, G. A.; Bzesovsky, V. M.; Gusarova, N. K.; Kushnarev, D. F.; Amosova, S. V. *Reakts. Sposobn. Org. Soedin.* 1974, *11*, 365.
41. Taskinen, E. *Tetrahedron* 1978, *34*, 353.
42. Taskinen, E. *Tetrahedron* 1978, *34*, 433.
43. Guthrie, R. D.; Irvine, R. W. *Aust. J. Chem.* 1980, *33*, 1037.
44. Kalabin, G. A.; Kushnarev, D. F.; Valeyev, R. B.; Trofimov, B. A.; Fedotov, M. A. *Org. Magn. Reson.* 1982, *18*, 1.
45. Bouchoux, G.; Hanna, I.; Houriet, R; Rolli, E. *Can. J. Chem.* 1986, *64*, 1345.
46. Bouchoux, G.; Djazi, F.; Hoppilliard, Y.; Houriet, R.; Rolli, E. *Org. Mass Spectrom.* 1986, *21*, 209.
47. Taskinen, E.; Kukkamaki, E.; Kotilainen, H. Tetrahedron 1978, 34, 1203.
48. Samdal, S.; Seip, H. M. *J. Mol. Struct.* 1975, *28*, 193.
49. Zacheslavskaya, R. Kh.; Rappoport, L. Y.; Petrov, G. N.; Trofimov, B. A. *Reakts. Sposobn. Org. Soedin.* 1978, *15*, 163.
50. Bou, W.; Radom, L. *J. Mol Struct.* 1978, *43*, 267.
51. Chalmers, A. A.; Hall, R. H. *J. Chem. Soc., Perkin Trans.* 2 1974, 728.
52. Brown, R. S.; Marcinko, R. W. *J. Am. Chem. Soc.* 1978, *100*, 5721.
53. Deslongchamps, P. "Stereoelectronic Effects in Organic Chemistry"; Pergamon Press: New York, 1983.
54. Houk, K. N.; Paddon-Row, M. N.; Rondan, N. G.; Wu, Y. D.; Brown, F. K.; Spellmeyer, D. C.; Metz, J. T.; Li., Y.; Loncharich, R. J. *Science* 1986, *231*, 1108.
55. Kirby, A. J. "The Anomeric Effect and Related Stereoelectronic Effects at Oxygen"; Springer-Verlag: New York, 1983.
56. Curran, D. P.; Suh, Y. G. *J. Am. Chem. Soc.* 1984, *106*, 5002.
57. Denmark, S. E.; Dappen, M. S. *J. Org. Chem.* 1984, *49*, 798.
58. Wakatsuki, Y.; Nozakura, S.; Murahashi, S. *Bull. Chem. Soc. Jpn.* 1972, *45*, 3426.
59. McCrindle, R.; Ferguson, G.; Khan, M. A.; McAlees, A. J.; Ruhl, B. L. *J. Chem. Soc., Dalton Trans.* 1981, 986.
60. Fleming, I. "Frontier Orbitals and Organic Chemical Reactions"; John Wiley: New York, 1976.
61. Houk, K. N. *J. Am. Chem. Soc.* 1973, *95*, 4092.
62. Collman, J. P.; Hegedus, L. S.; Norton, J. R.; Finke, R. G. "Principles and Applications

of Organotransition Metal Chemistry"; University Science Books: Mill Valley, CA, 1987.
63. Dehand, J.; Mauro, A.; Ossor, H.; Pfeffer, M.; Santos, R. H. DeA.; Lechat, J. R. *J. Organomet. Chem.* 1983, *250*, 537.
64. Ito, Y.; Aoyama, H.; Hirao, T.; Mochizuki, A.; Saegusa, T. *J. Am. Chem. Soc.* 1979, *101*, 494.
65. Ito, Y.; Nakatsuka, M.; Kise, N.; Saegusa, T. *Tetrahedron Lett.* 1980, *21*, 2873.
66. Herberhold, M. "Metal π-Complexes"; Elsevier: Amsterdam, 1974, Vol. II, Part 2, p. 119.
67. Busse, P.; Pesa, F.; Orchin, M. *J. Organomet. Chem.* 1977, *140*, 229.
68. Chang, T. C. T.; Rosenblum, M.; Samuels, S. B. *J. Am. Chem. Soc.* 1980, *102*, 5930.
69. Arai, I; Daves, G. D., Jr. *J. Org. Chem.* 1978, *43*, 4110.
70. Roth, W.; Pigman, W. "Methods in Carbohydrate Chemistry"; Academic Press: New York, 1963, Vol. II, pp. 405–408.
71. Cheng, J. C. Y.; Daves, G. D., Jr. *J. Org. Chem.* 1987, *52*, 3083.
72. Hacksell, U.; Kalinoski, H. T.; Barofsky, D. F.; Daves, G. D., Jr. *Acta Chem. Scand. B* 1985, *39*, 469.
73. Arai, I.; Lee, T. D.; Hanna, R.; Daves, G. D., Jr. *Organometallics* 1982, *1*, 742.
74. Arai, I.; Daves, G. D., Jr. *J. Am. Chem. Soc.* 1981, *103*, 7683.
75. Dunkerton, L. V.; Brady, K. T.; Mohamed, F. *Tetrahedron Lett.* 1982, *23*, 599.
76. Czernecki, S.; Dechavanne, V. *Can. J. Chem.* 1983, *61*, 533.
77. Czernecki, S.; Gruy, F. *Tetrahedron Lett.* 1981, *22*, 437.
78. Kwok, D. I.; Daves, G. D., Jr., unpublished results.
79. Hacksell, U.; Daves, G. D., Jr. *J. Org. Chem.* 1983, *48*, 2870.
80. Cheng, J. C. Y.; Hacksell, U.; Daves, G. D., Jr. *J. Org. Chem.* 1986, *51*, 3093.
81. Bäckvall, J. E. *Acc. Chem. Res.* 1983, *16*, 335.
82. Bäckvall, J. E.; Björkman, E. E.; Pettersson, L.; Siegbahn, P. J. *Am. Chem. Soc.* 1984, *106*, 4369.
83. Bäckvall. J. E.; Björkman, E. E.; Pettersson, L.; Siegbahn, P. *J. Am. Chem. Soc.* 1985, *107*, 7265.
84. McKeon, J. E.; Fitton, P; Griswold, A. A. *Tetrahedron* 1972, *28*, 227.
85. McKeon, J. E.; Fitton, P. *Tetrahedron* 1972, *28*, 233.
86. Dunkerton, L. V.; Serino, A. J. *J. Org. Chem.* 1982, *47*, 2812.
87. Arai, I.; Daves, G. D., Jr. *J. Org. Chem.* 1979, *44*, 21.
88. Danno, S.; Moritani, I.; Fujiwara, T. *Tetrahedron* 1969, *25*, 4819.
89. Heck, R. F. *Organomet. Chem. Syn.* 1972, *1*, 1455.
90. Kasahara, A.; Izumi, T.; Fukuda, T. *Bull. Soc. Chem. Jpn.* 1977, *50*, 551.
91. Arai, I.; Daves, G. D., Jr. *J. Heterocycl. Chem.* 1978, *15*, 351.
92. Hallberg, A.; Westfelt, L.; Holm, B. *J. Org. Chem.* 1981, *46*, 5414.
93. Heck, R. F. *J. Am. Chem. Soc.* 1971, *93*, 6896.
94. Dieck, H. A.; Heck, R. F. *J. Am. Chem. Soc.* 1974, *96*, 1133.
95. Melpolder, J. B.; Heck, R. F. *J. Org. Chem.* 1976, *41*, 265.
96. Zeigler, C. B., Jr.; Heck, R. F. *J. Org. Chem.* 1978, *43*, 2941.
97. Zeigler, C. B., Jr.; Heck, R. F. *J. Org. Chem.* 1978, *43*, 2949.
98. Hallberg, A.; Westfelt, L. *J. Chem. Soc., Perkin Trans. 1* 1984, 933.
99. Hallberg, A.; Westfelt, L.; Andersson, C. M. Synth. Commun. 1985, *15*, 1131.
100. Lee, T. D.; Daves, G. D., Jr. *J. Org. Chem.* 1983, *48*, 399.
101. Echavarren, A. M.; Stille, J. K. *J. Am. Chem. Soc.* 1987, *109*, 5478.
102. RajanBabu, T. V.; Reddy, G. S. *J. Org. Chem.* 1986, *51*, 5458.
103. Scott, W. J.; Pena, M. R.; Swärd, K.; Stoessel, S. J.; Stille, J. K. *J. Org. Chem.* 1985, *50*, 2302.
104. Fujimoto, H.; Yamasaki, T. *J. Am. Chem. Soc.* 1986, *108*, 578.
105. Koga, N.; Obara, S.; Kitaura, K.; Morokuma, K. *J. Am. Chem. Soc.* 1985, *107*, 7109.
106. Doherty, N. M.; Bercaw, J. E. *J. Am. Chem. Soc.* 1985, *107*, 2670.
107. Bryndza, H. E. *J. Chem. Soc., Chem. Commun.* 1985, 1696.

108. Eisenstein, O.; Hoffmann, R. *J. Am. Chem. Soc.* 1981, *103*, 4308.
109. Thorn, D. L.; Hoffmann, R. *J. Am. Chem. Soc.* 1978, *100*, 2079.
110. Kasahara, A.; Izumi, T.; Yodono, M.; Saito, R.; Takeda, T.; Sugawara, T. *Bull. Soc. Chem. Jpn.* 1973, *46*, 1220.
111. Saito, R.; Izumi, T.; Kasahara, A. *Bull. Soc. Chem. Jpn.* 1973, *46*, 1776.
112. Bäckvall, J. E.; Heumann, A. *J. Am. Chem. Soc.* 1986, *108*, 7107.
113. Arai, I.; Hanna, R.; Daves, G. D., Jr. *J. Am. Chem. Soc.* 1981, *103*, 7684.
114. Hartley, F. R. *Coord. Chem. Rev.* 1981, *35*, 143.
115. Newcome, G. R.; Puckett, W. E.; Crupta, V. K.; Fronczek, F. R. *Organometallics* 1983, *2*, 1247.
116. Cheng, J. C. Y.; Hacksell, U.; Daves, G. D., Jr. *J. Org. Chem.* 1985, *50*, 2778.
117. Hacksell, U.; Daves, G. D., Jr. *Organometallics* 1983, *2*, 772.
118. Sharpless, K. B.; Teranishi, A. Y.; Bäckvall, J. E. *J. Am. Chem. Soc.* 1977, *99*, 3120.
119. Andrews, M. A.; Cheng, C. W. F. *J. Am. Chem. Soc.* 1982, *104*, 4268.
120. Hudrlik, P. F.; Peterson, D.; Rona, R. *J. Org. Chem.* 1975, *40*, 2263.
121. Kauffmann, T.; Kriegesmann, R.; Hamsen, A. *Chem. Ber.* 1982, *115*, 1818.
122. Cheng, J. C. Y.; Daves, G. D., Jr. Organometallics 1986, 5, 1753.
123. Ireland, R. E.; Thiasrivongs, S.; Vanier, N.; Wilcox, C. S. *J. Org. Chem.* 1980, *45*, 48.
124. Lesyng, B.; Saenger, W. *Carbohydr. Res.* 1984, *133*, 187.
125. Wiorkiewicz-Kuczera, J.; Rabczenko, A. *J. Chem. Soc., Perkin Trans.* 2 1985, 789.
126. Daves, G. D., Jr. *Acc. Chem. Res.* 1990, *23*, 201.
127. Daves, G. D., Jr.; Hallberg, A. *Chem. Rev.* 1989, *89*, 1433.

TRANSITION METAL CATALYZED SILYLMETALATION OF ACETYLENES AND Et_3B-INDUCED RADICAL ADDITION OF Ph_3SnH TO ACETYLENES
SELECTIVE SYNTHESIS OF VINYLSILANES AND VINYLSTANNANES

Koichiro Oshima

Advances in Metal-Organic Chemistry, Volume 2, pages 101–141
Copyright © 1991 JAI Press Ltd
All rights of reproduction in any form reserved
ISBN: 0-89232-948-3

I. INTRODUCTION

During the past decade, we have developed synthetically useful reactions mediated by reagents containing aluminium as a key atom.[1] The reagents perform combined acid–base attack on substrates with lower activation energies. The aluminium Lewis acid center serves to bind the substrate and subsequently the base directly attached to aluminium is excited by the coordination and attacks the substrate in the subsequent rate- and product-determining step. Three typical examples supporting this idea are shown below. (1) The highly specific isomerization has been affected by diethylaluminium 2,2,6,6-tetramethylpiperidide (DATMP) (Scheme 1).[2]

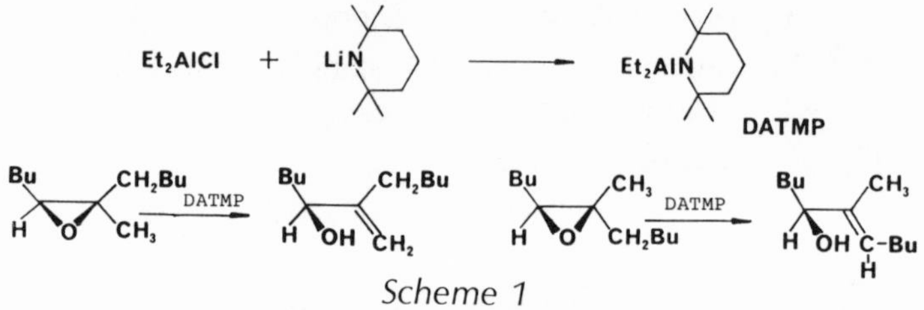

Scheme 1

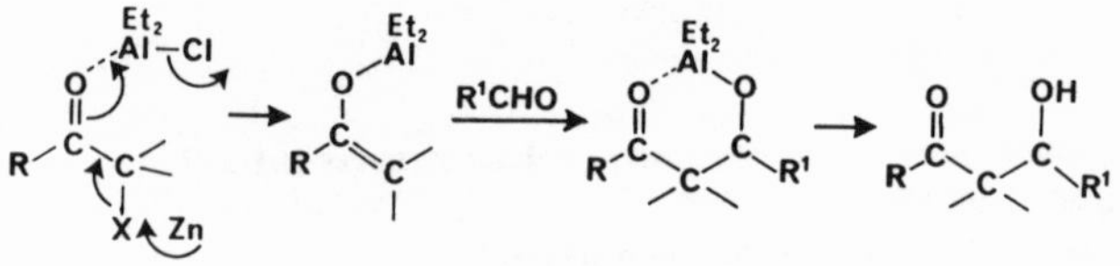

Scheme 2

(2) Coexistence of diethylaluminium chloride facilitates the zinc reduction of α-halocarbonyl compounds to generate aluminium enolates effectively.[3] The enolates thus produced are sufficiently reactive to attack the carbonyl components to provide the β-ketolates, workup of which yields the β-hydroxy carbonyl compounds (Scheme 2). (3) Most recently we have found another effective method for the regioselective formation of aluminium enolates. A novel reagent which is believed to have an aluminium–tin single bond is produced by treatment of n-Bu$_3$SnLi with an equimolar amount of Et$_2$AlCl. Treatment of an α-bromo carbonyl compound with this reagent has provided an enolate which reacted with a ketone or aldehyde to afford a β-hydroxy carbonyl compound in good yield (Scheme 3).[4]

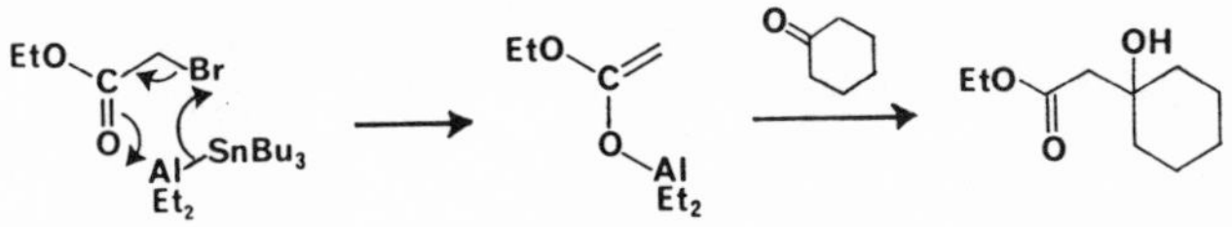

Scheme 3

A few years ago we started an investigation on the stereoselective synthesis of the side chain of a plant growth steroidal hormone, brassinolide. Our synthetic route which consists of three key reactions is shown in Scheme 4.

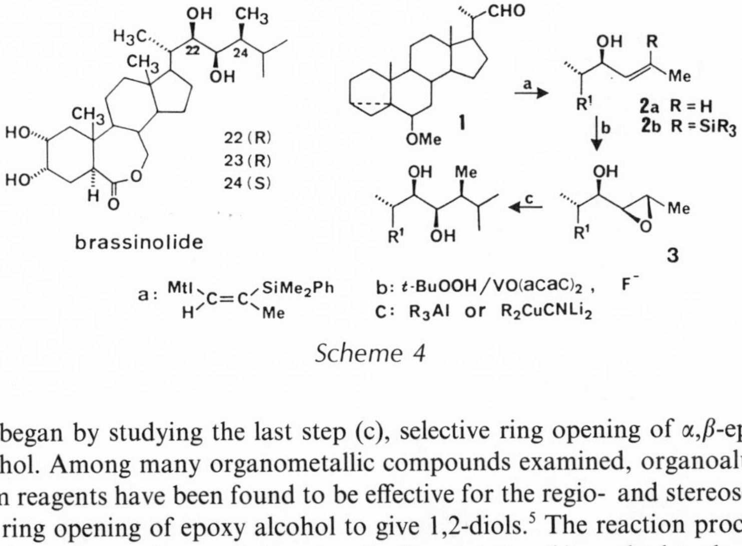

Scheme 4

We began by studying the last step (c), selective ring opening of α,β-epoxy alcohol. Among many organometallic compounds examined, organoaluminium reagents have been found to be effective for the regio- and stereoselective ring opening of epoxy alcohol to give 1,2-diols.[5] The reaction proceeds with inversion at the reacting center. In the next step (b), we had to develop a new method of preparing *threo* epoxy alcohol 3 from (*E*)-allylic alcohol 2a. The vanadium-catalyzed epoxidation of (*E*)-allylic alcohol gave the *erythro* isomer as a main product. On the other hand, *m*-CPBA epoxidation gave the *threo* isomer as a major product, but the selectivities of these reactions were rather low and not enough for our purpose. This problem was solved by substitution of an appropriate hydrogen on the double bond by a bulky Me$_3$Si group.[6] Treatment of **2b** with VO(acac)$_2$-t-BuOOH or m-CPBA gave *threo* epoxy alcohol **3** with high stereoselectivity (99%). The Me$_3$Si group was easily removed by treatment with n-Bu$_4$NF or CsF in DMSO and the desilylation of epoxysilane proceeded with retention of configuration at the oxiranyl carbon.

We then have faced the difficult problem (step a) of how to get an alkenyl-metal species, (Z)-R$_3$SiC(Me)=CHMtl in order to have the desired silyl-substituted allylic alcohol **2b** from the readily available aldehyde **1**. Regio- and stereoselective addition of Si-metal compounds to propyne was a possibility to obtain this alkenylmetal compound. Thus, the combined acid–base reagents such as Si–Al or Si–Zn were examined to solve the problem. This was the basis of our studies on the silylmetalation of acetylenes.

II. SILYLMETALATION OF ACETYLENES AND ITS APPLICATION TO THE STEREOSELECTIVE SYNTHESIS OF STEROIDAL SIDE CHAINS

A. Silylmetalation of Acetylenes

In connection with an investigation of the stereoselective synthesis of the side chain of brassinolide, we had to develop a method for the formation of the alkenyl metal species (**5**, R = CH$_3$) (Scheme 5). Although exclusive formation of 2-metallo-1-silyl-1-alkenes (**4**) by silylcupration,[7] silyltitanation[8] or silylalumination[9] was previously reported, the selective generation of 1-metallo-2-silyl isomer **5** has not been described to our knowledge. First we investigated simultaneous addition of the silyl group and metal to acetylenes with regioselectivity using PhMe$_2$SiLi and several metal compounds such as MeMgI, Et$_2$AlCl and ZnBr$_2$ in the presence of a couple of transition metal catalysts.[10]

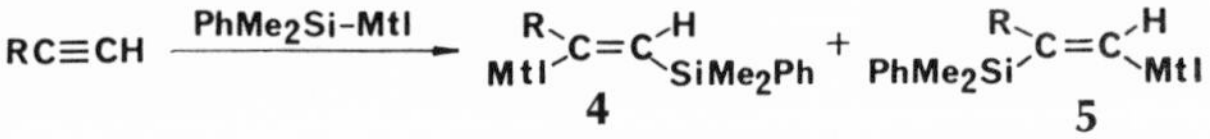

$$RC\equiv CH \xrightarrow{\text{PhMe}_2\text{Si-Mtl}} \underset{\textbf{4}}{\overset{R}{\underset{Mtl}{}}C=C\overset{H}{\underset{SiMe_2Ph}{}}} + \underset{\textbf{5}}{\overset{R}{\underset{PhMe_2Si}{}}C=C\overset{H}{\underset{Mtl}{}}}$$

Scheme 5

Platinum- or copper-catalyzed silylmagnesation followed by aqueous quenching provided exclusively (E)-1-silyl-1-alkenes, which have previously been produced by stoichiometric silylcupration[7] or silyltitanation.[8] In contrast, the use of a Pd catalyst resulted in the formation of a mixture of two regio-isomers. The reagent prepared from PhMe$_2$SiLi and ZnBr$_2$ also added to acetylenes in *cis* fashion to give isomeric mixtures. In the presence of a Pt

Table 1. Transition Metal Catalyzed Silylmetalation[a]

$$RC{\equiv}CH \xrightarrow[\text{2. H}_3\text{O}^+]{\text{1. PhMe}_2\text{SiLi--MX cat}} \underset{\mathbf{6}}{\overset{R}{\underset{H}{>}}C{=}C\overset{H}{\underset{SiMe_2Ph}{<}}} + \underset{\mathbf{7}}{\overset{R}{\underset{PhMe_2Si}{>}}C{=}C\overset{H}{\underset{H}{<}}}$$

R	MX	Catalyst	Yield (%)	**6:7**
a: n-$C_{10}H_{21}$	MeMgI	cis-$PtCl_2(P$-n-$Bu_3)_2$	90	> 99 : < 1
		CuI^b	86	> 99 : < 1
		$PdCl_2(PPh_3)_2$	76	60 : 40
	Et_2AlCl	—^{c}	60	97 : 3
		$RhCl(PPh_3)_3$	70	91 : 9
		CuI^b	78	55 : 45
		$PdCl_2(P(o$-$CH_3C_6H_4)_3)_2$	85	15 : 85
	$ZnBr_2{}^d$	cis-$PtCl_2(P$-n-$Bu_3)_2$	55	70 : 30
		$PdCl_2(PPh_3)_2$	71	30 : 70
		$RuCl_2(PPh_3)_3$	75	20 : 80
b: n-BuCH(OH)	MeMgI	CuI^b	85	95 : 5
	Et_2AlCl	$Pd(OAc)_2$–$P(o$-$CH_3C_6H_4)_3$	80	30 : 70
c: $PhCH_2OCH_2CH_2$	MeMgI	CuI^b	90	> 99 : < 1
	Et_2AlCl	$PdCl_2(PPh_3)_2$	88	30 : 70

[a]A mixture of acetylene substrate, $PhMe_2SiLi$–MX reagent and a catalyst (1:2:0.01 mol ratio) was employed. The reactions were performed at 25°C in THF unless otherwise stated and completed within 1 h. [b]CuI (5 mol%) was used. [c]Pt or Ru catalyst did not accelerate the silylalumination reactions. The reaction mixture was heated at reflux for 8 h without catalyst. [d]A reagent was produced by mixing the silyllithium with $ZnBr_2$ in a 2:1 ratio.

catalyst, 1-silyl-1-alkene was the main product. However, the Ru or Pd catalyzed reaction gave the 2-silyl isomer as the major product. The reagent derived from $PhMe_2SiLi$ and Et_2AlCl was added to 1-dodecyne without any catalyst after heating at 80°C for 8 h to give 1-(dimethylphenylsilyl)-1-dodecene almost exclusively. The Cu catalyzed reaction provided two isomers in 50:50 ratio. Palladium-catalyzed silylalumination afforded 2-silyl-1-alkene with high regioselectivity. The results are summarized in Table 1.

The regiochemistry of silylmetalation depends heavily on the nature of the transition metal catalysts as well as metallic species of silyl-metal reagents employed. Ligands of the palladium catalysts also play the important role of controlling product distribution. The reaction of 1-dodecyne with $PhMe_2$-$SiAlEt_2$ in the presence of a variety of palladium catalysts was studied, ligand L of $PdCl_2L_2$ and the ratio of **6/7** being as follows: (m-$MeOC_6H_4)_3P$, 90/10; $Ph_2PCH_2CH_2PPh_2$, 65/35; (2-(di-phenylphosphino)ferrocenyl)methyl-dimethylamine, 50/50; n-Bu_3P, 35/65; PPh_3, 30/70; (o-$CH_3C_6H_4)_3P$, 15/85.

The *cis* addition of a silyl-metal component was confirmed by the examination of the ^{1}H-NMR spectrum of the product **6a** ($J = 18.9$ Hz) and also by the GLPC comparison with the authentic (Z) sample prepared from hydro-alumination of 1-(dimethylphenylsilyl)-1-dodecyne.[11] Silylalumination has been also proved to proceed in *cis* fashion by the examination of ^{1}H-NMR of (E)-1-deuterio-1-dodecene prepared by palladium-catalyzed silylalumination of 1-dodecyne followed by quenching with D_2O and successive desilylation (Scheme 6). Deuterolysis of the intermediate derived from silylmetalation gave the monodeuteriated alkenylsilanes. Thus, silyl reagents do not cause the acetylenic proton–metal exchange (Scheme 7).

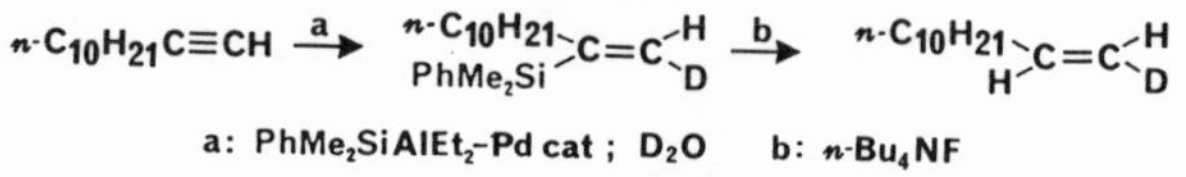

$$n\text{-}C_{10}H_{21}C\equiv CH \xrightarrow{a} \underset{PhMe_2Si}{n\text{-}C_{10}H_{21}}C=C\underset{D}{\overset{H}{}} \xrightarrow{b} \underset{H}{n\text{-}C_{10}H_{21}}C=C\underset{D}{\overset{H}{}}$$

a: PhMe$_2$SiAlEt$_2$-Pd cat ; D$_2$O   b: n-Bu$_4$NF

Scheme 6

$$n\text{-}C_{10}H_{21}C\equiv CH \xrightarrow{a,b} \underset{D}{n\text{-}C_{10}H_{21}}C=C\underset{SiMe_2Ph}{\overset{H}{}}$$

$$n\text{-}C_{10}H_{21}C\equiv CD \xrightarrow{a,c} \underset{H}{n\text{-}C_{10}H_{21}}C=C\underset{SiMe_2Ph}{\overset{D}{}}$$

a: PhMe$_2$SiMgMe/ PtCl$_2$ (P^{n}Bu$_3$)$_2$ b: D$_2$O   c: H$_2$O

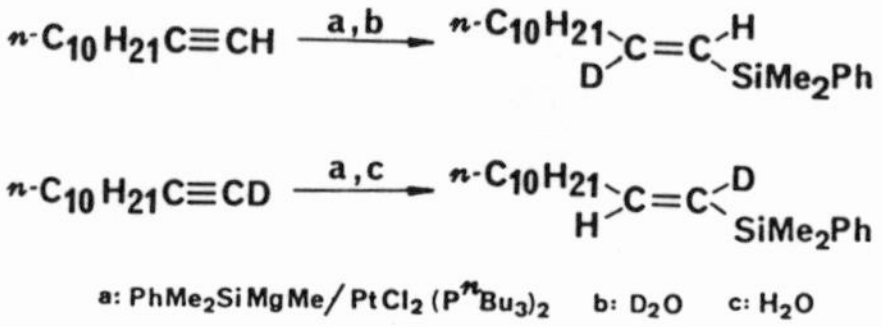

Scheme 7

The present method provides not only simple silyl-substituted alkenes but also functionalized alkenylsilanes. Some electrophiles react with the alkenyl-metal species without difficulty. For instance, treatment of the intermediate derived from PtCl$_2$(P-n-Bu$_3$)$_2$-catalyzed silylmagnesation of 1-dodecyne with iodine, methyl iodide and valeraldehyde gave the corresponding silylalkenes carrying the electrophilic partner E (Scheme 8). All these electrophiles reacted at 25°C within 1 h.

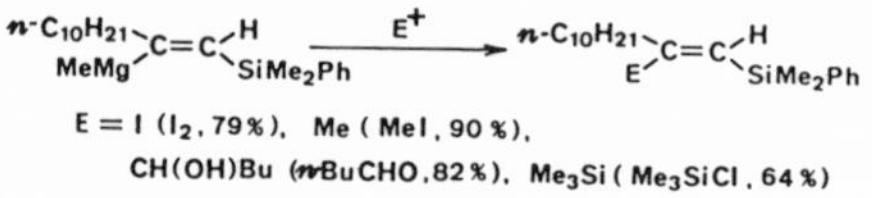

$$\underset{MeMg}{n\text{-}C_{10}H_{21}}C=C\underset{SiMe_2Ph}{\overset{H}{}} \xrightarrow{E^+} \underset{E}{n\text{-}C_{10}H_{21}}C=C\underset{SiMe_2Ph}{\overset{H}{}}$$

E = I (I$_2$, 79%), Me (MeI , 90%),
CH(OH)Bu (nBuCHO,82%), Me$_3$Si (Me$_3$SiCl , 64%)

Scheme 8

An ethereal solution of methylmagnesium iodide (1.1 M, 1.8 mL, 2.0 mmol) was added to a THF solution of dimethylphenylsilyllithium (0.52 M, 3.8 mL, 2.0 mmol) at 0°C under an argon atmosphere. After stirring for 15 min, *cis*-PtCl$_2$(P-n-Bu$_3$)$_2$ (6.7 mg, 0.01 mmol) was added, and the resulting solution was stirred for an additional 15 min. A solution of 1-dodecyne (0.17 g, 1.0 mmol) in THF (3.0 mL) was added, and the whole was stirred for 30 min at 25°C. Workup (ether, 1 M HCl) and purification by silica gel column chromatography gave (E)-1-(dimethylphenylsilyl)-1-dodecene (0.27 g) in 90% yield.

B. Application of Silylmetalation of Acetylenes to the Synthesis of Brassinolide

Palladium-catalyzed silylalumination of 1-dodecyne provided a mixture of two regioisomers in an 85/15 (**7a/6a**) ratio. This ratio was the best so far and could not be improved in spite of various attempts at that moment. (This problem was solved later, see Section III). A combination of this method with our previous findings[5,6] provided us with an easy route to the stereoselective synthesis of the side chain of brassinolide.[12] Palladium-catalyzed silylalumination of propyne followed by addition of iodine provided an 88:12 mixture of 2-(dimethylphenylsilyl)-1-iodo-1-propene (**8a**) and its regioisomer. The desired 2-silylalkene **8a** was obtained in pure form by silica gel column chromatography.

Treatment of a mixture of an aldehyde **1**[13] and iodoalkene **8a** with butyllithium at −78°C gave, after chromatography, the (22*R*)-allylic alcohol **9** in 48% yield along with the (22*S*) isomer (16% yield, Scheme 9). Silyl-group-assisted stereoselective epoxidation[6] ($VO(acac)_2$-t-BuOOH) followed by elimination of $PhMe_2Si$ group with n-Bu_4NF gave the key intermediate *threo*-α,β-epoxy alcohol **10a** exclusively (65% yield). Regio- and stereoselective ring opening[5] of epoxy alcohol with the organoaluminium compound $Et_2AlC{\equiv}CSiMe_3$ proceeded with inversion at the reacting center to give 1,2-diol **11a** at a 60% yield. Removal of the Me_3Si group (KF, DMSO) and hydrogenation (H_2, PtO_2) provided **11b**. Reaction of benzyl ether **10b** with the higher order mixed cuprate[14] $(Me_2CH)_2Cu(CN)Li_2$ afforded **11c** (63% yield), which was transformed into **11d** (Li in liquid NH_3).[12b]

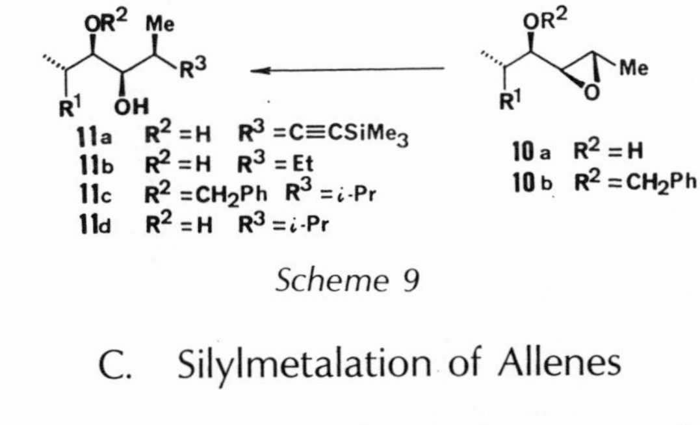
CHO
X C=C SiMe2Ph H Me
8a X = I
8b X = Li
OMe
1
OH SiMe2Ph
Me
R1
9
OR2
Me
O
R1
10 a R2 = H
10 b R2 = CH2Ph
OR2 Me
R3
R1 OH
11a R2 = H R3 = C≡CSiMe3
11b R2 = H R3 = Et
11c R2 = CH2Ph R3 = i-Pr
11d R2 = H R3 = i-Pr

Scheme 9

C. Silylmetalation of Allenes

Silylmagnesation of 1,2-cyclononadiene in the presence of a CuI catalyst, followed by aqueous quenching, produces exclusively 1-(dimethylphenylsilyl)-1-cyclononene (**12**) in which the PhMe$_2$Si group is attached to the central carbon of the allene.[15] Among many combinations examined, see Table 2 for examples, the PhMe$_2$SiAlEt$_2$–PdCl$_2$(P(o-MeC$_6$H$_4$)$_3$)$_2$ system was found to afford the isomeric 3-(dimethylphenylsilyl)-1-cyclononene (**13**) predominantly. Ligands of palladium catalysts play an important role in controlling product distribution. The reaction of 1,2-cyclononadiene with PhMe$_2$SiAlEt$_2$ in the presence of Pd(PPh$_3$)$_4$ gave a mixture of **12** and **13** in a 55:45 ratio. Acyclic allene gave similar results as cyclic ones. For instance, CuI catalyzed silylmagnesation of 1,2-tridecadiene provided 2-(di-methyl-phenylsilyl)-1-tridecene (**14**) exclusively (68%). In contrast, PdCl$_2$(PPh$_3$)$_2$ promoted silylalumination gave an isomeric mixture (**14**:**15**:**16** = 40:35:25, 55%) and silylzincation in the presence of PdCl$_2$(PPh$_3$)$_2$ afforded the mixture **14**:**15**:**16** = 30:22:48, 81% (Scheme 10).

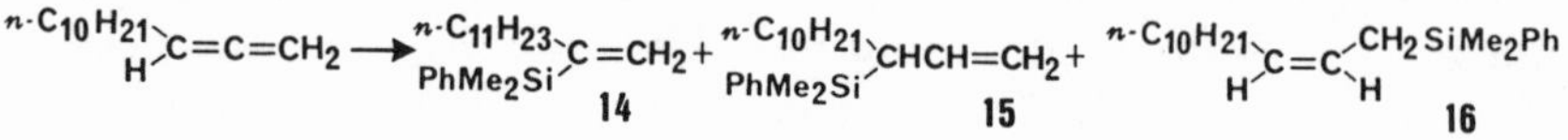

$$
\underset{H}{\overset{n\text{-}C_{10}H_{21}}{>}}C=C=CH_2 \longrightarrow \underset{PhMe_2Si}{\overset{n\text{-}C_{11}H_{23}}{>}}C=CH_2 + \underset{PhMe_2Si}{\overset{n\text{-}C_{10}H_{21}}{>}}CHCH=CH_2 + \underset{H}{\overset{n\text{-}C_{10}H_{21}}{>}}C=C\overset{CH_2SiMe_2Ph}{\underset{H}{<}}
$$

$$\mathbf{14} \qquad\qquad \mathbf{15} \qquad\qquad \mathbf{16}$$

Scheme 10

Some electrophiles smoothly reacted with the allylic Grignard intermediates derived from CuI catalyzed silylmagnesation of 1,2-cyclononadiene (Table 3). Thus, methyl iodide, allyl bromide, trialkylchlorosilanes, and aldehydes gave the corresponding silylalkenes carrying the electrophilic partners E.

This new method has provided us with a simple route to *dl*-muscone.[16]

Table 2.   Silylmetalation of 1,2-Cyclononadiene[a]

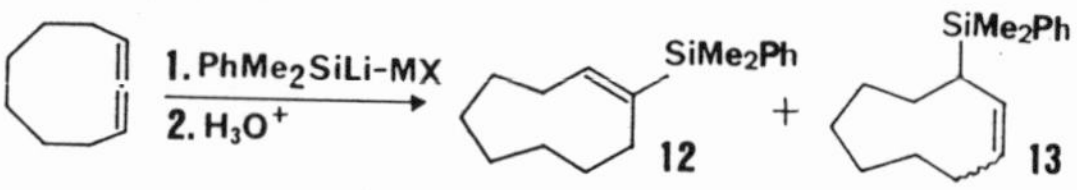

Reagent	Catalyst	Yield (%)	Ratio of **12**:**13** 12	13
PhMe$_2$SiMgMe	CuI	74	> 95	< 5
(PhMe$_2$Si)$_2$Zn[b]	PdCl$_2$(P-n-Bu$_3$)$_2$	70	50	50
(PhMe$_2$Si)$_2$Zn	PdCl$_2$(P(o-MeC$_6$H$_4$)$_3$)$_2$	40	30	70
PhMe$_2$SiAlEt$_2$	Pd(PPh$_3$)$_4$	41	55	45
PhMe$_2$SiAlEt$_2$	PdCl$_2$(P(o-MeC$_6$H$_4$)$_3$)$_2$	36	10	90

[a] 2 mmol of PhMe$_2$Si–Mtl reagent, 1 mmol of 1,2-cyclononadiene and 5 mol% of catalyst were employed. [b] A reagent was produced by mixing the silyllithium with ZnBr$_2$ in a 2:1 ratio.

Table 3. Silylmetalation of 1,2-Cyclononadiene Followed by Reaction with Electrophiles

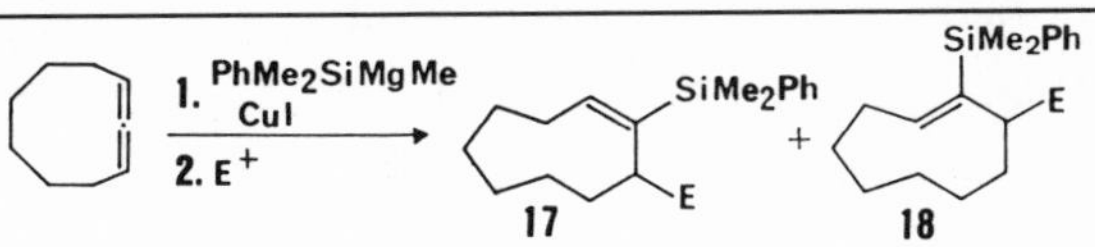

	Conditions		Yield	Ratio[a] of **17**:**18**	
Electrophile	Temperature (°C)	Time (h)	(%)	**17**	**18**
MeI	0	0.1	88	100	0
$CH_2 = CHCH_2Br$	0	0.1	94	100	0
$PhMe_2SiCl$	0	0.2	91	15	85
Me_3SiCl	0	0.2	84	0	100
PhCHO	−78	0.2	56	23	77
CH_3CHO	−78	0.5	67	100	0
$CH_2 = CHCHO$	−78	0.5	71	—[b]	
$CH_3CH = CHCHO$	−78	1.0	91	—[b]	

[a]The ratios were determined by the examination of ^{1}H-NMR spectra of the mixtures according to the reported data. [b]Major products were **17**. Exact isomeric ratio could not be determined because of the complexity of ^{1}H-NMR spectra. 1,2-Adduct was obtained exclusively.

Silylmagnesation of 1,2-cyclopentadecadiene followed by quenching with MeI gave 3-methyl-2-(dimethylphenylsilyl)-1-cyclopentadecene (77% yield). Epoxidation (97%) and oxirane ring opening[17] (47%) gave hydroxysilane. Final oxidation[18] of the hydroxysilane produced *dl*-muscone (96%, Scheme 11).

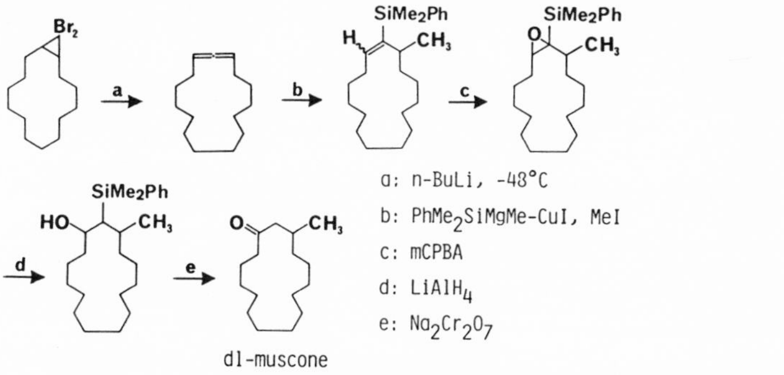

a: n-BuLi, −48°C
b: $PhMe_2SiMgMe-CuI$, MeI
c: mCPBA
d: $LiAlH_4$
e: $Na_2Cr_2O_7$

dl-muscone

Scheme 11

D. Intramolecular Cyclization Mediated by Silylmetalation

The silylmagnesium compound prepared from $PhMe_2SiLi$ and MeMgI reacts with terminal acetylenes to give vinylsilanes in the presence of various transition metal catalysts as described in Section II.A. Among them, the copper(I) iodide catalyzed reaction affords 1-silyl-1-alkene with high regio- and stereoselectivity. Further application of this reaction to substrates carrying a terminal acetylenic moiety and a leaving group such as OTs, OMs or Br has provided a novel synthesis of cycloalkanes with a silylmethylene substituent.[19]

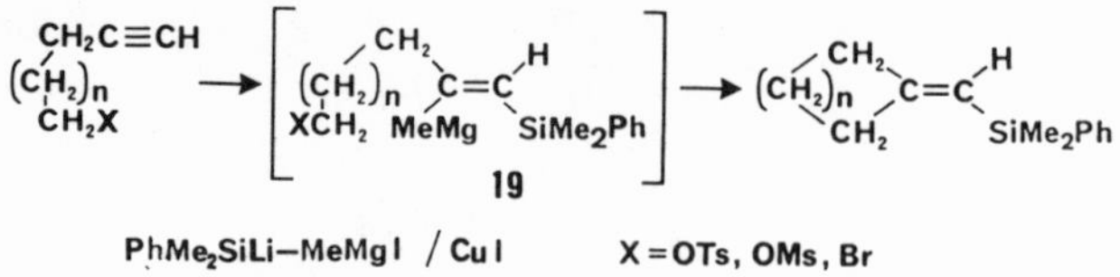

$$
\begin{array}{c}
CH_2C{\equiv}CH \\
(CH_2)_n \\
CH_2X
\end{array}
\longrightarrow
\left[
\begin{array}{c}
CH_2 \\
(CH_2)_n \quad C{=}C \quad H \\
XCH_2 \quad MeMg \quad SiMe_2Ph
\end{array}
\right]
\longrightarrow
\begin{array}{c}
CH_2 \\
(CH_2)_n \quad C{=}C \quad H \\
CH_2 \qquad SiMe_2Ph
\end{array}
$$

19

PhMe₂SiLi—MeMgI / CuI   X = OTs, OMs, Br

Scheme 12

Treatment of a tosylate of 5-hexyn-1-ol **20a** with PhMe₂SiMgMe in the presence of CuI catalyst gave (dimethylphenylsilyl)methylenecyclopentane **21a** (91% yield). Acetylenes **20b** and **20c** gave the corresponding cyclopentane derivatives **21b** and **21c** (46 and 43% yield). Mesylates of 6-heptyn-1-ol and 4-pentyn-1-ol (**22** and **24**) afforded the corresponding six-membered ring compound **23** and four-membered one **25**, respectively. These reactions proceed as follows. The addition of PhMe₂SiMgMe to the terminal acetylene linkage takes place at 0°C to provide alkenyl Grignard compound **19**. In a second step the organomagnesium reagent **19** attacks the carbon bearing the leaving group intramolecularly at 25°C to give the cyclized products. The intermediary alkenyl metal species **19** (n = 3, X = OMs) could be trapped by quenching the reaction mixture with D₂O at 0°C to give the mesylate of (E)-6-deuterio-7-(dimethylphenylsilyl)-6-hepten-1-ol quantitatively.

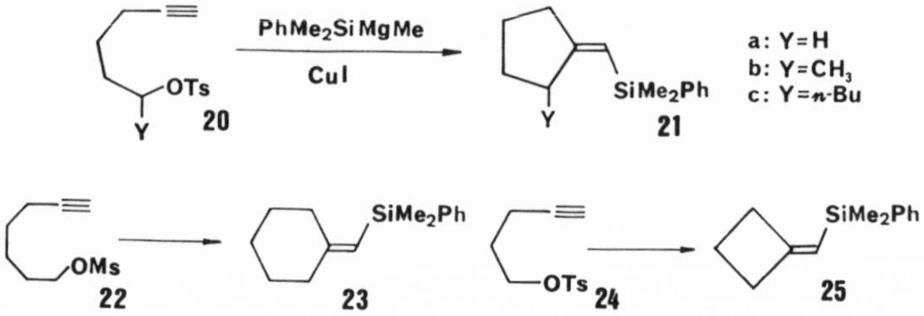

a: Y=H
b: Y=CH₃
c: Y=n-Bu

Scheme 13

Bromides such as 7-bromo-1-heptyne or 6-bromo-1-hexyne also gave the cyclized products but the yields were rather low as compared with the corresponding mesylate or tosylate. For instance, treatment of 7-bromo-1-heptyne with PhMe₂SiMgMe in the presence of CuI gave cyclohexylidene compound **23** (25% yield) along with (E)-1-(dimethylphenylsilyl)-1-heptene

(15%) which is apparently generated by the replacement of the Br group by hydrogen followed by silylmagnesation of the resulting 1-heptyne. The latter bromide gave a mixture of 1-(dimethylphenylsilyl)methyl-1-cyclopentene (40%) and (*E*)-1-(dimethylphenylsilyl)-1-hexene (20%).

The reaction of prosaic tosylate or alkyl halides provided evidence for the radical nature of the reagent, PhMe$_2$SiMgMe. The tosylate of dodecyl alcohol gave a mixture of 1-dodecyldimethylphenylsilane (20%), tridecane (3%) and dodecane (36%) on treatment with PhMe$_2$SiMgMe in the presence of CuI catalyst. 1-Bromododecane also provided a mixture of these three compounds in 22, 10 and 24% yields, respectively. The silane and tridecane are ordinary coupling reaction products between the magnesium reagent and the dodecyl group. However, the formation of dodecane could be ascribed to the intermediary dodecyl radical generated by a single electron transfer from the reagent.

Extension of this new method to mesylate (**26a, 26b**) derived from homopropargyl alcohols provided 1-(dimethylphenylsilyl)-1-cyclobutenes **28a** and **28b** in addition to the expected three membered ring products **27a** and **27b**. The structure of **28a** was confirmed by the comparison with a sample prepared from 4-bromo-1-(dimethylphenylsilyl)-1-butyne and i-Bu$_2$AlH according to the reported procedure.[20] The reaction of **26a** proceeded fast at 0°C and intermediacy of alkenyl metal compound could not be detected by TLC. Cyclopropylidene compounds **27** could be formed by the same addition–substitution sequence as the formation of four- to six-membered ring products. Meanwhile, we are tempted to assume that the cyclobutene **28** arises from a radical intermediate **29** cyclizing to give the four-membered ring. Baldwin's rule prefers 4-*Endo*-Dig mode rather than 3-*Exo*-Dig.[21]

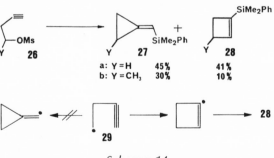

a: Y = H   45%   41%
b: Y = CH$_3$ 30%   10%

Scheme 14

The new cyclization technique could be applied to the allenic compound bearing leaving group and provided us with a simple route to bicyclic compounds which has been exemplified by the synthesis of *cis*-bicyclo [4.3.0]nonan-3-one.[22]

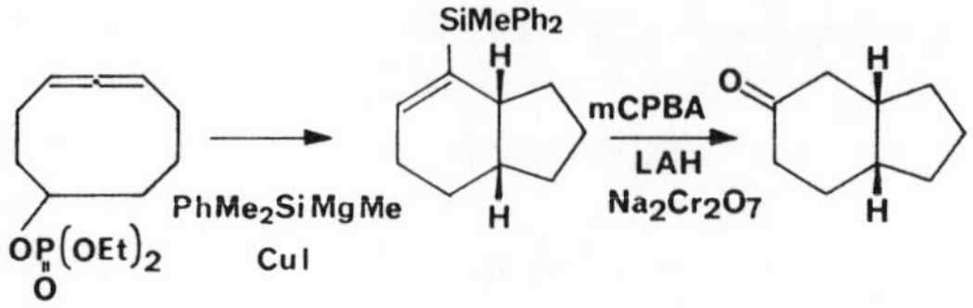

Scheme 15

A THF solution of PhMe$_2$SiLi (0.5 M, 12 mL, 6.0 mmol) was treated with MeMgI (1.0 M ether solution, 6.0 mL, 6.0 mmol) in THF (5 mL) at 0°C. CuI (19 mg, 0.1 mmol) was added and the resulting mixture was stirred for 5 min. A THF solution of the tosylate of 5-hexyn-1-ol **20a** (0.51 g, 2.0 mmol) was added and the whole was stirred for 30 min at 0°C, then for an additional 1 h at 25°C. Extractive workup (ether, 1 M HCl) followed by purification by silica gel column chromatography gave (dimethylphenylsilyl)methylenecyclopentane **21a** (91% yield).

E.   Stannylmetalation of Acetylenes

The synthetic utility of vinylstannanes[23] is critically dependent on their availability. The hydrostannation (see Section VI) of acetylenes is the simplest and most direct route to vinylstannanes. This reaction, however, is generally not highly stereoselective.[24] *trans*-Addition products predominate under kinetic conditions, whereas *cis*-addition products tend to predominate under thermodynamic conditions. Although several other preparative routes[25] are now available, there still exists a need for new methods which have generality and high regio- and stereoselectivities. Here we describe a new method for the selective synthesis of vinylstannanes.

Piers and Chong[26] have reported that treatment of ω-substituted 1-alkynes with trimethylstannylcopper–dimethyl sulfide in THF in the presence of methanol affords good to excellent yields of the corresponding 2-(trimethylstannyl)-1-alkenes (Scheme 16). Without methanol the reaction is quite slow and a large amount of starting material is recovered because of the reversible nature of the reaction. In contrast, the reagents derived from n-Bu$_3$SnLi[27]

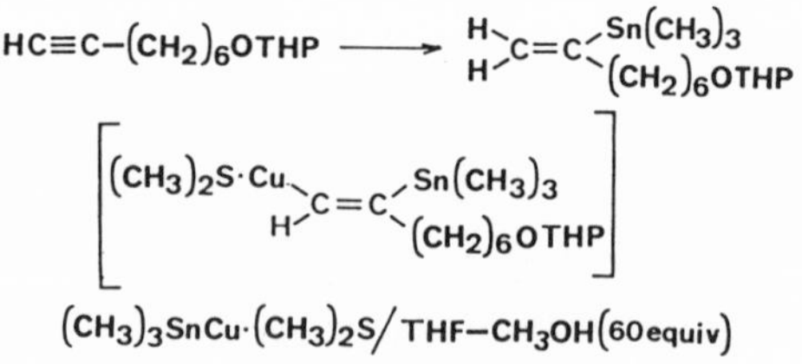

Scheme 16

Table 4. Transition Metal Catalyzed Stannylmetalation of Acetylenes[a]

$$RC{\equiv}CH \xrightarrow[\text{2.H}_3\text{O}^+]{\text{1.}n\text{-Bu}_3\text{Sn-Mtl cat}} \underset{\textbf{30}}{\overset{R}{\underset{H}{}}C{=}C\overset{H}{\underset{\text{Sn-}n\text{-Bu}_3}{}}} + \underset{\textbf{31}}{\overset{R}{\underset{n\text{-Bu}_3\text{Sn}}{}}C{=}C\overset{H}{\underset{H}{}}}$$

Entry	Reagent	Catalyst	Yield[b] (%)	Ratio of **30**:**31**[c]	
				30	**31**
R = PhCH$_2$OCH$_2$CH$_2$					
1	(n-Bu$_3$Sn)$_2$CuCN[d]	—	75	36	64
2	n-Bu$_3$SnMgMe[e]	CuCN	88	100	0
3	n-Bu$_3$SnMgMe	CuBr·SMe$_2$	23	34	66
4	n-Bu$_3$SnAlEt$_2$[e]	CuCN	86	81	19
5	(n-Bu$_3$Sn)$_2$Zn[f]	CuCN	63	26	74
6	(n-Bu$_3$Sn)$_2$Zn	Pd(PPh$_3$)$_4$	81	14	86
R = Ph					
7	n-Bu$_3$SnMgMe	CuCN	89	> 95	< 5
8	n-Bu$_3$SnAlEt$_2$	CuCN	88	79	21
9	(n-Bu$_3$Sn)$_2$Zn	Pd(PPh$_3$)$_4$	93	60	40
10	(n-Bu$_3$Sn)$_2$Zn	PdCl$_2$(PPh$_3$)$_2$	89	> 95	< 5
11	n-Bu$_3$SnMgMe	CuCN	70	70	30
12	n-Bu$_3$SnAlEt$_2$	CuCN	87	38	62
13	(n-Bu$_3$Sn)$_2$Zn	Pd(PPh$_3$)$_4$	70	< 5	> 95

[a]3 mmol of n-Bu$_3$Sn-Mtl reagent, 1 mmol of acetylenic compound and 5 mol% of catalyst were employed. [b]Isolated yields. [c]The ratios were determined by GLPC and ^{1}H-NMR spectra. [d]A reagent was produced by mixing the stannyllithium with CuCN in a 2/1 ratio. [e]Prepared from the stannyllithium and MeMgI (or Et$_2$AlCl) in a 1/1 ratio. [f]Prepared from the stannyllithium and ZnBr$_2$ in a 2/1 ratio.

and MeMgI, Et$_2$AlCl or ZnBr$_2$ have provided us with a novel route to vinylstannanes under good control of regio- and stereoselectivity.[28]

Treatment of 4-benzyloxy-1-butyne with n-Bu$_3$SnMgMe, generated from n-Bu$_3$SnLi and MeMgI, in the presence of a catalytic amount of CuCN, gave (*E*)-4-benzyloxy-1-tributylstannyl-1-butene as a single product.[29] Representative results are summarized in Table 4. Many combinations of n-Bu$_3$Sn–metal and transition metal catalysts were examined. Whereas the n-Bu$_3$SnAlEt$_2$ and CuCN system provided a mixture of (*E*)-1-tributylstannyl-1-alkene (**30**) and its regioisomer (**31**) in an 81/19 ratio (entry 4 in Table 4), (n-Bu$_3$Sn)$_2$Zn and Pd(PPh$_3$)$_4$ gave **31** predominantly (entry 6 in Table 4).

The reaction of n-Bu$_3$SnMgMe and CuCN with an internal acetylene such as 5-benzyloxy-2-pentyne gave no addition product and the starting compound was recovered unchanged. As shown in Scheme 17, stannyl-metalation proceeds in a *cis*-fashion. Treatment of phenylacetylene with n-Bu$_3$SnAlEt$_2$ in the presence of CuCN gave two isomers which were separated by preparative GLPC (Silicone OV 17 3% on Uniport HP, 2 m, 200°C). The product **32** (t_r = 4 min) showed ^{1}H-NMR (CDCl$_3$) signals at δ 5.43 (d, J = 2.7 Hz, H$_b$), 6.03 (d, J = 2.7 Hz, H$_a$). The other isomer, **33** having a longer retention time (t_r = 7 min) gave ^{1}H-NMR absorption at δ 6.87 (bs, 2H, H$_c$ and H$_d$). Assignment of the stereochemistry of H$_a$ and H$_b$ was based on the ^{1}H-NMR spectral data of the hydrostannation products.[30] Quenching the reaction mixture with D$_2$O provided an isomeric mixture whose ^{1}H-NMR spectrum showed only two signals in the olefinic region at δ 5.98 and 6.82. The disappearance of the higher field signal at δ 5.43 in the spectrum of compound **32** is also consistent with a *cis*-addition process.

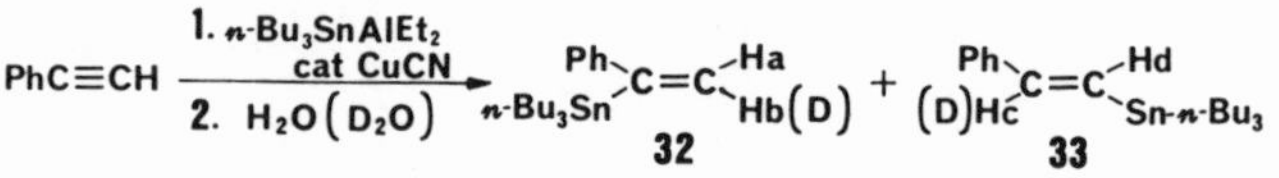

$$PhC{\equiv}CH \xrightarrow[\text{2. H}_2\text{O (D}_2\text{O)}]{\substack{\text{1. } n\text{-Bu}_3\text{SnAlEt}_2 \\ \text{cat CuCN}}} \underset{\textbf{32}}{\overset{Ph}{\underset{n\text{-Bu}_3\text{Sn}}{>}}C{=}C\overset{H_a}{\underset{H_b(D)}{<}}} + \underset{\textbf{33}}{\overset{Ph}{\underset{(D)H_c}{>}}C{=}C\overset{H_d}{\underset{Sn\text{-}n\text{-Bu}_3}{<}}}$$

Scheme 17

The new reaction has provided not only simple vinylstannanes but also functionalized alkenylstannanes on treatment of an intermediary alkenyl-metal species with various electrophiles. For instance, stannylmagnesation of 4-benzyloxy-1-butyne catalyzed by CuCN followed by the addition of MeI (large excess) gave **34b** (69% yield; Scheme 18).

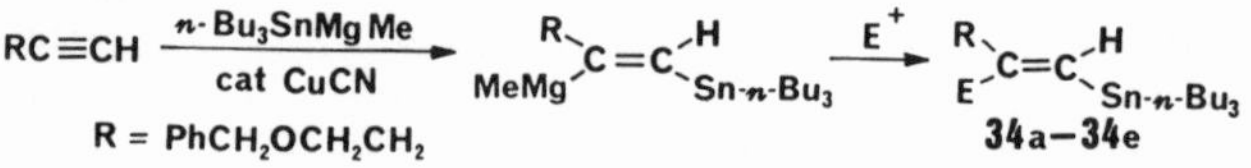

$$RC{\equiv}CH \xrightarrow[\substack{\text{cat CuCN} \\ R = PhCH_2OCH_2CH_2}]{n\text{-Bu}_3\text{SnMg Me}} \overset{R}{\underset{MeMg}{>}}C{=}C\overset{H}{\underset{Sn\text{-}n\text{-Bu}_3}{<}} \xrightarrow{E^+} \underset{\textbf{34a–34e}}{\overset{R}{\underset{E}{>}}C{=}C\overset{H}{\underset{Sn\text{-}n\text{-Bu}_3}{<}}}$$

a: E = H (H$_2$O, 88%) b: E = Me (MeI, 69%) c: E = Et (EtI, 73%)
d: E = Allyl (CH$_2$=CHCH$_2$Br, 78%) e: E = PhCH(OH) (PhCHO, 65%)

Scheme 18

We have developed a synthetic method with a reagent which is believed to have a Zn–Sn single bond as described above. In relation to this reagent, the isolable complex (Ph$_3$Sn)$_2$Zn·TMEDA[31] has been prepared and examined. The complex easily added to acetylenic bonds in the same manner as (Ph$_3$Sn)$_2$Zn derived from Ph$_3$SnLi and ZnBr$_2$ *in situ*. Treatment of 1-dodecyne with (Ph$_3$Sn)$_2$Zn·TMEDA in THF in the presence of CuCN catalyst gave a mixture of 2-(triphenylstannyl)-1-dodecene and the (*E*)-1-triphenylstannyl isomer (83:17) in 87% combined yield (Scheme 19).[32]

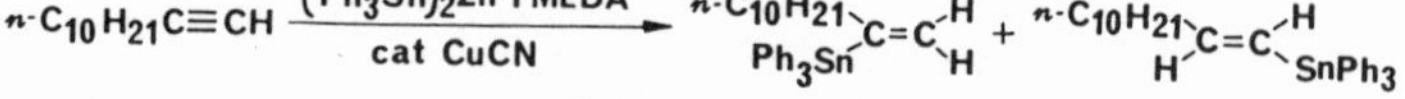

$$n\text{-C}_{10}\text{H}_{21}C{\equiv}CH \xrightarrow[\text{cat CuCN}]{(Ph_3Sn)_2Zn\cdot TMEDA} \overset{n\text{-C}_{10}\text{H}_{21}}{\underset{Ph_3Sn}{>}}C{=}C\overset{H}{\underset{H}{<}} + \overset{n\text{-C}_{10}\text{H}_{21}}{\underset{H}{>}}C{=}C\overset{H}{\underset{SnPh_3}{<}}$$

Scheme 19

Table 5. ^{119}Sn-NMR Spectral Data of Vinylstannanes[a]

R	R'	$\begin{matrix}R'_3Sn\\R\end{matrix}C=C\begin{matrix}H\\H\end{matrix}$	$\begin{matrix}H\\R\end{matrix}C=C\begin{matrix}SnR'_3\\H\end{matrix}$	$\begin{matrix}H\\R\end{matrix}C=C\begin{matrix}H\\SnR'_3\end{matrix}$
PhCH$_2$OCH$_2$CH$_2$	n-Bu	− 45.2	− 51.8	—
	Ph	− 132.8	− 137.6	—
n-C$_6$H$_{13}$	Me	− 35.7	− 40.7	− 58.3
	Ph	—	− 136.4	− 150.2
n-C$_5$H$_{11}$	Ph	− 133.5	—	—
n-C$_{10}$H$_{21}$	Ph	− 133.3	− 136.7	− 150.4
Ph	Me	− 28.0	− 32.4	− 51.3
	n-Bu	− 39.7	− 44.6	− 57.3
	Ph	− 130.3	− 131.6	− 149.8
Cyclohexyl	Me	− 35.9	− 38.2	− 57.9

[a]All spectra were measured in CDCl$_3$. The chemical shifts are given in δ (ppm) with tetramethylstannane as the internal standard.

The ^{119}Sn-NMR spectra are very useful for the determination of stereo-chemistry.[33] The chemical shifts of vinylstannanes which have different substitution patterns are summarized in Table 5.[28,32] The ratios of isomeric mixtures can be easily measured by integrating the peak area of the corresponding signals.

A hexane solution of butyllithium (1.5 M, 6.0 mL, 9.0 mmol) was added to a suspension of anhydrous tin(II) chloride (0.58 g, 3.0 mmol) in THF (4 mL) at 0°C. After stirring for 20 min, the reaction mixture was treated with an ethereal solution of methylmagnesium iodide (1.0 M, 3.0 mL, 3.0 mmol) at 0°C. After the resulting mixture had been stirred for 15 min, CuCN (4 mg, 5 mol%) and 4-benzyloxy-1-butyne (0.16 g, 1.0 mmol) in THF (5 mL) were added successively, and the whole was stirred for 30 min at 0°C. The reaction mixture was poured into a saturated NH$_4$Cl aqueous solution (20 mL) and extracted with ether. The combined organic layers were washed with brine, dried over anhydrous sodium sulfate, and concentrated. Purification by alumina column chromatography gave pure 4-benzyloxy-1-(tributylstannyl)-1-butene (0.38 g, 88% yield) as a colorless oil.

F.   Germylcupration of Acetylenes

The reaction of triphenylgermylcopper reagent generated from Ph$_3$GeLi and CuCN with 1-dodecyne in the presence of n-BuOH gave a mixture of 2-(triphenylgermyl)-1-dodecene (**35a**) and 1-(triphenylgermyl)-1-dodecene (**36a**) (**35a**:**36a** = 80:20) in 71% combined yield. The representative results are shown in Table 6.

Treatment of 1-dodecyne with the triphenylgermylcopper reagent without

Table 6. Germylcupration of Acetylenes

$$R^1C{\equiv}CH \xrightarrow{(R_3^2Ge)_2Cu(X)Li} \underset{35}{\overset{R^1}{\underset{R_3^2Ge}{>}}C{=}C\overset{H}{\underset{H}{<}}} + \underset{36}{\overset{R^1}{\underset{H}{>}}C{=}C\overset{H}{\underset{GeR_3^2}{<}}}$$

Reagent			Product	
R_3^2GeLi	Cu salt	Additive	Yield (%)[a]	Ratio of **35:36**
$R^1 = n\text{-}C_{10}H_{21}$				
Ph$_3$GeLi[b]	CuCN	n-BuOH	71	80:20
	CuCN	n-BuCHO	99	70:30
	CuI	n-BuOH	78	40:60
	CuBr·SMe$_2$	n-BuCHO	81	75:25[c]
Et$_3$GeLi[d]	CuBr·SMe$_2$	—	(81)[e]	70:30[f]
$R^1 = PhCH_2OCH_2CH_2$				
Ph$_3$GeLi	CuCN	n-BuOH	46	5:95
	CuBr·SMe$_2$	n-BuCHO	82	5:95
Et$_3$GeLi	CuBr·SMe$_2$	—	91	65:35[f]
$R^1 = Ph$				
Ph$_3$GeLi	CuCN	n-BuOH	85	<5:>95
Et$_3$GeLi	CuBr·SMe$_2$	—	91	15:85[f]
$R^1 = CH_2{=}C(CH_3)$				
Ph$_3$GeLi	CuBr·SMe$_2$	n-BuOH	98	1:99
Et$_3$GeLi	CuBr·SMe$_2$	—	95	13:87[f]

[a]Isolated yields unless otherwise noted. [b](Ph$_3$Ge)$_2$CuX (0.75 mmol), n-BuOH (or n-BuCHO, 2.3 mmol) and acetylene substrate (0.5 mmol) were employed. [c]Ether was used as solvent instead of THF. [d](Et$_3$Ge)$_2$Cu(SMe$_2$)$_{Li}$ (0.5 mmol) and acetylene substrate (0.5 mmol) were employed. [e]GLPC yield using octadecane as an internal standard (Silicone OV-17 3%, 2 m, 170°C). [f]MeOH was used for quenching the reaction.

n-BuOH gave no trace of triphenylgermylalkenes **35a** and **36a**. The starting acetylenic compound was recovered unchanged (>90%). Coexistence of a proton donor such as n-BuOH or water is essential for the formation of **35a** and **36a**. These weakly acidic proton donors react easily only with the intermediary vinylcopper compound and are reluctant to react with tri-phenylgermylcopper reagents. Benzenethiol and aniline were ineffective because the former reacted with germylcopper reagents and the latter was inert to both copper species due to its lower acidity. These results indicated that the reaction is reversible and the equilibrium favours the original system more than the intermediary vinylcopper species (Scheme 20). Similar results have been reported in the reaction of the stannylcopper reagent with acetylenic substrates by Piers and Chong (see Section I.E). Valeraldehyde,

3-pentanone and acetophenone were also effective to complete the reaction. The adducts between the intermediary vinylcopper species and carbonyl compounds could not be observed in the reaction mixture. The function of the carbonyl compounds is unclear.

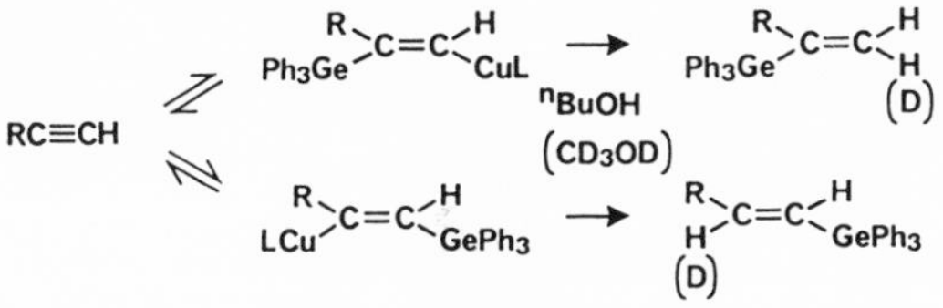

Scheme 20

The nature of the additives slightly affects the distribution of the isomeric vinylgermanes. The isomeric ratio of **35a/36a** = 80/20 was obtained with water and alcohols (MeOH, n-BuOH, s-BuOH and t-BuOH), and 70/30 was obtained with carbonyl compounds (nonanal, 3-pentanone and aceto-phenone). The type of copper salts with or without ligands also affects the product distributions. Isomeric ratios (**35a/36a**) in the reaction between 1-dodecyne and the copper reagents derived from Ph_3GeLi and various kinds of copper salts in the presence of n-BuCHO are given: CuCN (70/30), CuI (50/50), CuSCN (40/60), CuI/P(OPh)$_3$ (60/40), CuI/P-n-Bu$_3$ (40/60).

The reagent $(Et_3Ge)_2Cu(SMe_2)Li$ **37** prepared from Et_3GeLi and $CuBr \cdot SMe_2$ was also effective for the germylmetalation. In contrast to the reagent $(Ph_3Ge)_2Cu(CN)Li_2$, the addition of reagent **37** proceeded essentially to completion in the absence of proton donors. Thus, the intermediary alkenylcopper species could be trapped by various electrophiles. For instance, germylcupration of phenylacetylene followed by the addition of MeI gave the corresponding alkenylgermane in 82% combined yield (Scheme 21).

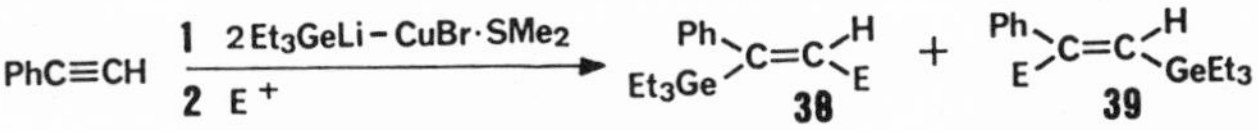

a: E = D (D$_2$O, **38/39** = 7/93, 89%)   b: E = Me (MeI, **38/39** = 17/83, 82%)
c: E = CH$_2$=CHCH$_2$ (CH$_2$=CHCH$_2$Br, **38/39** = 10/90, 96%)

Scheme 21

The regiochemistry heavily depends not only on the nature of the copper reagent but also on the structure of the acetylenic compounds. Treatment of propyne or 1-dodecyne with the reagent $(Ph_3Ge)_2Cu(CN)Li_2$ or $(Et_3Ge)_2$-$Cu(SMe_2)Li$ provided 2-germyl-1-alkenes as the main products. Meanwhile, the germylcupration of the substrate bearing unsaturated groups α,β to the acetylenic linkage proved to give 1-germyl isomer preferentially.

cis-Addition of germylcopper reagents was confirmed by the examination of the products with the authentic samples prepared from the corresponding vinyl bromide and t-BuLi/Et$_3$GeCl.

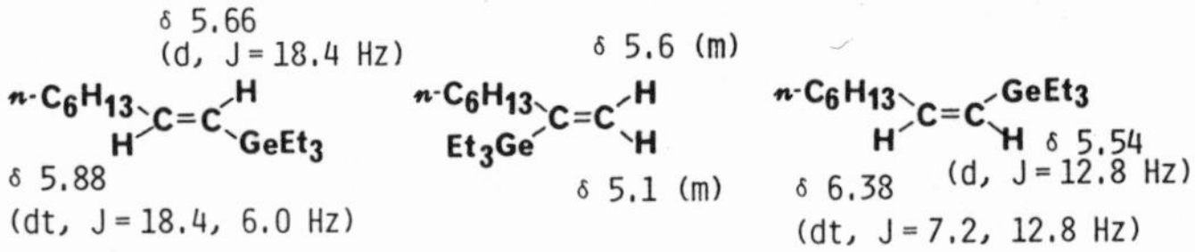

Scheme 22

The copper reagent derived from R$_3$GeLi and CuX (1:1) was as effective as the digermylcuprates (Ph$_3$Ge)$_2$Cu(CN)Li$_2$ and (Et$_3$Ge)$_2$Cu(SMe$_2$)Li. The use of catalytic amounts of CuX (5 mol%) instead of stoichiometric amounts decreased the yield of the corresponding germylalkenes (<5%). Many combinations of organometallic compounds and transition metal catalysts were examined for the germylmetalation of 1-dodecyne. In contrast to silylmetalation and stannylmetalation, almost all combinations failed to give vinylgermanes. Only Ph$_3$Ge–AlEt$_2$ derived from Ph$_3$GeLi and Et$_2$AlCl provided the corresponding vinylgermanes in the presence of various kinds of transition metal catalysts. For instance, the PdCl$_2$(PPh$_3$)$_2$ (2 mol%) catalyzed reaction of 1-dodecyne at 25°C for 14 h gave a mixture of **35a** and **36a** (**35a/36a** = 40/60) in 65% yield after hydrolysis.

III.   DEVELOPMENT OF PhMe$_2$SiZnR$_2$Li REAGENT AND ITS CHARACTERISTICS

The previously reported reaction of Si–Mg, Si–Al or Si–Zn reagents with an acetylenic linkage affords a simple and general method of vinylsilane synthesis (see Section II.A). The method, however, has two major drawbacks: (1) whereas the terminal acetylenes react with these reagents very easily, the internal acetylenes are reluctant to react, (2) regioselective preparation of 1-silyl-1-alkenes from 1-dodecyne was easily performed with the combination of PhMe$_2$SiMgMe-CuI or PhMe$_2$SiMgMe-PtCl$_2$(P-n-Bu$_3$)$_2$. In contrast, it is difficult to obtain 2-silyl-1-alkenes with high regioselectivity. The combination of PhMe$_2$SiAlEt$_2$–PdCl$_2$(P(o-tolyl)$_3$)$_2$ gave the best results so far and provided 2-silyl-1-dodecene as the main product (85%) along with the 1-silyl isomer (15%) upon treatment of 1-dodecyne. Here we describe the new silylmetalation reactions which solve these problems.[34]

Extensive studies have been explored with the reactions of cuprates such as conjugate addition or substitution.[35] In contrast, few examples are known for the synthetic utility of organozincate reagents.[36] It is well known that ate complexes such as R_4BLi and R_4AlLi are much more reactive than the corresponding organometallic reagents, R_3B and R_3Al.[37] Thus, the ate complexes, $PhMe_2SiZnR_2Li$ and $PhMe_2SiAlR_3Li$, were hoped to react with internal acetylenes as well as terminal ones. This was indeed the case and the representative results are shown in Tables 7 and 8.[38] The reaction has following characteristics:

(1) CuI, CuCN and $Pd(PPh_3)_4$ were effective catalysts for these silylmetalation reactions, while $RhCl(PPh_3)_3$ and $RuCl_2(PPh_3)_3$ were not efficient.[39] The uncatalyzed reaction of silylzinc and silylaluminium compounds with acetylenes proceeded very slowly to provide silylated olefins in low yield after prolonged reaction time. For instance, the reaction of $PhMe_2SiZnEt_2Li$ with 1-dodecyne gave a mixture of 1-(dimethylphenylsilyl)-1-dodecene and the 2-silyl isomer (5:3) in 13% combined yield after stirring at 25°C for 20 h. (2) The regioselectivity of the reaction was strongly affected by the nature of the dialkylzinc and trialkylaluminium employed. The use of bulky alkyl group favours the formation of 2-silyl-1-alkenes. Silylzinc reagent, $PhMe_2$ $SiZn$-t-Bu_2Li[40] gave higher regioselectivity than the silylalanate, $PhMe_2SiAl$-t-Bu_3Li.[41] In the case of the former reagent, high selectivity ($> 95\%$) was achieved for all examined substrates except entry 16 in Table 7. The selective preparation of 1-(dimethylphenylsilyl)-1-alkenes has already been achieved with $PhMe_2SiMgMe$ in the presence of CuI or $PtCl_2(P$-n-$Bu_3)_2$ catalyst. Thus, we have succeeded to obtain both regioisomers with high selectivity ($> 95\%$). (3) Regioselective silylmetalation was performed for propargylic and homopropargylic alcohols. As shown in Table 7 (entry 18 and 19) and Table 8 (entry 10), 2-butyn-1-ol or 3-pentyn-1-ol gave (E)-3-(dimethylphenylsilyl)-2-buten-1-ol or (E)-4-(dimethylphenylsilyl)-3-penten-1-ol as a single product. The silylanion attacks the remote acetylenic carbon from hydroxyl group exclusively.

The *cis* mode of the addition was confirmed by the comparison of the tetrahydropyranyl ether of (Z)-3-penten-1-ol derived from the silylzincation product of 3-pentyn-1-ol (entry 19 in Table 7) with an authentic sample (Scheme 23).

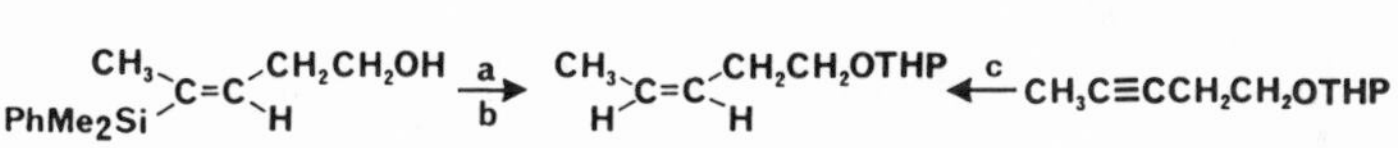

a: dihydropyran, $TsOH/CH_2CH_2$ b: n-$Bu_4NF/HMPA$
c: H_2/Pd-C, 5% $BaSO_4$, quinoline

Scheme 23

As an extension of this technique, we have examined the reaction of silylated acetylenes with the silylzinc reagents. The results are summarized in Table 9. 1,2-Disilylated alkenes were obtained exclusively regardless of the nature of the other substituent on acetylenic carbon. However, the stereo-

Table 7. Silylzincation of Acetylenes[a]

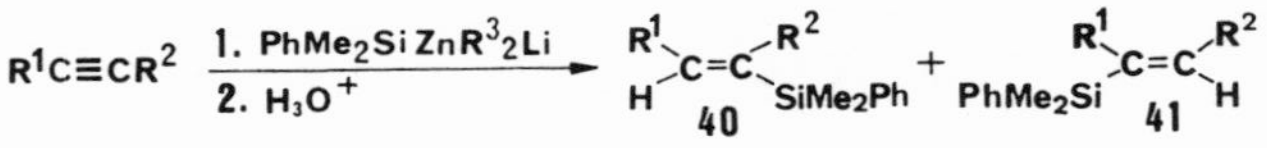

$$R^1C{\equiv}CR^2 \xrightarrow[\text{2. } H_3O^+]{\text{1. } PhMe_2SiZnR^3_2Li} \underset{\mathbf{40}}{R^1{\diagup}C{=}C{\diagdown}{}^{R^2}_{SiMe_2Ph}} + \underset{\mathbf{41}}{{}_{PhMe_2Si}{\diagdown}^{R^1}C{=}C{\diagup}{}^{R^2}_{H}}$$

| Entry | Substrate | | Reagent[b] | Yield (%) | Ratio of 40:41 | |
	R¹	R²			40	41
1	n-C₁₀H₂₁	H	Ph₃SiZnEt₂Li	90[c]	100	0
2			PhMe₂SiZnEt₂Li[d]	80	75	25
3			PhMe₂SiZnEt₂Li	81	58	42
4			PhMe₂SiZnEt₂Li[e]	60	30	70
5			PhMe₂SiZn–t-Bu₂Li	92	1	99
6	THPOCH₂CH₂	H	(PhMe₂Si)₃ZnMgMe[f]	97	100	0
7			PhMe₂SiZnEt₂Li	80	67	33
8			PhMe₂SiZn–i-Pr₂Li	97	30	70
9			PhMe₂SiZn–t-Bu₂Li	87	1	99
10	PhCH₂OCH₂CH₂	H	Ph₃SiZnEt₂Li	87[c]	100	0
11			PhMe₂SiZnEt₂Li	78	67	33
12			PhMe₂SiZn–i-Pr₂Li	91	33	67
13			PhMe₂SiZn–t-Bu₂Li	98	5	95
14	HOCH₂	H	PhMe₂SiZnEt₂Li[g]	82	100	0
15	HOCH₂CH₂	H	PhMe₂SiZnEt₂Li[g]	84	75	25
16			PhMe₂SiZn–t-Bu₂Li[g]	83	47	53

120

Entry	R	R'	Reagent	Yield		
17	n-C₅H₁₁	n-C₅H₁₁	PhMe₂SiZnEt₂Li	90		—
18	HOCH₂	CH₃	PhMe₂SiZnEt₂Li[g]	85	100	0
19	HOCH₂CH₂	CH₃	(PhMe₂Si)₃ZnLi[g]	89	100	0
20	THPOCH₂CH₂	n-C₃H₇	(PhMe₂Si)₃ZnLi	96	50	50
21	PhCH₂OCH₂CH₂	CH₃	PhMe₂SiZnEt₂Li	78	80	20
22	t-BuMe₂SiOCH₂CH₂	CH₃	PhMe₂SiZnEt₂Li	62	83	17

[a] The reactions were performed at 25°C in THF. Reagent (2.0 mmol), acetylene (1.0 mmol) and catalyst (CuCN, 2 mol%) were employed. [b] Prepared from PhMe₂SiLi and the corresponding dialkylzinc (1:1) at 25°C. [c] The corresponding triphenylvinylsilane was obtained. [d] Pd(PPh₃)₄ was used as a catalyst. [e] CoCl₂(PPh₃)₂ was used as a catalyst. [f] Prepared from PhMe₂SiMgMe (3.0 mmol) and ZnBr₂ (1.0 mmol). [g] Reagent (3.0 mmol) and acetylene (1.0 mmol) were employed.

Table 8. Silylalumination of Acetylenes[a]

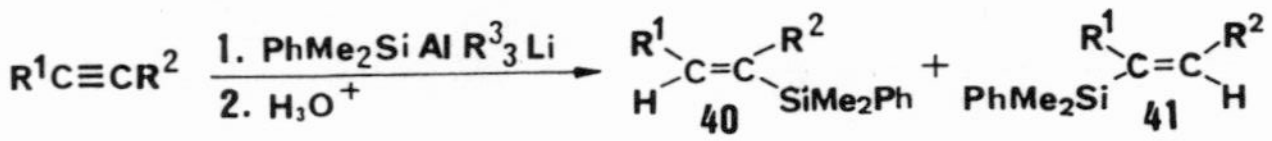

$$R^1C{\equiv}CR^2 \xrightarrow[\text{2. H}_3\text{O}^+]{\text{1. PhMe}_2\text{Si Al R}^3_3\text{ Li}} \underset{\mathbf{40}}{R^1{\diagdown}C{=}C{\diagup}^{R^2}_{\text{SiMe}_2\text{Ph}}_{\;H}} + \underset{\mathbf{41}}{^{R^1}_{\text{PhMe}_2\text{Si}}{\diagdown}C{=}C{\diagup}^{R^2}_{H}}$$

Entry	Substrate R¹	Substrate R²	Reagent[b]	Yield (%)	Ratio of 40:41 — 40	41
1	n-C₁₀H₂₁	H	PhMe₂SiAlMe₃Li[c]	78	67	33
2			PhMe₂SiAlEt₃Li[d]	73	64	36
3			PhMe₂SiAl–t-Bu₃Li[d]	65	40	60
4			PhMe₂SiAl–t-Bu₃Li[c]	45	17	83
5	PhCH₂OCH₂CH₂	H	PhMe₂SiAlEt₃Li[c]	72	83	17
6	HOCH₂	H	PhMe₂SiAlEt₃Li[c]	89	100	0
7			PhMe₂SiAlEt₃Li[d]	90	100	0
8			PhMe₂SiAl–t-Bu₃Li[d]	90	67	33
9			PhMe₂SiAl–t-Bu₃Li[c]	90	14	86
10	HOCH₂	CH₃	PhMe₂SiAlEt₃Li[c,e]	90	100	0
11	n-Bu(HO)CH	n-C₃H₇	PhMe₂SiAlEt₃Li[d,e]	63	100	0
12	HOCH₂CH₂	H	PhMe₂SiAlEt₃Li[d,e]	90	64	36
13			PhMe₂SiAlEt₃Li[c,e]	40	83	17

[a]The reactions were performed at 25°C in THF. Reagent (2.0 mmol), acetylene (1.0 mmol) and catalyst (2 mol%) were employed. [b]Reagent (3.0 mmol) and acetylene (1.0 mmol) were employed. [c]CuCN was used as a catalyst. [d]Pd(PPh₃)₄ was used as a catalyst. [e]Reagent (3.0 mmol) and acetylene (1.0 mmol) were employed.

Table 9. Silylzincation of Timethylsilylacetylenes in the Presence of CuCN Catalyst[a]

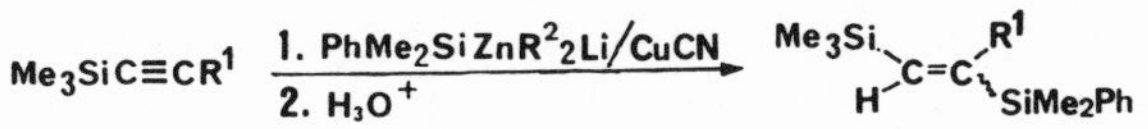

$$Me_3SiC{\equiv}CR^1 \xrightarrow[\text{2. } H_3O^+]{\text{1. } PhMe_2SiZnR^2{}_2Li/CuCN} \underset{H}{\overset{Me_3Si}{>}}C{=}C\underset{SiMe_2Ph}{\overset{R^1}{<}}$$

	Substrate		Yield	Ratio of E/Z	
Entry	R^1	Reagent	(%)	E	Z
1	H	$PhMe_2SiZnEt_2Li$	38	100	0
2		$PhMe_2SiZn{-}t{-}Bu_2Li$	78	100	0
3	$n{-}C_6H_{13}$	$(PhMe_2Si)_3ZnLi$	42	100	0
4		$PhMe_2SiZn{-}t{-}Bu_2Li$	53	100	0
5	CH_2OH	$PhMe_2SiZn{-}t{-}Bu_2Li$[b]	75	0	100
6	CH_2CH_2OH	$PhMe_2SiZn{-}t{-}Bu_2Li$[b]	44	0	100
7	CH_2CH_2OTHP	$PhMe_2SiZn{-}t{-}Bu_2Li$	72	100	0
8	$CH_2CH_2CH_2OH$	$PhMe_2SiZn{-}t{-}Bu_2Li$[b]	74	100	0

[a]Reagent (2.0 mmol), acetylene (1.0 mmol) and catalyst (2 mol%) were employed. [b]3 mmol of reagent was used per 1.0 mmol of substrate.

chemistry of the reaction depends on the structure of the substrates. 1-Trimethylsilyl-1-octyne and 4-(2-tetrahydropyranyloxy)-1-trimethylsilyl-1-butyne gave the corresponding *cis* addition products, (*E*)-1,2-disilylated alkenes. Meanwhile, silyl-acetylenes having free hydroxyl group such as 3-trimethylsilyl-2-propyn-1-ol and 4-trimethylsilyl-3-butyn-1-ol provided (*Z*)-2-(dimethylphenylsilyl)-3-trimethylsilyl-2-propen-1-ol and (*Z*)-3-(dimethylphenylsilyl)-4-trimethylsilyl-3-buten-1-ol, respectively (Table 9, entry 5 and 6). Selective desilylation (PhMe₂Si) of these products in the presence of a Me₃Si group by n-Bu₄NF[42] in THF provided (*E*)-3-trimethyl-silyl-2-propen-1-ol and (*E*)-4-trimethyl-silyl-3-buten-1-ol, respectively (Scheme 24).

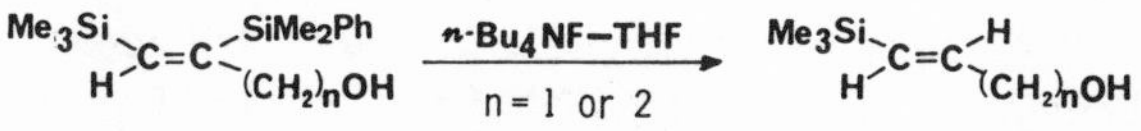

$$\underset{H}{\overset{Me_3Si}{>}}C{=}C\underset{(CH_2)_nOH}{\overset{SiMe_2Ph}{<}} \xrightarrow[n=1 \text{ or } 2]{n\text{-}Bu_4NF-THF} \underset{H}{\overset{Me_3Si}{>}}C{=}C\underset{(CH_2)_nOH}{\overset{H}{<}}$$

Scheme 24

Apparently, the reaction proceeded in *trans* fashion in contrast to other substrates. These exceptional results could be explained by assuming the isomerization of primary *cis* adducts into (*Z*)-isomers which are stabilized by the chelation of neighbouring oxygen atom forming the five- or six-membered rings depicted in Scheme 25. Alternatively, *trans* addition of the silyl group and zinc might occur by the intramolecular assistance of the zinc-alkoxy group. However, no *trans* adducts were observed in the reaction mixture of 2-butyn-1-ol or 3-pentyn-1-ol with $PhMe_2SiZnEt_2Li$. The coexistence of the silyl group and the free hydroxyl group is essential for the *trans* addition. Thus, we prefer the former isomerization process to the direct *trans* addition mechanism for the formation of (Z)-disilylated alkenes. This was confirmed by the fact that (Z)-3-(dimethylphenylsilyl)-4-iodo-4-trimethyl-silyl-3-buten-1-ol gave (Z)-3-(dimethylphenylsilyl)-4-trimethylsilyl-3-buten-1-ol exclusively on treatment with n-BuLi followed by aqueous workup (Scheme 26).[43]

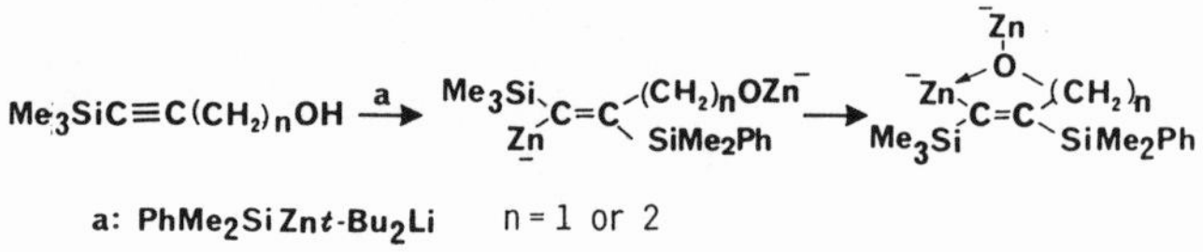

$$Me_3SiC{\equiv}C(CH_2)_nOH \xrightarrow{a} \underset{Zn}{\overset{Me_3Si}{\diagdown}}C{=}C\overset{(CH_2)_nOZn^-}{\underset{SiMe_2Ph}{\diagup}} \longrightarrow \underset{Me_3Si}{\overset{^-Zn\overset{Zn}{\overset{O}{\diagdown}}}{\diagdown}}C{=}C\overset{(CH_2)_n}{\underset{SiMe_2Ph}{\diagup}}$$

a: $PhMe_2SiZn\,t\text{-}Bu_2Li$ n = 1 or 2

Scheme 25

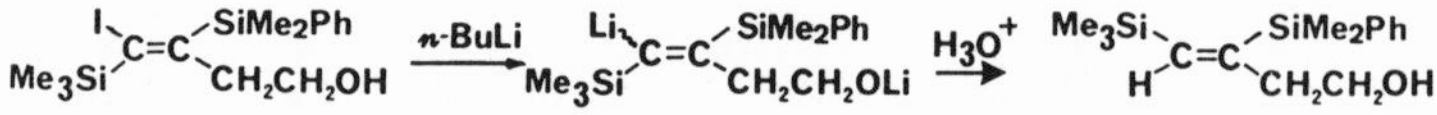

$$\underset{Me_3Si}{\overset{I}{\diagdown}}C{=}C\overset{SiMe_2Ph}{\underset{CH_2CH_2OH}{\diagup}} \xrightarrow{n\text{-BuLi}} \underset{Me_3Si}{\overset{Li}{\diagdown}}C{=}C\overset{SiMe_2Ph}{\underset{CH_2CH_2OLi}{\diagup}} \xrightarrow{H_3O^+} \underset{H}{\overset{Me_3Si}{\diagdown}}C{=}C\overset{SiMe_2Ph}{\underset{CH_2CH_2OH}{\diagup}}$$

Scheme 26

The authentic samples of tetrahydropyranyl ethers were prepared by hydrosilylation followed by treatment with PhMgBr (Scheme 27).

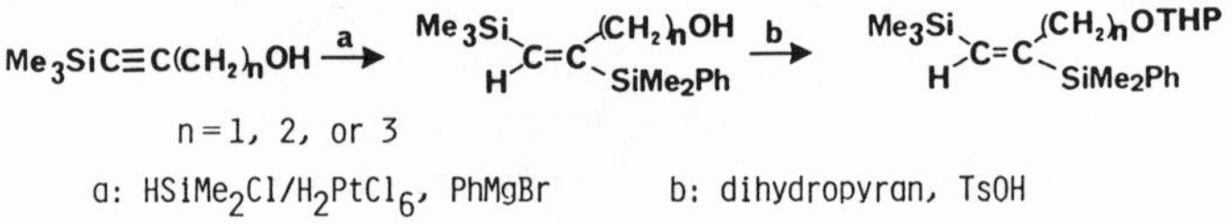

$$Me_3SiC{\equiv}C(CH_2)_nOH \xrightarrow{a} \underset{H}{\overset{Me_3Si}{\diagdown}}C{=}C\overset{(CH_2)_nOH}{\underset{SiMe_2Ph}{\diagup}} \xrightarrow{b} \underset{H}{\overset{Me_3Si}{\diagdown}}C{=}C\overset{(CH_2)_nOTHP}{\underset{SiMe_2Ph}{\diagup}}$$

n = 1, 2, or 3

a: $HSiMe_2Cl/H_2PtCl_6$, PhMgBr   b: dihydropyran, TsOH

Scheme 27

A THF solution of $PhMe_2SiLi$ (0.59 M, 1.7 mL, 1.0 mmol) was added to an ethereal solution of $t\text{-}Bu_2Zn$ (1.0 M, 1.0 mL, 1.0 mmol) at 0°C under an argon atmosphere. After stirring for 20 min, the reaction mixture was diluted with THF (2.0 mL). Then a catalytic amount of CuCN (2.0 mg, 0.02 mmol) and a THF solution of 1-dodecyne (85 mg, 0.5 mmol) were added at 0°C, and the whole was stirred at 0°C for 30 min and at 25°C for an additional 50 min. The resulting mixture was poured into sat. NH_4Cl and extracted with ethyl acetate. The combined organic layers were dried and concentrated *in vacuo*. Purification of the residual oil by preparative TLC on silica gel gave 2-(dimethylphenylsilyl)-1-dodecene (0.14 g) in 92% yield.

IV. DISILYLATION OF ACETYLENES WITH Si–Mn REAGENTS

Extensive studies on the chemistry of the combined acid–base reagents such as Si–Al, Si–Zn, Sn–Al and Sn–Zn compounds with many substrates (acetylenes, epoxides or α-halocarbonyl compounds) have shown us that the counter cations of silylanion or stannylanion play a critical role for the determination of the products and also for the distribution of the isomeric ratios of the products. Thus, we investigated many reagents containing various kinds of metal cations as a gegen ion of silylanion or stannylanion and found that the reaction of Si–Mn with acetylenes resulted in rather unexpected formation of disilylated products.[44,45]

(Trimethylsilyl)lithium[46] was treated with methylmagnesium iodide and anhydrous manganese(II) chloride.[47] Addition of 1-alkyne to the resulting reagent provided an *E*, *Z* mixture of 1,2-bis(trimethylsilyl)-1-alkene. The reaction proceeded smoothly with silylacetylenes as well as terminal acetylenes. Terminal acetylenes gave mixtures of *E* and *Z* isomers of disilylated products. Although the role of MeMgI is not clear, the presence is essential for the formation of disilylated products. Without MeMgI, monosilylated products were obtained predominantly after aqueous workup. For instance, treatment of 4-(benzyloxy)-1-butyne with $3Me_3SiLi-MnCl_2$ gave a mixture of 4-(benzyloxy)-2-(trimethylsilyl)-1-butene, 4-(benzyloxy)-1-(trimethylsilyl)-1-butene and the disilylated product in a 1:1:1 ratio (65% combined yield). The reaction has been extended to distannylation of acetylenes. The results are summarized in Table 10.

Treatment of 5-(benzyloxy)-2-pentyne with the reagent which is believed to be $(Me_3Si)_3MnMgMe$ gave 2,3-bis(trimethylsilyl)-2-alkene **44a** in 78% yield. Meanwhile, the addition of H_2O (or D_2O, MeI) to the reaction mixture after stirring at 0°C for 20 min without warming up to room temperature gave monosilylated product **44b** (71%) (or **44c** (71%), **44d** (70%)) along with the disilylated product **44a** (13–20%, Scheme 29).[48] Thus, the formation of the disilylated product may be explained as follows: (1) addition of the reagent $(Me_3Si)_3MnMgMe$ to triple bond in *cis* fashion giving silylated alkenyl-manganese **43** and (2) reductive elimination of manganese affording disilylated olefin. The *cis* addition of Si–Mn component was confirmed by the following experiment. A monosilylated alkene, 5-(benzyloxy)-2-(dimethylphenylsilyl)-2-pentene was prepared from $(PhMe_2Si)_3MnMgMe$ and 5-(benzyloxy)-2-pentyne in the same way as **44b**. Protodesilylation with n-Bu_4NF[42] gave 5-(benzyloxy)-2-pentene. The examination of the ^1H-NMR spectra proved that the olefin had *Z* configuration (> 95%. *J* = 11 Hz, Scheme 30). Similar desilylation of **44b** failed.

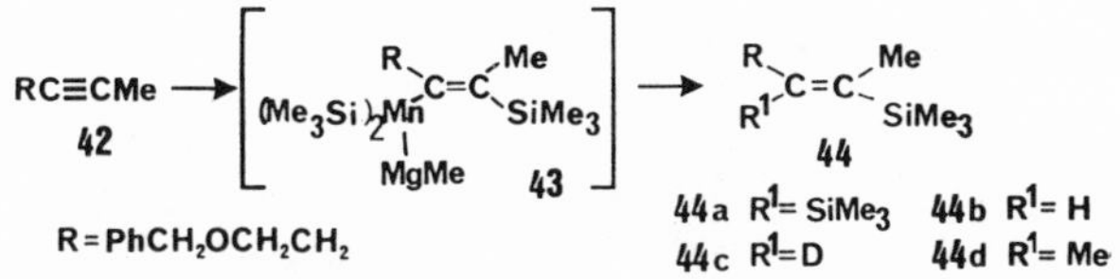

$R = PhCH_2OCH_2CH_2$

44a $R^1 = SiMe_3$ 44b $R^1 = H$
44c $R^1 = D$ 44d $R^1 = Me$

Scheme 29

Table 10.   Disilylation or Distannylation of Acetylenes

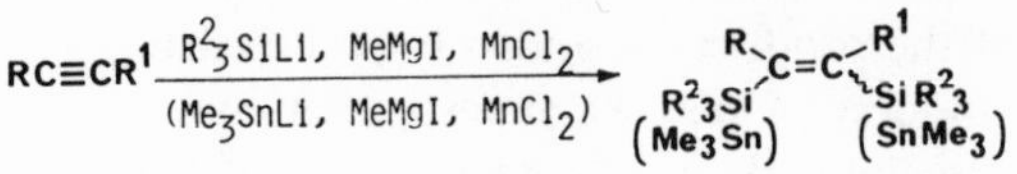

$$RC{\equiv}CR^1 \xrightarrow[\text{(Me}_3\text{SnLi, MeMgI, MnCl}_2)]{R^2_3\text{SiLi, MeMgI, MnCl}_2} \quad \underset{\substack{R^2_3\text{Si} \\ (\text{Me}_3\text{Sn})}}{\overset{R}{\diagdown}}C=C\underset{\substack{\text{SiR}^2_3 \\ (\text{SnMe}_3)}}{\overset{R^1}{\diagup}}$$

Acetylene	Reagent	Product	
$RC{\equiv}CR^1$	$R^2_3\text{SiLi}$	Yield (%)	Z/E
$HC{\equiv}CSiMe_3$	Me_3SiLi	72	—
$n\text{-}C_6H_{13}C{\equiv}CH$	Me_3SiLi	66	33/67
$PhC{\equiv}CH$	Me_3SiLi	65	50/50
$n\text{-}C_6H_{13}C{\equiv}CSiMe_3$	Me_3SiLi	80	—
$PhCH_2OCH_2CH_2C{\equiv}CH$	Me_3SiLi	55	50/50
$PhCH_2OCH_2CH_2C{\equiv}CH$	$PhMe_2SiLi$	51	42/58
$PhCH_2OCH_2CH_2C{\equiv}CD$	$PhMe_2SiLi$	55	42/58[a]
$PhCH_2OCH_2CH_2C{\equiv}CSiMe_3$	Me_3SiLi	59	—
$THPOCH_2CH_2C{\equiv}CH$	Me_3SiLi	63	65/35
$THPOCH_2CH_2C{\equiv}CH$	$PhMe_2SiLi$	70	65/35
$THPOCH_2CH_2C{\equiv}CSiMe_3$	Me_3SiLi	83	—
$HOCH_2CH_2C{\equiv}CSiMe_3$	Me_3SiLi[b]	58	—
$n\text{-}C_{10}H_{21}C{\equiv}CH$	Me_3SnLi[c]	56	100/0
$n\text{-}C_{10}H_{21}C{\equiv}CD$	Me_3SnLi	55	100/0[a]
$THPOCH_2CH_2C{\equiv}CH$	Me_3SnLi	54	100/0
$THPOCH_2CH_2C{\equiv}CH$	$n\text{-}Bu_3SnLi$	48	90/10
$PhCH_2OCH_2CH_2C{\equiv}CH$	Me_3SnLi	52	90/10

[a]Deuterium remained completely, thus, reagents did not cause the acetylenic proton–metal exchange. [b]3 mmol of manganese reagent and 1.0 mmol of substrate were employed. [c]Prepared from $SnCl_2$ and 3 equivalents of alkyllithium.

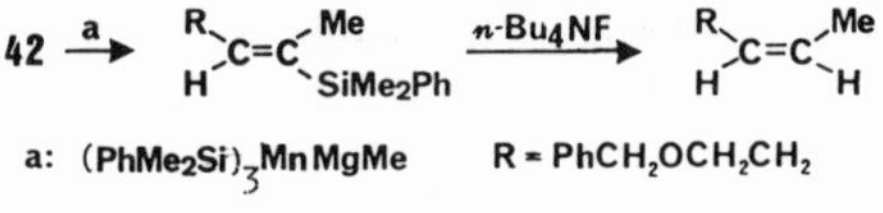

$$42 \xrightarrow{a} \underset{H}{\overset{R}{\diagdown}}C=C\underset{SiMe_2Ph}{\overset{Me}{\diagup}} \xrightarrow{n\text{-}Bu_4NF} \underset{H}{\overset{R}{\diagdown}}C=C\underset{H}{\overset{Me}{\diagup}}$$

a: $(PhMe_2Si)_3MnMgMe$ $R = PhCH_2OCH_2CH_2$

Scheme 30

It is worth noting that the reaction can be successfully applied to the synthesis of highly strained tetrakis(trimethylsilyl)ethene which is not readily available by any known method.[49]

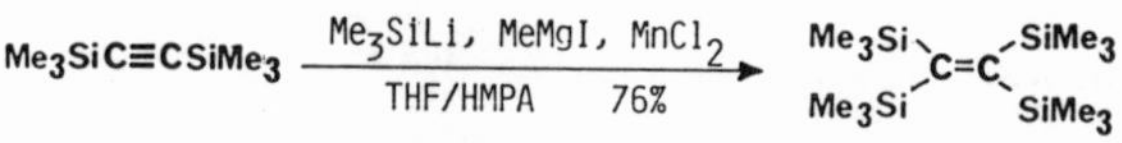

$$Me_3SiC{\equiv}CSiMe_3 \xrightarrow[\text{THF/HMPA}\quad 76\%]{Me_3SiLi, MeMgI, MnCl_2} \underset{Me_3Si}{\overset{Me_3Si}{\diagdown}}C=C\underset{SiMe_3}{\overset{SiMe_3}{\diagup}}$$

Scheme 31

Methyllithium (1.4 M, 3.2 mL, 4.5 mmol) was added to a solution of hexamethyldisilane (0.93 mL, 4.5 mmol) in THF (8 mL)–HMPA (2 mL) at 0°C. After the mixture was stirred for 15 min, methylmagnesium iodide (1.5 M, 3.0 mL, 4.5 mmol) was added to the resulting solution of (trimethylsilyl)lithium. The mixture was stirred for another 15 min and anhydrous manganese(II) chloride (0.19 g, 1.5 mmol) was added in one portion. Then a solution of tetrahydropyranyl ether of 4-(trimethylsilyl)-3-butyn-1-ol (0.23 g, 1.0 mmol) in THF (3 mL) was added and the resulting mixture was stirred for 3 h at 0°C. The mixture was diluted with ether and poured into saturated NH_4Cl. Purification by preparative TLC on silica gel gave tetrahydropyranyl ether of 3,4,4-tris (trimethylsilyl)-3-buten-1-ol (0.31 g) in 81% yield as a colorless oil.

V. TRANSITION METAL CATALYZED ADDITION OF PhMe₂SiBEt₃Li TO ACETYLENES IN THE PRESENCE OF METHANOL

The reactions of organoboron compounds such as hydroboration or haloboration have been explored in detail.[50] Several compounds containing the B–Si bond have been synthesized;[51] nevertheless, no examples are known of their synthetic utility. Here we wish to disclose that silylboron and stannylboron compounds, $PhMe_2SiBEt_3Li$ and $n-Bu_3SnBEt_3Li$, add to triple bonds effectively to give vinylsilanes or vinylstannanes, respectively, under good control of the regio- and stereoselectivity.[52]

The reaction of 1-dodecyne with 2 equivalents of $PhMe_2SiBEt_3Li$ in the presence of a catalytic amount of CuCN afforded a mixture consisting of the starting material (35% recovery) and the desired vinylsilanes (1-(dimethylphenylsilyl)-1-dodecene:2-silyl isomer = 64:36, 50% combined yield) after quenching the reaction mixture with 1 M HCl. Prolonging the reaction time could not improve the conversion. In contrast, treatment of propargyl alcohol under the same reaction conditions provided (E)-3-(dimethylphenylsilyl)-2-propen-1-ol as a single product in 87% yield. These results indicated that (1) the reaction of acetylenic compounds with $PhMe_2SiBEt_3Li$ is reversible and the equilibrium favors only marginally the intermediate vinylborate, and (2) in the presence of a proton source, the equilibrium is shifted to the right by protonation of the intermediate. The reversible nature of the reaction of $PhMe_2SiBEt_3Li$ with acetylenic compounds has been clearly demonstrated by the following result. Treatment of the reaction mixture derived from 4-benzyloxy-1-butyne and $PhMe_2SiBEt_3Li$ with allyl bromide (5 equivalents) gave allyldimethylphenylsilane in 90% yield along with the recovered starting material (85%).

Thus, it was anticipated that the addition of methanol to the reaction mixture of 1-dodecyne and $PhMe_2SiBEt_3Li$ would force the reaction go to completion, and this was indeed the case. Water, diethyl malonate and phenol were also effective to complete the reaction. We are tempted to assume the reaction scheme shown in Scheme 32. Weak acids such as water, methanol and phenol react easily only with the intermediary vinylcopper compound or vinylborate, but are reluctant to react with the silylborate or with dimethylphenylsilylcopper reagents. Moderate stability of silylborates to water has been reported.[51b] The use of acetic acid instead of methanol as a proton source resulted in the formation of a mixture containing the vinylsilane and starting material, 1-dodecyne (1:1). Similar results have been observed in the reaction of stannylcopper and germylcopper reagents with acetylenic substrates (see Section II.F).

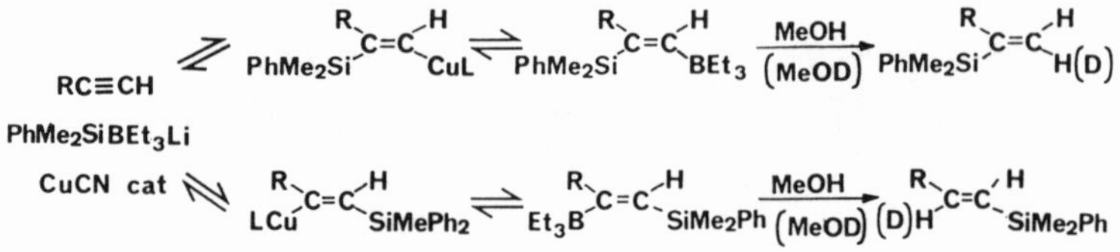

Scheme 32

Representative results are summarized in Table 11. CuI, CuBr·SMe$_2$ and CoCl$_2$(PPh$_3$)$_2$ also catalyze the reaction, while Pd(PPh$_3$)$_4$, RuCl$_2$(PPh$_3$)$_3$ and NiCl$_2$(PPh$_3$)$_2$ are not effective. The addition proceeded in *cis* fashion exclusively. The regioselectivity of the reaction heavily depends on the nature of the catalyst employed. CoCl$_2$(PPh$_3$)$_2$ gives exclusively the 1-silyl-1-alkene starting from 1-alkynes. Acetylenic compounds containing a hydroxyl group provide vinylsilanes selectively which have the silyl group on the carbon remote from the hydroxyl group.

The reaction has been extended to stannylboration of acetylenes. The results are also shown in Table 11. The uncatalyzed reaction of stannylboron compounds with acetylenes proceeded very slowly to give vinylstannanes in low yield (entry 14 in Table 11). Catalysts such as CuCN and CoCl$_2$(PPh$_3$)$_2$ were effective for this reaction.

The addition of a hexane solution of triethylborane (1.0 M, 2.0 mL, 2.0 mmol) to a greenish black THF solution of PhMe$_2$SiLi (0.75 M, 2.7 mL, 2.0 mmol) at 0°C gave a decolorized pale brown solution. Its ^{11}B-NMR spectrum showed a signal at -36.8 ppm, the high field shift of which (Et$_3$B: 68.5 ppm) reveals the formation of the ate complex, PhMe$_2$SiBEt$_3$Li. A catalytic amount of CuCN (0.1 mmol) was added and the color of the solution immediately changed to black. Then a solution of 1-dodecyne (0.17 g, 1.0 mmol) and methanol (0.41 mL, 10 mmol) in THF (2 mL) was added and the whole was stirred at 25°C for 1 h. Extractive workup (AcOEt, 1 M HCl) and purification by preparative TLC on silica gel gave a mixture of 1- and 2-(dimethylphenylsilyl)-1-dodecene in 89% combined yield (1-silyl isomer : 2-silyl isomer = 61:39).

Table 11. Silylboration and Stannylboration in the Presence of Methanol[a]

$$R^1C{\equiv}CR^2 \xrightarrow[\text{(n-Bu}_3\text{SnBEt}_3\text{Li)}]{\text{PhMe}_2\text{SiBEt}_3\text{Li}} \underset{\textbf{45}\ \left(\text{Sn-}n\text{-Bu}_3\right)}{\overset{R^1}{\underset{H}{>}}C{=}C\overset{R^2}{\underset{\text{SiMe}_2\text{Ph}}{<}}} + \underset{\left(n\text{-Bu}_3\text{Sn}\right)\ \textbf{46}}{\overset{R^1}{\underset{\text{PhMe}_2\text{Si}}{>}}C{=}C\overset{R^2}{\underset{H}{<}}}$$

Entry	Substrate R^1	R^2	Catalyst	Yield[b] (%)	Ratio[c] of **45:46**
PhMe$_2$SiBEt$_3$Li					
1	n-C$_{10}$H$_{21}$	H	CuCN	89	61:39
2			CoCl$_2$(PPh$_3$)$_2$	78	100:0
3	PhCH$_2$OCH$_2$CH$_2$	H	CuCN	91	68:32
4			CoCl$_2$(PPh$_3$)$_2$	57	100:0
5	n-C$_5$H$_{11}$	n-C$_5$H$_{11}$	CuCN	73	—
6	HOCH$_2$	H	CuCN	95	100:0
7	HOCH$_2$CH$_2$	H	CuCN	95	78:22
8			CoCl$_2$(PPh$_3$)$_2$	96	100:0
9	HOCH$_2$CH$_2$CH$_2$	H	CuCN	96	66:34
10			CoCl$_2$(PPh$_3$)$_2$	81	100:0
11	HOCH$_2$	CH$_3$	CuCN	85	100:0
12	HOCH$_2$CH$_2$	CH$_3$	CuCN	97	91:9
13	SiMe$_3$	n-C$_5$H$_{11}$	CuCN	66	100:0
n-Bu$_3$SnBEt$_3$Li					
14	PhCH$_2$OCH$_2$CH$_2$	H	—	24[d]	100:0
15			CuCN	40[d]	65:35
16			CoCl$_2$(PPh$_3$)$_2$	58[d]	80:20
17	HOCH$_2$CH$_2$	H	CoCl$_2$(PPh$_3$)$_2$	47	100:0

[a] 2 mmol of PhMe$_2$SiBEt$_3$Li (or n-Bu$_3$SnBEt$_3$Li) reagent, 1 mmol of acetylenic compound, 10 mol% of catalyst and 10 mmol of methanol were employed. [b] Isolated yields unless otherwise noted. [c] The ratios were determined by GLPC and ^{1}H-NMR. [d] GLPC yields using n-C$_{28}$H$_{58}$ as an internal standard (Silicone OV-17 5% on Uniport HP 60–80 mesh, 2 m, 250°C).

VI. Et$_3$B INDUCED RADICAL ADDITION OF R$_3$SnH AND R$_3$GeH TO ACETYLENES

In a former section (V), the behavior of n-Bu$_3$SnBEt$_3$Li was discussed. The addition of the ate complex to acetylenes required a coexistence of a proton source such as methanol, so that the intermediary vinylborate species could not be used for further transformation. Next we studied the reaction of acetylenes with the reagent which was prepared from Et$_3$B and Ph$_3$SnH and believed to be Ph$_3$SnBEt$_2$. Treatment of 1-dodecyne with the reagent gave a mixture of (*E*)- and (*Z*)-1-(triphenylstannyl)-1-dodecene. However, the expected intermediary vinylborane could not be trapped by any electrophiles such as D_2O, MeI or allyl bromide. The ^{11}B-NMR spectrum of a solution of Et_3B showed no change upon treatment with Ph_3SnH. These facts indicated that the reagent did not have the structure of Ph_3SnBEt_2 and it turned out that the reaction proceeded via free radical chain mechanism.

A. Et_3B Induced Radical Addition of R_3SnH to Acetylenes and its Application to Cyclization Reactions

The cyclization of vinyl acetylene to methylene-substituted five-membered rings has been described by Stork and Mook.[53] We have studied this reaction further and report that trialkylborane mediates a facile addition of R_3SnH to an acetylenic bond to give vinylstannane regioselectively and that this new method is applied to vinyl radical cyclization reactions[54,55] effectively.

The hydrostannation of acetylenes[56] takes place readily either in the absence of a catalyst or in the presence of a catalytic amount of free radical initiator such as azobisisobutyronitrile (AIBN),[57] but these reaction conditions (without solvent, heat to 80–100°C) may not always be suitable for an intramolecular radical cyclization reaction.[55f]

Table 12. Et_3B-Induced Hydrostannation of Acetylenes

$$R^1C{\equiv}CR^2 \longrightarrow \underset{47}{\overset{R^1}{\underset{H}{>}}C{=}C\overset{R^2}{\underset{SnR_3}{<}}} + \underset{48}{\overset{R^1}{\underset{H}{>}}C{=}C\overset{SnR_3}{\underset{R^2}{<}}}$$

Acetylene	Reagent	Reaction time (h)	Yield (%)	Ratio of **47**:**48**
$n\text{-}C_{10}H_{21}C{\equiv}CH$	Ph_3SnH	0.3	80	79:21
	$n\text{-}Bu_3SnH$	2.0	40	80:20
$PhCH_2OCH_2CH_2C{\equiv}CH$	Ph_3SnH	0.3	79	69:31
	$n\text{-}Bu_3SnH$	10	71	90:10
$THPOCH_2CH_2C{\equiv}CH$	Ph_3SnH	0.3	81	80:20
	$n\text{-}Bu_3SnH$	2.0	49	90:10
$HOCH_2CH_2C{\equiv}CH$	Ph_3SnH	0.3	87	82:18
	$n\text{-}Bu_3SnH$	2.0	40	69:31
$PhC{\equiv}CH$	Ph_3SnH	0.3	75	100:0
$Me_3SiC{\equiv}CH$	Ph_3SnH	0.3	83[b]	100:0
$n\text{-}C_5H_{11}C{\equiv}CC_5H_{11}$	Ph_3SnH	10	86[c]	0:100
$PhC{\equiv}CCH_3$	Ph_3SnH	1.0	74	25:75

[a]Acetylene (1.0 mmol), R_3SnH (1.2 mmol) and Et_3B (0.1 mmol) were employed. [b]Excess of (trimethylsilyl)acetylene (5.0 mmol) and Ph_3SnH (1.0 mmol) were employed and the yield was based on Ph_3SnH. [c]Excess of Ph_3SnH (5.0 mmol) was used.

We have found that an addition of a catalytic amount of Et_3B to a solution of acetylenic compound and Ph_3SnH (or $n-Bu_3SnH$) in toluene promotes the formation of vinylstannanes effectively.[58] The representative results are summarized in Table 12. The triphenylstannyl group adds to terminal acetylenic carbon regioselectively but non-stereoselectively to give a mixture of (E)- and (Z)-1-(triphenylstannyl)-1-alkenes. The E/Z ratios of double bonds were generally 8/2–7/3 and were not affected by solvents and reaction temperature. The ratios of (E)-1-(triphenylstannyl)-1-dodecene and the Z isomer were 79/21, 80/20, 77/23 and 63/37 in toluene, benzene, Et_2O and THF, respectively. In contrast, Corey *et al.* have reported[57] that the E/Z ratios depend on the reaction temperature in the case of uncatalyzed hydrostannylation. Heating a mixture of 1-dodecyne and Ph_3SnH at 80°C for 1.5 h gave a mixture of (E)- and (Z)-1-(triphenylstannyl)-1-dodecene ($E/Z = 22/78$) in 53% combined yield. A mixture of the E and Z isomer ($E/Z = 75/25$, 65% yield) was obtained after 5 h at 150°C. Phenylacetylene and (trimethylsilyl)acetylene provided (E)-vinylstannanes exclusively. An addition of $n-Bu_3SnH$ required a longer reaction time and gave the corresponding vinylstannanes in poor yields.

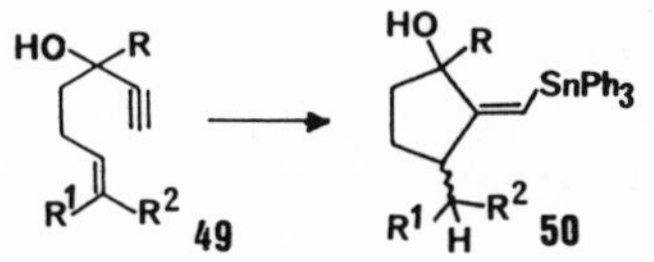

a: R = Me $R^1 = R^2 = Me$ 75% (78/22)
b: R = Me $R^1 = R^2 = H$ 50% (80/20)
c: R = Me $R^1 = Me$ $R^2 = CH_2CH_2CH=CMe_2$ 78% (79/21)
d: R = n-C_5H_{11} $R^1 = H$ $R^2 = Ph$ 87% (63/37)

Scheme 33

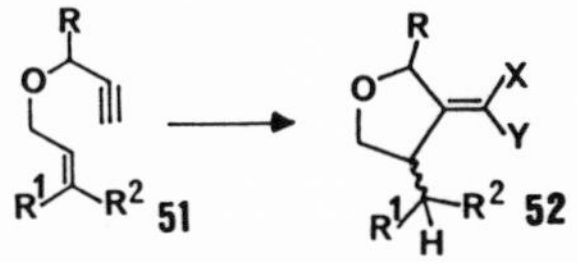

a: R = H $R^1 = R^2 = Me$ X = $SnPh_3$ Y = H   (78%)
b: R = H $R^1 = H$ $R^2 = n-C_3H_7$ X = $SnPh_3$ Y = H   (85%)
c: R = n-Bu $R^1 = R^2 = Me$ X = $SnPh_3$ Y = H   (69%, 64/36)
d: R = H $R^1 = R^2 = Me$ X = Ha   Y = Hb
e: R = H $R^1 = R^2 = Me$ X = $SnPh_3$ Y = D
f: R = H $R^1 = R^2 = Me$ X = H   Y = D

Scheme 34

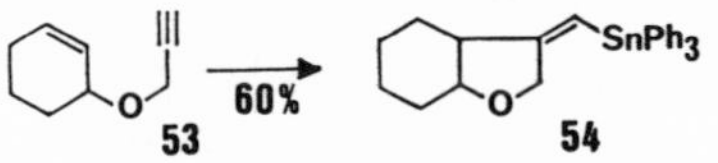

60%

Scheme 35

96%

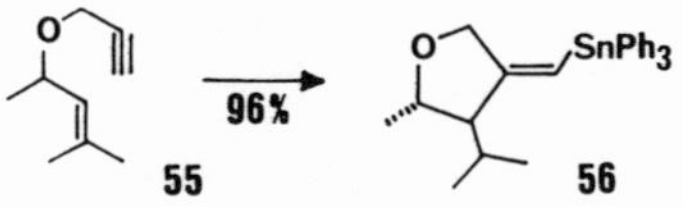

Scheme 36

The reaction was successfully applied to the radical cyclization reactions shown in Schemes 33–36. The concentration of Ph_3SnH affected the yield and distribution of the products. Uncyclized product was obtained in addition to the cyclized desired compound in a higher concentration. For instance, the compound **49a** gave cyclized product **50a** exclusively at 0.02 M concentration of Ph_3SnH, while, at 0.30 M concentration, **50a** and uncyclized product $Me_2C=CHCH_2CH_2C(OH)MeCH=CHSnPh_3$ were obtained in 60 and 15% yield, respectively. Heating a mixture of **49a** and Ph_3SnH without solvent at 80°C for 15 h gave a complex mixture consisting of (E)- and (Z)-vinylstannanes $(Me_2C=CHCH_2CH_2C(OH)MeCH=CHSnPh_3$, 46%), regioisomer $(Me_2C=CHCH_2CH_2C(OH)MeC(SnPh_3)=CH_2$, 9%) and the desired cyclized product **50a** (38% yield). It is worth noting that the serious limitation, i.e. the non-stereoselectivity shown in Table 12, was overcome in these cyclization reactions and the cyclized products consist of only the Z isomer without contamination by the other stereoisomer. The formation of a single isomer may be explained by assuming the rapid cyclization of the intermediary radical **A** which is generated by the kinetically favoured *trans* addition of the triphenylstannyl radical. Isomerization of **A** to **B** can be slow compared to cyclization. The compound **52d** derived from **52a** by destannylation(n-BuLi/THF, H_2O)[59] showed ^1H-NMR signals at δ 5.00 (m, H_a) and 4.95 (m, H_b). Treatment of the deuteriated acetylene **51a** $(DC\equiv CCH_2OCH_2CH=CMe_2)$ with Ph_3SnH followed by destannylation provided **52f**, whose ^1H-NMR spectrum showed only one signal in the olefinic region at δ 4.99. The complete disappearance of the higher field signal is consistent with a formation of single stereoisomer **52e**.

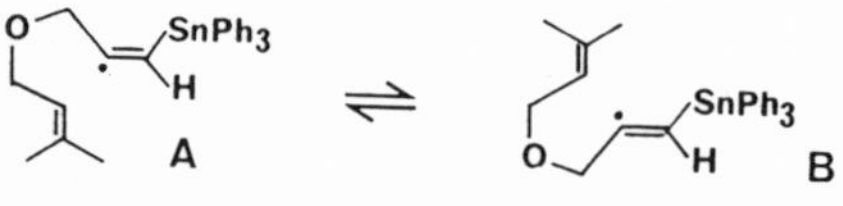

Scheme 37

The structure of the cyclized product was also confirmed as follows. Treatment of **51** ($R = R^1 = R^2 = H$) with our new method provided **52** (32% yield) along with the six-membered ring product 3-(triphenylstannyl)methylenetetrahydropyran (45%). The vinylstannane **52** was converted into vinylsilane by treatment with n-BuLi and Me_3SiCl, which was identical with the sample prepared from allyl (trimethylsilyl)propargyl ether following Negishi's procedure.[60]

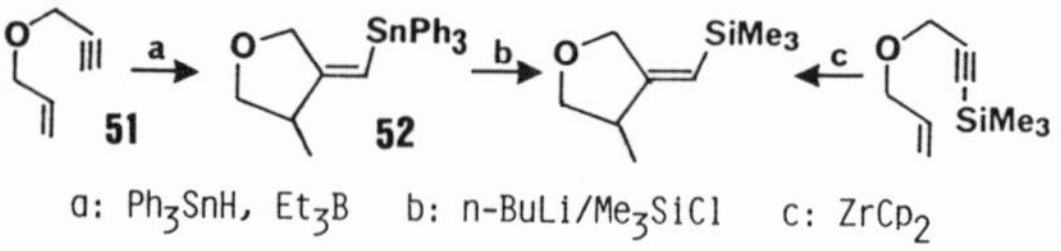

a: Ph_3SnH, Et_3B b: n-BuLi/Me_3SiCl c: $ZrCp_2$

Scheme 38

Compounds **49a–d** and **51c** provided *cis–trans* stereoisomeric mixtures concerning the substituents on a five-membered ring. In contrast, compound **55** gave *trans* isomer **56** as a single product. This stereoselective cyclization reaction was applied[61] to the synthesis of α-methylene-γ-butyrolactones, which represent a major class of known natural products and possess wide-ranging biological activities.[62] The results are summarized in Table 13. The cyclized products, **58a–d** consist of only (*Z*)-*trans*-isomers, independently of the stereochemistry of the double bond in the starting enynes (**57c** and **57d**). In contrast, treatment of **57e** with Ph₃SnH gave *cis*-fused oxolane **58e** exclusively, which is thermodynamically more stable than the *trans*-isomer. Destannylation followed by oxidation with $CrO_3 \cdot 2py$[63] gave the desired α-methylene-γ-butyrolactones **59**.

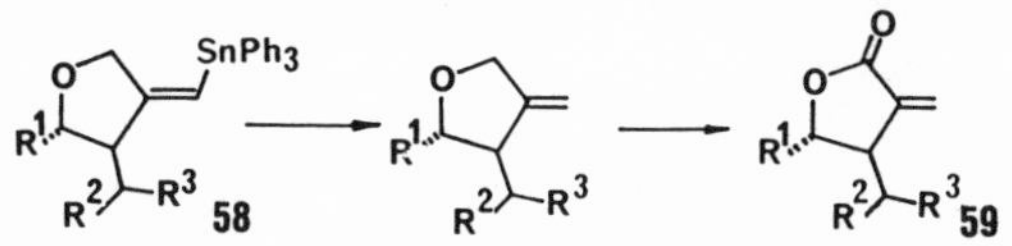

Scheme 39

Scheme 40 illustrates the synthesis of dehydroiridodiol and isodehydroiridodiol. The triethylborane-induced triphenyltin radical addition–cyclization process provided vinylstannane **62** (84%) starting from readily available propargylic alcohol **61**. Collins oxidation of **62** gave **63** (54%). Diisobutylaluminium hydride reduction followed by treatment with *p*-TsOH provided a mixture of dehydroiridodiol (3*R**, 8*S**) and isodehydroiridodiol (3*R**, 8*R**)

Table 13.   Synthesis of α-Methylene-γ-butyrolactones

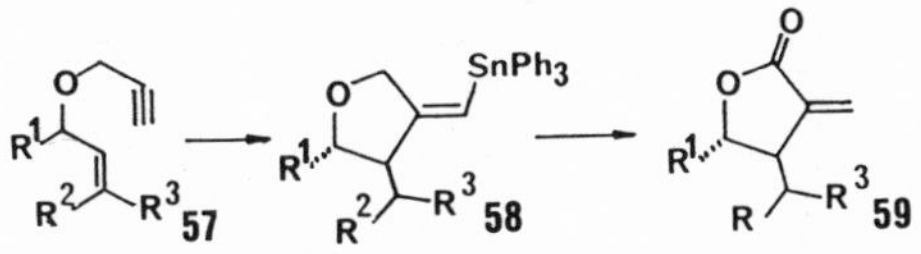

57	R¹	R²	R³	Yield (%) 58	Yield (%) 59[a]
a	Me	Me	Me	84	57
b	Ph	Me	Me	70	39
c	*n*-C₄H₉	H	*n*-C₃H₇	83	41
d	Me	*n*-C₄H₉	H	75	59
e	—(CH₂)₄—		H	71[b]	31

[a]Overall yield from **58**. [b]*Cis* product was obtained.

(26/74, 58% overall yield from **63**),[64] which was easily separated by preparative TLC on silica gel.

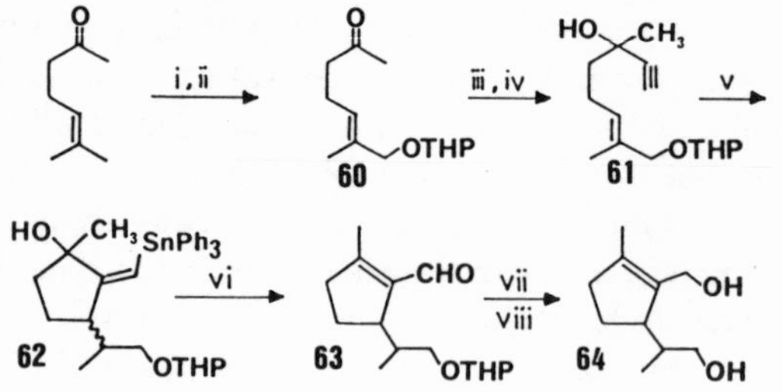

i) SeO$_2$/EtOH-H$_2$O   ii) Dihydropyran, TsOH   iii) Me$_3$SiC≡CLi
iv) KF/DMSO   v) Ph$_3$SnH, Et$_3$B   vi) CrO$_3$·2Py   vii) i-Bu$_2$AlH
viii) TsOH/MeOH

Scheme 40

The reaction was not so effective for the formation of a six-membered ring. For instance, treatment of $HC≡CCH_2OCH_2CH_2CH=CHCH_2CH_3$ gave the desired cyclized product in only 28% yield along with uncyclized vinyl-stannane (49%). An addition of galvinoxyl to a reaction mixture of 1-dodecyne, Ph$_3$SnH and Et$_3$B resulted in a recovery of the acetylene (73%). The organoboranes are known to be excellent sources of free radicals in the presence of oxygen.[65] Thus, we are tempted to assume a radical chain mechanism for the reaction. A trace of oxygen could be in a reaction mixture and initiate the free-radical reaction, although the reactions have been achieved under an argon atmosphere.

A hexane solution of Et$_3$B (1.0 M, 0.2 mL, 0.2 mmol) was added to a solution of Ph$_3$SnH (0.42 g, 1.2 mmol) and the acetylene **49a** (0.15 g, 1.0 mmol) in toluene (100 mL) at 25°C under an argon atmosphere. After stirring for 3 h at 25°C, the reaction mixture was poured into water and extracted with ethyl acetate. Purification by preparative TLC on silica gel gave the cyclized product **50a** (0.37 g, 75% yield) as a stereoisomeric mixture (78/22).

B.   Et$_3$B Induced Stereoselective Radical Addition of Ph$_3$GeH to Acetylenes and its Application to the Isomerization of Olefins

Free radical reactions have been used increasingly in recent years for the synthesis of organic molecules. The hydrogermylation[66] or hydrostannation of acetylenes takes place readily either in the absence of a catalyst or in the presence of catalytic amount of free radical initiator such as azobisisobutyronitrile (AIBN). These reactions producing the corresponding alkenyltrialkyl-

Table 14. Stereoselective Hydrogermylation of Acetylenes

$$RC{\equiv}CH \xrightarrow[\text{Et}_3\text{B}]{\text{Ph}_3\text{GeH}} \underset{H}{\overset{R}{>}}C=C\underset{H}{\overset{\text{GePh}_3}{<}} + \underset{H}{\overset{R}{>}}C=C\underset{\text{Ge Ph}_3}{\overset{H}{<}}$$

Entry	Acetylene R	Reaction conditions Temperature (°C)	Time (h)	Product Yield (%)[a]	Z/E[b]
1	n-C$_{10}$H$_{21}$	−78[c]	3	76	> 20/1
2		−20[d]	2	78	2/1
3		25[d]	2	77	1/9
4		60[d]	2	99	< 1/20
5		0[d]	(THF)2	84	8/1
6		0[d]	(PhCH$_3$–MeOH)2	80	10/1
7	CH$_3$	−78[e]	2	65	> 20/1
8	HOCH$_2$CH$_2$	−78[c]	5	80	> 20/1
9		60[d]	15	75	< 1/20
10	HOCH$_2$CH$_2$CH$_2$CH$_2$	−78[c]	5	80	> 20/1
11	EtOOC(CH$_2$)$_9$	−78[c]	12	64	> 10/1
12		60[d]	15	93	< 1/20
13	6-dodecyne	−78[c]	8	65	> 20/1

[a]Isolated yields. [b]Determined by GC and/or NMR. [c]Acetylene (1.1 mmol), Ph$_3$GeH (1.0 mmol) and Et$_3$B (1.0 mmol) were employed. Toluene was used as solvent. [d]Acetylene (1.0 mmol), Ph$_3$GeH (1.1 mmol) and Et$_3$B (1.0 mmol) were employed. Benzene was used as solvent unless otherwise noted. [e]Propyne (3.0 mmol), Ph$_3$GeH (1.0 mmol) and Et$_3$B (1.0 mmol) were employed.

germane or alkenyltrialkylstannane are of particular synthetic interest; however, they have a serious limitation. Thus, the reactions are generally not highly regio- and stereoselective. Moreover, the mechanism of the reactions does not appear to have been well established, mainly because the products can undergo isomerization under the hydrogermylation or hydrostannation reaction conditions. Here we want to describe that trialkylborane facilitates the addition of Ph$_3$GeH to acetylenes to give (*E*)- or (*Z*)-alkenyltriphenylgermanes, respectively, under excellent control of regio- and stereoselectivities.[67]

The representative results are summarized in Table 14. The isomeric ratios of the products heavily depend on the reaction temperature and the ratio of [acetylene]/[Ph$_3$GeH]. This is a big difference from the Et$_3$B induced addition reaction of Ph$_3$SnH to acetylenes. In the case of Ph$_3$SnH, the ratios of the products, (*E*)-alkenyltriphenylstannane and its (*Z*)-isomer, were always 8/2–7/3 and not affected by the reaction temperature and the ratio of [acetylene]/[Ph$_3$SnH]. In contrast, the reaction of Ph$_3$GeH at −78°C in toluene in the presence of slight excess of the acetylene provides (*Z*)-alkenyltriphenylgermane exclusively, whereas the reaction at 60°C in benzene with slight excess of Ph_3GeH gives (*E*)-alkenyltriphenylgermane as a single product. Solvent also affects the isomeric ratio of the products. In polar solvents, the (*Z*)-isomer was obtained as the major product. For instance, treatment of 1-dodecyne with Ph_3GeH–Et_3B in THF at 0°C for 2 h gave a mixture of (*Z*)-1-triphenylgermyl-1-dodecene and (*E*)-isomer ($Z/E = 8/1$) in 84% yield. Addition of methanol (10 mmol per 1.0 mmol of substrate) to toluene is also effective for the selective formation of (*Z*)-isomer (entry 6 in Table 14).

Et_3B-induced addition of n-Pr_3GeH to acetylenes did not give high stereoselectivity as compared to the addition of Ph_3GeH. For instance, the reaction of 1-dodecyne with Pr_3GeH at 60°C in the presence of Et_3B gave a isomeric mixture of (*E*)-1-tripropylgermyl-1-dodecene and (*Z*)-isomer in 79% yield ($E/Z = 2/1$). The amount of Et_3B could be reduced to 0.1 mol per 1.0 mol of acetylene without any decrease of the yield and the reaction rate at the temperature above 0°C. However, the reaction rate drops considerably at low temperature such as $-78°C$. Thus the use of stoichiometric amounts of Et_3B is recommended in these cases. i-Pr_3B and (n-C_8H_{17})$_3B$ were as effective as Et_3B. Et_3B initiates the radical reaction at low temperature such as $-78°C$ which is a great advantage. The ordinary radical initiators such as AIBN and t-BuOO–t-Bu requires the heating of the reaction mixture (80–130°C) to promote the reaction, so that the isomerization of the produced alkenylgermanes easily takes place under such conditions.

It was anticipated that the *trans* addition products (i.e. (*Z*)-isomers) were kinetic-controlled products and isomerized into (*E*)-isomers under thermodynamic conditions. This was indeed the case as demonstrated by the isomerization of (*Z*)-1-triphenylgermyl-1-dodecene into the (*E*)-isomer. Heating a benzene solution of (*Z*)-1-triphenylgermyl-1-dodecene at 60°C in the presence of catalytic amounts of Ph_3GeH and Et_3B gave (*E*)-isomer exclusively. The isomerization is explained by addition–elimination sequences of the triphenylgermyl radical (Scheme 41). The germyl radical, $Ph_3Ge\cdot$, attacks the olefin to give a radical intermediate **A**. Free rotation scrambles the stereochemistry, so that the composition of the mixture reaches the thermodynamic equilibrium.[68] This mechanism is supported by the following facts that treatment of (*Z*)-1-triphenylgermyl-1-dodecene (1.0 mmol) with n-Pr_3GeH-Et_3B (1.0 mmol each) at 60°C gave a mixture of (*E*)-1-tripropylgermyl-1-dodecene (**65**) and (*E*)-1-triphenylgermyl-1-dodecene (**66**) (**65/66** = 2/5) and that treatment of (*Z*)-1-triethylgermyl-1-dodecene with Ph_3GeH-Et_3B gave (*E*)-1-triethylgermyl-1-dodecene (**67**) and (*E*)-1-triphenylgermyl-1-dodecene (**66**) (**67/66** = 2/5, Scheme 42).

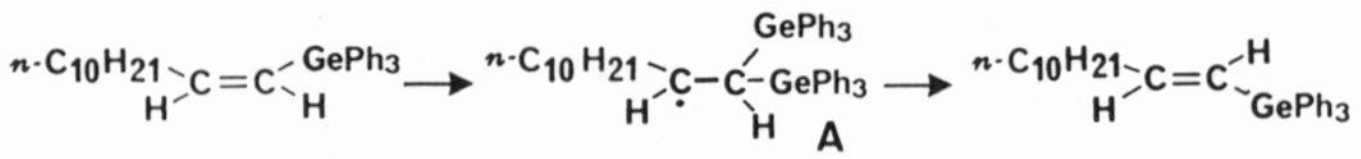

Scheme 41

Table 15. Isomerization of Olefins by means of $Ph_3GeH-Et_3B$

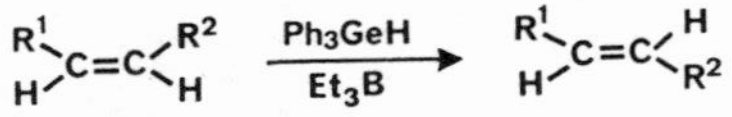

$$R^1{\diagup}C=C{\diagdown}R^2 \quad \xrightarrow[Et_3B]{Ph_3GeH} \quad R^1{\diagup}C=C{\diagdown}H$$
(H) (H) (H) (R²)

Entry	Substrate			Reaction time (h)	Product	
	R^1	R^2	Z/E		Y(%)	Z/E
1	$n\text{-}C_5H_{11}$	$n\text{-}C_5H_{11}$	> 20/1	10	90	15/85
2	$t\text{-}C_4H_9$	$n\text{-}C_8H_{17}$	> 20/1	10	91	0/100
3	$n\text{-}C_6H_{13}$	Ph	100/0	5	96	0/100
4	Ph	Ph	> 20/1	2	81	< 1/20
5	$n\text{-}C_6H_{13}$	$SiMe_2Ph$	> 20/1	10	84	< 1/20
6	$n\text{-}C_{10}H_{21}$	$GePh_3$	10/1	4	88	< 1/20
7	CH_3	$GePh_3$	> 20/1	10	75	< 1/20
8	$HOCH_2CH_2$	$GePh_3$	> 20/1	10	70	< 1/20
9	$EtOOC(CH_2)_9$	$GePh_3$	7/1	10	95	0/100
10	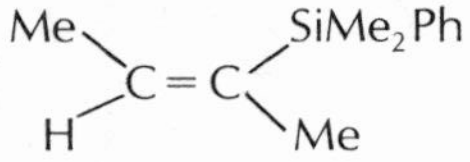 $Me{\diagup}C=C{\diagdown}^{SiMe_2Ph}_{Me}$ (H)		100/0	10 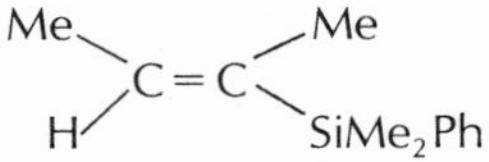	$Me{\diagup}C=C{\diagdown}^{Me}_{SiMe_2Ph}$ (H) 71		< 1/20
11	Me_3Si	$GeEt_3$	100/0	10[a]	89	0/100

[a] Et_3GeH was used instead of Ph_3GeH.

137

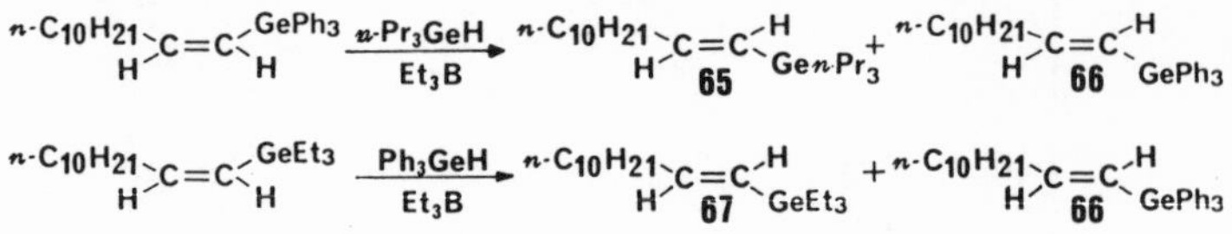

$$n\text{-}C_{10}H_{21} \diagdown C = C \diagup GePh_3 \quad \xrightarrow{\text{n-Pr}_3\text{GeH}} \quad n\text{-}C_{10}H_{21} \diagdown C = C \diagup H \quad + \quad n\text{-}C_{10}H_{21} \diagdown C = C \diagup H$$

$$n\text{-}C_{10}H_{21} \diagdown C = C \diagup GeEt_3 \quad \xrightarrow{\text{Ph}_3\text{GeH}} \quad n\text{-}C_{10}H_{21} \diagdown C = C \diagup H \quad + \quad n\text{-}C_{10}H_{21} \diagdown C = C \diagup H$$

Scheme 42

The new reaction was successfully applied to the isomerization of various kinds of olefins[69] and typical results are summarized in Table 15.

A hexane solution of Et_3B (1.0 M, 1.0 mL, 1.0 mmol) was added to a solution of 1-dodecyne (0.18 g, 1.1 mmol) and Ph_3GeH (0.30 g, 1.0 mmol) in toluene (8 mL) at $-78°C$ under an argon atmosphere. After stirring for 3 h at $-78°C$, the reaction mixture was poured into ice-cooled water and extracted with ethyl acetate three times. Combined organic layers were washed with brine, dried over Na_2SO_4 and concentrated *in vacuo*. The residual oil was submitted to preparative TLC on silica gel to give (Z)-1-triphenylgermyl-1-dodecene exclusively (0.36 g, 76% yield, $Z/E = >20/1$).

ACKNOWLEDGEMENTS

The author would like to thank Professors Hitosi Nozaki and Kiitiro Utimoto for their encouragement and advice throughout this work. We also thank the undergraduate and graduate students for their enthusiastic collaboration. Their names are given in the list of references. Financial support by the Ministry of Education, Science and Culture, Japanese Government (Grant-in-Aid for Special Project Research) is acknowledged.

REFERENCES AND NOTES

1. Yamamoto, H.; Nozaki, H. *Angew. Chem. Int. Ed.* 1978, *17*, 169; Oshima, K.; Yamamoto, H.; Nozaki, H. Selective synthetic reactions by means of organoaluminum amphoteric reagents, in "New Synthetic Methodology and Biologically Active Substrates"; (Yoshida, Z., Ed.); Elsevier-Kodansha: Tokyo, 1981, p. 19.

2. Yasuda, A.; Tanaka, S.; Oshima, K.; Yamamoto, H.; Nozaki, H. *J. Am. Chem. Soc.* 1974, *96*, 6513; Yasuda, A.; Yamamoto, H.; Nozaki, H. *Bull. Chem. Soc. Jpn.* 1979, *52*, 1705; Yasuda, A.; Tanaka, S.; Yamamoto, H.; Nozaki, H. *ibid.* 1979, *52*, 1752.

3. Maruoka, K.; Hashimoto, S.; Kitagawa, Y.; Yamamoto, H.; Nozaki, H. *ibid.* 1980, *53*, 3301.

4. Matsubara, S.; Tsuboniwa, N.; Morizawa, Y.; Oshima, K.; Nozaki, H. *ibid.* 1984, *57*, 3242.

5. Suzuki, T.; Saimoto, H.; Tomioka, H.; Oshima, K.; Nozaki, H. *Tetrahedron Lett.* 1982, *23*, 3597.

6. Tomioka, H.; Suzuki, T.; Oshima, K.; Nozaki, H. *ibid.* 1982, *23*, 3387.

7. Fleming, I.; Newton, T. W.; Rossler, F. *J. Chem. Soc., Perkin Trans. 1* 1981, 2528.

8. Tamao, K.; Akita, M.; Kanatani, R.; Ishida, N.; Kumada, M. *J. Organomet. Chem.* 1982, *226*, C9.

9. Altnau, G.; Rösch, L.; Bohlmann, F.; Lonitz, M. *Tetrahedron Lett.* 1980, *21*, 4069.

10. Hayami, H.; Sato, M.; Kanemoto, S.; Morizawa, Y.; Oshima, K.; Nozaki, H. *J. Am. Chem. Soc.* 1983, *105*, 4491.

11. Uchida, K.; Utimoto, K.; Nozaki, H. *J. Org. Chem.* 1976, *41*, 2215.

12. (a) Grove, M.D.; Spencer, G.F.; Rohwedder, W.K.; Mandava, N.; Worley, J.F. Warthen, J.D., Jr.; Steffens, G.L.; Flippen-Anderson, J.L.; Cook, J.C. *Nature (London* 1979, *281*, 216. (b) Fung, S.; Siddall, J.B. *J. Am. Chem. Soc.* 1980, *102*, 6591. (c) Ishiguro M.; Takatsuto, S.; Morisaki, M.; Ikekawa, N. *J. Chem. Soc., Chem. Commun.* 1980, 962 (d) Wada, K.; Maruno, S.; Ikekawa, N.; Morisaki, M.; Mori, K. *Plant Cell Physiol.* 1981 *22*, 323. (e) Singh, H.; Bhardwaj, T.R. *Indian J. Chem.* 1986, *25B*, 989.

13. Wiersig, J.R.; Waespe-Sarcevic, N.; Djerassi, C. *J. Org. Chem.* 1979, *44*, 3374.

14. Lipshutz, B.H.; Kozlowski, J.; Wilhelm, R.S. *J. Am. Chem. Soc.* 1982, *104*, 2305.

15. Morizawa, Y.; Oda, H.; Oshima, K.; Nozaki, H. *Tetrahedron Lett.* 1984, *25*, 1163.

16. Taguchi, H.; Yamamoto, H.; Nozaki, H. *ibid*, 1972, 4661; Hiyama, T.; Mishima, T. Kitatani, K.; Nozaki, H. *ibid.* 1974, 3297; Utimoto, K.; Tanaka, M.; Kitai, M.; Nozaki, H *ibid.* 1978, 2301; Tsuji, J.; Yamada, M.; Kaito, M.; Mandai, T. *ibid.* 1979, 2257.

17. Eisch, J.J.; Trainor, J.T. *J. Org. Chem.* 1963, *28*, 2870.

18. Fristad, W.E.; Bailey, T.R.; Paquette, L.A. *ibid.* 1980, *45*, 3028.

19. Okuda, Y.; Morizawa, Y.; Oshima, K.; Nozaki, H. *Tetrahedron Lett.* 1984, *25*, 2483.

20. Negishi, E.; Boardman, L.D.; Tour, J.M.; Sawada, H.; Rand, C.L. *J. Am. Chem. Soc.* 1983, *105*, 6344.

21. Baldwin, J.E. *J. Chem. Soc., Chem. Commun.* 1976, 734; Beckwith, A.L.J. *Tetrahedron* 1981, *37*, 3073.

22. Augustine, R.L.; Broom, A.D. *J. Org. Chem.* 1960, *25*, 802; Macdonald, T.L.; Mahalingam, S. *J. Am. Chem. Soc.* 1980, *102*, 2113.

23. Kosugi, M.; Shimizu, Y.; Migita, T. *Chem. Lett.* 1977, 1423; Milstein, D.; Stille, J.K. *J. Org. Chem.* 1979, *44*, 1613; idem *J. Am. Chem. Soc.* 1979, *101*, 4992; idem *ibid.* 1978, *100*, 3636.

24. For a general discussion, see Negishi, E. "Organometallics in Organic Synthesis," Vol. 1, John Wiley: New York 1980, pp. 410–412.

25. Seyferth, D.; Stone, F.G. *J. Am. Chem. Soc.* 1957, *79*, 515; Labadie, J.W.; Stille, J.K. *ibid.* 1983, *105*, 6129; Shibasaki, M.; Torisawa, Y.; Ikegami, S. *Tetrahedron Lett.* 1982, *23*, 4607.

26. Piers, E.; Chong, J.M. *J. Chem. Soc., Chem. Commun.* 1983, 934; *J. Org. Chem.* 1982, *47*, 1604.

27. Tamborski, C.; Ford, F.E.; Soloski, E.J. *ibid.* 1963, *28*, 237; Still, W.C. *J. Am. Chem. Soc.* 1977, *99*, 4836; idem *ibid.* 1978, *100*, 1481; Kitching, W.; Olszowy, H.A.; Harvey, K. *J. Org. Chem.* 1982, *47*, 1893.

28. Matsubara, S.; Hibino, J.; Morizawa, Y.; Oshima, K.; Nozaki, H. *J. Organomet. Chem.* 1985, *285*, 163.

29. For other examples using n-Bu$_3$SnCu or n-Bu$_3$SnCu(L)Li, see Cox, S.D.; Wudl, F. *Organometallics* 1983, *2*, 184; Westmijze, H.; Ruitenberg, K.; Meijer, J.; Vermeer, P. *Tetrahedron Lett.* 1982, *23*, 2797; Seitz, D.E.; Lee, S.-H. *ibid.* 1981, *22*, 4909; R$_3$SnMgR′, see Quintard, J.-P.; Elissondo, B.; Pereyre, M. *J. Organomet. Chem.* 1981, *212*, C31.

30. Leusink, A.J.; Budding, H.A.; Marsman, J.W.; *ibid.* 1967, *9*, 285.

31. (a) des Tombe, F.J.A.; van der Kerk, G.J.M.; Creemers, H.M.J.C.; Noltes, J.G. *J. Chem. Soc., Chem. Commun.* 1966, 914. (b) des Tombe, F.J.A.; van der Kerk, G.J.M. *J. Organomet. Chem.* 1972, *43*, 323. (c) Vyazankin, N.S.; Razuvaev, G.A.; Kruglaya, O.A. *Organomet, Chem. Rev.* 1968, *A3*, 323.

32. Nonaka, T.; Okuda, Y.; Matsubara, S.; Oshima, K.; Utimoto, K.; Nozaki, H. *J. Org. Chem.* 1986, *51*, 4716.

33. Smith, P. J.; Tupciauskas, A. P. in Webb, G. A. Ed. Annual Reports on NMR Spectroscopy, Vol. 8, Academic Press, New York 1978, pp. 291–370.

34. Wakamatsu, K.; Nonaka, T.; Okuda, Y.; Tückmantel, W.; Oshima, K.; Utimoto, K.; Nozaki, H. *Tetrahedron* 1986, *42*, 4427.

35. Gmelin Handbook of Inorganic Chemistry (8th edn.), Springer, Berlin, 1983, "Organocopper Compounds Part 2."

36. (a) Isobe, M.; Kondo, S.; Nagasawa, N.; Goto, T. *Chem. Lett.* 1977, 679. (b) Seebach, D.; Langer, W. *Helv. Chim. Acta* 1979, *62*, 1701, 1710. (c) Watson, R. A.; Kjonaas, R. A. *Tetrahedron Lett.* 1986, *27*, 1437. (d) Tückmantel, W.; Oshima, K.; Nozaki, H. *Chem. Ber.* 1986, *119*, 1581.

37. Negishi, E. "Organometallics in Organic Synthesis"; Vol. 1, Wiley: New York, 1980, pp. 286–393.

38. The reagents are also effective for 1,4-addition to α,β-unsaturated carbonyl compounds. For instance, the addition of $PhMe_2SiZnEt_2Li$ to 2-cyclohexenone gave the 1,4-adduct, 3-(dimethylphenylsilyl)cyclohexanone in 80% yield, while $PhMe_2SiLi$ provided 1-(dimethylphenylsilyl)-2-cyclohexen-1-ol (60% yield) as the main product. See Ref. 36(d).

39. $CoCl_2(PPh_3)_2$ was also effective catalyst for silylzinc reagent, but not for silylaluminium reagent.

40. Dialkylzincs were prepared according to the following references. $t\text{-}Bu_2Zn$: Abraham, M. H. *J. Chem. Soc.* 1960, 4130. $i\text{-}Pr_2Zn$: Rathke, M. W.; Yu, H. *J. Org. Chem.* 1972, *37*, 1732.

41. $t\text{-}Bu_3Al$ was prepared from $t\text{-}BuMgCl$ and $AlCl_3$ *in situ* and used directly without further purification.

42. Oda, H.; Sato, M.; Morizawa, Y.; Oshima, K.; Nozaki, H. *Tetrahedron* 1985, *41*, 3257.

43. This kind of isomerization has been reported. See Ref. 20.

44. Pd(0) catalyzed double silylation of acetylenes with disilanes of special substituents such as hydro, fluoro and methoxy group has given the corresponding silyl olefins in good yields. In contrast, hexamethyldisilane provided very poor yields of double silylated products (Scheme 28).

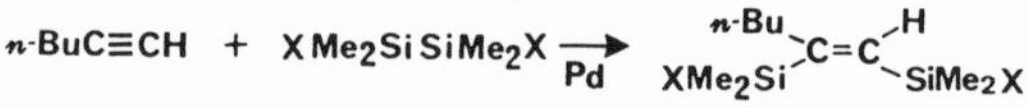

$$n\text{-}BuC{\equiv}CH \; + \; XMe_2SiSiMe_2X \xrightarrow[Pd]{} \underset{XMe_2Si}{\overset{n\text{-}Bu}{>}}C{=}C\underset{SiMe_2X}{\overset{H}{<}}$$

$$X = OMe \; (70\%) \quad X = Me \; (7\%)$$

Scheme 28

Watanabe, H.; Kobayashi, M.; Sato, M.; Nagai, Y. *J. Organomet. Chem.* 1981, *216*, 149; Watanabe, H.; Kobayashi, M,; Higuchi, K.; Nagai, Y. *ibid.* 1980, *186*, 51; Matsumoto, H.; Matsubara, I.; Kato, T.; Shono, H.; Watanabe, H.; Nagai, Y. *ibid.* 1980, *199*, 43; Sakurai, H.; Kamiyama, Y.; Nakadaira, Y. *J. Am. Chem. Soc.* 1975, *97*, 931; Okinoshima, H.; Yamamoto, K.; Kumada, M. *ibid.* 1972, *94*, 9263; Kusumoto, T.; Hiyama, T. *Tetrahedron Lett.* 1987, *28*, 1807.

45. Hibino, J.; Nakatsukasa, S.; Fugami, K.; Matsubara, S.; Oshima, K.; Nozaki, H. *J. Am. Chem. Soc.* 1985, *107*, 6416.

46. Still, W. C. *J. Org. Chem.* 1976, *41*, 3063.

47. We are tempted to assume that the active reagent in our method could be $(Me_3\text{-}Si)_3MnMgMe$ derived from 3 equivalents of $Me_3SiMgMe$ and $MnCl_2$. The addition of PhCHO to the reagent provided only phenyl(trimethylsilyl)carbinol and no trace of 1-phenylethanol.

48. In the case of terminal acetylenes and silylacetylenes in Table 10, the intermediary alkenylmanganese could not be trapped by the electrophiles such as D_2O and MeI.

49. Sakurai, H.; Nakadaira, Y.; Kira, M.; Tobita, H. *Tetrahedron Lett.* 1980, *21*, 3077; Chung,

C.; Lagow, R. J.; *J. Chem. Soc., Chem. Commun.* 1972, 1078; Sakurai, H.; Nakadaira, Y.; Tobita, H. *J. Am. Chem. Soc.* 1982, *104*, 300; idem, *Chem. Lett.* 1982, 771.

50. Brown, H. C. "Organic Synthesis via Boranes, "; John Wiley: New York, 1975; Mikhailov, B. M.; Bubnov, Yu. N. "Organoboron Compounds in Organic Synthesis, " Harwood, New York, 1984; Negishi, E. "Organometallics in Organic Synthesis "; Vol. 1, John Wiley: New York, 1980; Suzuki, A. *Topics in Current Chemistry*, 1983, *112*, 67.

51. (a) Biffar, W.; Nöth, H. *Angew. Chem.* 1980, *92*, 65 or *Int. Ed. Engl.* 1980, *19*, 58. (b) idem *Chem. Ber.* 1982, *115*, 934.

52. Nozaki, K.; Wakamatsu, K.; Nonaka, T.; Tückmantel, W.; Oshima, K.; Utimoto, K. *Tetrahedron Lett.* 1986, *27*, 2007.

53. Stork, G., in "Selectivity—A Goal for Synthetic Efficiency "; (Bartmann, W.; Trost, B. M. Eds.); Verlag Chemie: Weinheim, 1984, p. 281; Stork, G.; Mook, R. Jr., *Tetrahedron Lett.* 1986, *27*, 4529.

54. (a) Giese, B. "Radicals in Organic Synthesis, " (Baldwin, J. E., Ed.) Pergamon Press, Oxford, 1986. (b) Barton, D. H. R.; Crich, D.; Motherwell, W. B. *Tetrahedron* 1985, *41*, 3901. (c) Giese, B.; Horler, H. *ibid.* 1985, *41*, 4025. (d) Hart, D. *Science* 1984, *223*, 883. (e) Kraus, G. A.; Landgrebe, K. *Tetrahedron* 1985, *41*, 4039.

55. (a) Stork, G.; Baine, N. H. *J. Am. Chem. Soc.* 1982, *104*, 2321. (b) Stork, G.; Mook, R. Jr. *ibid.* 1983, *105*, 3720. (c) Stork, G.; Sher, P. M. *ibid.* 1983, *105*, 6765. (d) Porter, N. A.; Magnin, D. R.; Wright, B. Y. *ibid.* 1986, *108*, 2787. (e) Curran, D. P.; Chen, M.-H.; Kim, D. *ibid,* 1986, *108*, 2489. (f) High dilution favors the intramolecular radical cyclization. Ueno, Y.; Chino, K.; Okawara, M. *Tetrahedron Lett.* 1982, *23*, 2575. see, however, Stork, G.; Mook, R. Jr, *J. Am. Chem. Soc.* 1987, *109*, 2829.

56. (a) Leusink, A. J.; Budding, H. A.; Marsman, J. W. *J. Organomet. Chem.* 1967, *9*, 285. (b) Leusink, A. J.; Budding, H. A.; Drenth, W. *ibid.* 1967, *9*, 295.

57. Corey, E. J.; Ulrich, P.; Fitzpatrick, J. M.; *J. Am. Chem. Soc.* 1976, *98*, 222; Corey, E. J.; Wollenberg, R. H. *J. Org. Chem.* 1975, *40*, 2265.

58. Nozaki, K.; Oshima, K.; Utimoto, K. *J. Am. Chem. Soc.* 1987, *109*, 2547.

59. (Triphenylstannyl)alkenes were easily transformed into alkenyllithium as (trialkylstannyl) alkenes following the procedure described in Ref. 57.

60. Negishi, E.; Cederbaum, F. E.; Takahashi, T. *Tetrahedron Lett.* 1986, *27*, 2827; Negishi, E.; Holmes, S. J.; Tour, J. M.; Miller, J. A. *J. Am. Chem. Soc.* 1985, *107*, 2568.

61. Nozaki, K.; Oshima, K.; Utimoto, K. *Bull. Chem. Soc. Jpn.* 1987, *60*, 3465.

62. Hoffmann, H. M. R.; Rabe, J. *Angew. Chem. Int. Ed. Engl.* 1985, *24*, 44 and references cited therein.

63. Okabe, M.; Abe, M.; Tada, M. *J. Org. Chem.* 1982, *47*, 1775.

64. Sakai, T.; Nakajima, K.; Yoshihara, K.; Sakan, T.; Isoe, S. *Tetrahedron* 1980, *36*, 3115; Kimura, H.; Miyamoto, S.; Shinkai, H.; Kato, T. *Chem. Pharm. Bull.* 1982, *30*, 723.

65. Brown, H. C.; Midland, M. M. *Angew. Chem. Int. Ed. Engl.* 1972, *11*, 692; Suzuki, A.; Nozawa, S.; Itoh, M.; Brown, H. C.; Kabalka, G. W.; Holland, G. W. *J. Am. Chem. Soc.* 1970, *92*, 3503.

66. (a) Corriu, R. J. P.; Moreau, J. J. E. *J. Organomet. Chem.* 1972, *40*, 73. (b) Oda, H.; Morizawa, Y.; Oshima, K.; Nozaki, H. *Tetrahedron Lett.* 1984, *25*, 3221.

67. Ichinose, Y.; Nozaki, K.; Wakamatsu, K.; Oshima, K.; Utimoto, K. *Tetrahedron Lett.* 1987, *28*, 3709. Ichinose, Y.; Nozaki, K.; Wakamatsu, K.; Oshima, K.; Utimoto, K. *Bull. Chem. Soc. Jpn.* 1990, *63*, 2268.

68. Same mechanism has been proposed for the partial isomerization of (*Z*)-1-tributylgermyl-2-phenylethene into (*E*)-isomer. See Ref. 66(a).

69. Benzenethiyl radical initiated isomerization of double bonds has been reported. Moussebois, C.; Dale, J. *J. Chem. Soc. (C).* 1966, 260; Corey, E. J.; Hamanaka, E. *J. Am. Chem. Soc.* 1976, *89*, 2758.

DEVELOPMENT OF CARBENE COMPLEXES OF IRON AS NEW REAGENTS FOR SYNTHETIC ORGANIC CHEMISTRY

Paul Helquist

OUTLINE

Advances in Metal-Organic Chemistry, Volume 2, pages 143–194
Copyright © 1991 JAI Press Ltd
All rights of reproduction in any form reserved
ISBN: 0-89232-948-3

I. INTRODUCTION

Our entry into the area of transition metal carbene (or alkylidene)[1] complexes
several years ago may be traced back to considerations of reactivity of
carbenes themselves as classical organic intermediates.[2] Among the reactions
that carbenes are well-known to undergo are additions to alkenes to produce
cyclopropanes.[3] This reaction is an important one among synthetic organic
methods because of (1) the occurrence of a perhaps surprisingly large number
of natural products containing cyclopropane rings, (2) the presence of cyclopropanes among certain industrially important compounds (e.g. insec-
ticides), and (3) the use of cyclopropanes as synthetic intermediates leading
to several other types of compounds. The carbene addition reaction has been
shown to be useful in many cases, but all too often, complications arise
because of the inherently high reactivity of these species. Other pathways
such as insertion reactions and rearrangements may compete with the desired
cyclopropanation reactions (Eq. 1).

$$\begin{matrix} \diagdown \\ \diagup \end{matrix} C: \; + \; \begin{matrix} \diagdown \\ \diagup \end{matrix} C{=}C \begin{matrix} \diagup \\ \diagdown \end{matrix} \longrightarrow \quad \qquad \tag{1}$$

+ Insertion products

+ Rearrangement products

  Beginning with the initial development of modern transition metal organometallic chemistry a few decades ago, an increasingly commonly used
ploy for dealing with reactive, unstable organic species has been to allow
them to undergo coordination as novel ligands on various metals; a
compound that may be too unstable to isolate as a free species may thus be
stabilized in the coordination sphere of the metal. A now classic example of
this phenomenon is the case of cyclobutadiene, one of the examples of a
so-called anti-aromatic compound according to the Hückel formalism. Cyclobutadiene itself has been characterized in very low-temperature argon
matrices, but it cannot be isolated under normal laboratory conditions.
However, the complex containing cyclobutadiene as a ligand coordinated to
the tricarbonyliron unit is quite stable.[4] With this principal in mind, we
wondered whether the same approach could be employed with carbenes as a
means of controlling the reactivity of these intermediates (Eq. 2).

$$\begin{matrix} \diagdown \\ \diagup \end{matrix} C{:} \rightarrow M \longrightarrow \qquad \text{Control of reactivity?} \tag{2}$$

These considerations bring us to the general subject of transition metal carbene complexes. These compounds have the general formula **1** in which M is a transition metal, L represents various types of ligands, and most importantly for the present discussion, an alkylidene unit is bonded to the metal.

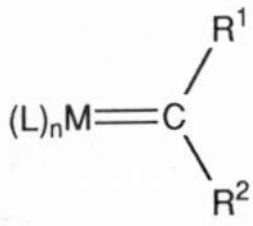

$$(L)_nM = C \begin{matrix} R^1 \\ \\ R^2 \end{matrix} \qquad 1$$

The basis for writing the structures of carbene complexes with a double bond between the metal and the carbon atom follows from simple molecular orbital considerations. In a formal sense, we can start with a carbene in the singlet spin state, and we can picture overlap of the filled sp^2 orbital of the carbene with an empty orbital on the metal to form a simple dative σ bond. We can also depict a means of back-bonding whereby a filled orbital of d-like symmetry on the metal overlaps with the empty p orbital of the carbene to form a retrodative π bond. The net effect is to form a double bond consisting of both σ and π components much as in the case of alkenes. Most importantly, however, this bonding scheme can lead to stabilization of the carbene. The high reactivity of a free, singlet carbene is due largely to the occurrence of the empty p orbital, but in carbene complexes, the electron deficiency of this orbital is at least partially satisfied by back donation of electron density from the metal. [The reader is cautioned that this representation is not meant to imply that carbene complexes are necessarily formed by reaction of free carbenes with metals, but rather this diagram is only meant to indicate the nature of the bonding in carbene complexes.]

Carbene complexes are now known for a large number of transition metals.[5] However, the first ones to be reported were of the Group 6 metals, i.e. chromium, molybdenum, and tungsten, and had the general structure **2**. These complexes, having an alkoxy substituent on the carbene center, were

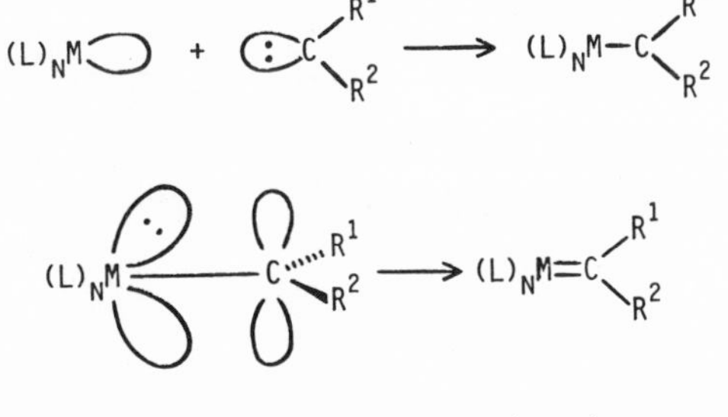

$$(L)_NM \,\bigcirc \;+\; \colon C \begin{matrix} R^1 \\ R^2 \end{matrix} \longrightarrow (L)_NM - C \begin{matrix} R^1 \\ R^2 \end{matrix}$$

$$(L)_NM \rule{1.2cm}{0.4pt} C \begin{matrix} R^1 \\ R^2 \end{matrix} \longrightarrow (L)_NM = C \begin{matrix} R^1 \\ R^2 \end{matrix}$$

Figure 1. Bonding in transition metal carbene complexes

first reported by E. O. Fischer in 1964.[6] They are easily prepared by the reaction of a Group 6 metal hexacarbonyl with an organolithium reagent followed by *O*-alkylation of the initially formed adduct (Eq. 3). For the most part, however, these complexes show limited carbene-like reactivity. For example, they have been reported to react with certain types of activated alkenes to give cyclopropanes, but this reaction has never been demonstrated to be applicable to alkenes in general.[7] Instead, these complexes show a range of other types of reactivity. Indeed, many of these other reactions have been developed in detail as important, synthetically useful methods that have been applied very effectively in a number of impressive syntheses of natural product by several workers.[8]

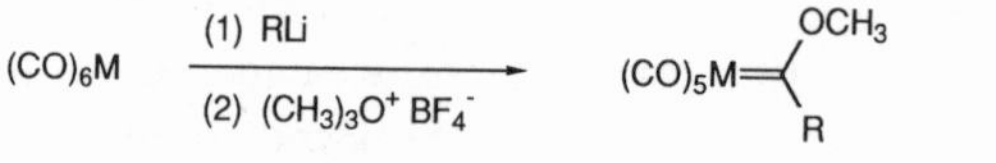

$$(CO)_6M \quad \xrightarrow[\text{(2) } (CH_3)_3O^+ BF_4^-]{\text{(1) RLi}} \quad (CO)_5M{=}\!\!\!<^{OCH_3}_{R} \qquad (3)$$

$$\textbf{2} \quad (M = Cr, Mo, W)$$

Yet, we were interested primarily in working with complexes that retain at least a certain degree of classical carbene reactivity. A key finding that has guided us in our work was reported by Rowland Pettit in 1966 (Scheme 1).[9] The binuclear complex, bis[cyclopentadienyldicarbonyliron] (3), which is a very stable, easily prepared,[10] and also commercially available compound, undergoes reductive cleavage to generate the sodium ferrate 4. This intermediate, which is not isolated, has been reported to be one of the most nucleophilic species known.[11] Alkylation with chloromethyl methyl ether gives the methoxymethyliron derivative 5. Of greatest interest is that when 5 is treated with an acid in the presence of an alkene, a cyclopropane is produced in at least modest yield. Very shortly after Pettit's initial report of these findings, Malcolm Green published quite similar results.[12]

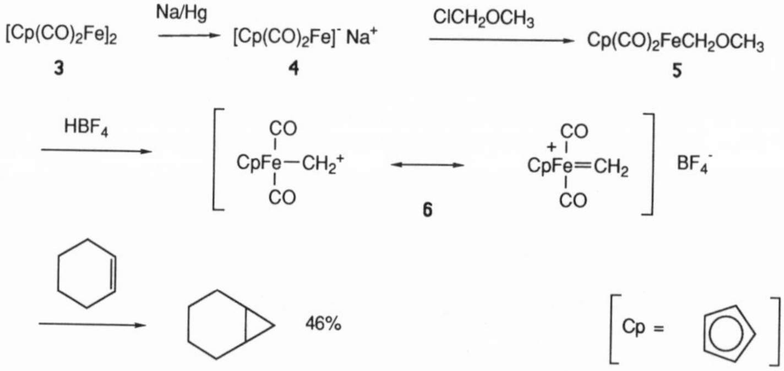

Scheme 1

The proposed pathway for the cyclopropane-producing reaction entails protolytic cleavage of the methoxy group to generate the cation **6** as the actual reactive intermediate. Participation of an appropriate filled orbital on the iron results in π back-bonding as described above for carbene complexes in general (Figure 1). The result is that one may write a resonance structure that allows us to recognize more readily the intermediate as an example of a carbene complex, or more specifically, a methylene complex. Due to the positive charge of this species, it may also be expected to be more electrophilic than the neutral complexes of the Group 6 metals discussed above. Parenthetically, the cyclopentadienyldicarbonyliron system has been studied extensively in several other contexts as well and has proven to be one of the most versatile and useful groups in organometallic chemistry. The work of Myron Rosenblum has been especially important in developing this area of organoiron chemistry.[13]

Many other investigators have taken a lead from Pettit's and Green's work and have gone on to study either the same methylene complex **6** or closely related species. Due to its very high reactivity and instability, the parent complex **6** itself has never actually been isolated or even characterized directly in the solution phase. However, in 1978 J. L. Beauchamp reported ion cyclotron resonance studies that provided direct evidence for the generation of **6** in the gas phase.[14] Also, in 1979 Roald Hoffmann published the results of theoretical studies probing the nature of the σ and π bonding in **6**. His calculations predicted a 6.2–kcal/mol barrier to rotation about the iron–carbon bond and a preferred "upright" conformation **6A** in which the methylene group lies in a plane that bisects the (CO)–Fe–(CO) angle and passes through the center of the cyclopentadienyl ligand.[15] An alternative "cross-wise" conformation **6B** was predicted to be less favorable.

Other investigators have made various stabilized derivatives of **6** in which other ligands are present in place of the cyclopentadienyl and carbonyl groups about the iron, and/or the hydrogen substituents on the carbene center are replaced by carbon or heteroatomic groups. An early example of a ligand substituted derivative from the work of Maurice Brookhart is $[\eta^5\text{-}C_5H_5(\text{diphos})Fe = CH_2]^+$ having 1,2-bis(diphenylphosphino) ethane (diphos or dppe) as a good donor ligand in place of the more electron-withdrawing carbonyls.[16] This complex is sufficiently stable to be characterized in solution by NMR spectrometry, but it has not been isolated as a pure substance. Maurice Brookhart has been especially active in this area, and this report is only one of several from his laboratory that will be cited in this chapter. More recently, Véronique Guerchais and Didier Astruc have described the pentamethylcyclopentadienyl analog of **6** which may also be observed by NMR but which decomposes after only a few minutes in solution at -50°C.[17]

Another early modification of Pettit's work was Alan Davison's preparation of an optically active analog of **5** having a menthyloxy group in place of the methoxy group and a triphenylphosphine ligand in place of one of the carbonyls. This derivative, η^5-$C_5H_5(CO)(PPh_3)FeCH_2O$-menthyl, undergoes acid-promoted reaction with alkenes to give cyclopropanes with modest enantiomeric excesses of 26–38.5%.[18] Values of ca. 90% ee have been obtained using the optically active ethylidene complex, $\{\eta^5$-$C_5H_5(CO)[Ph_2P$-(S)-2-methylbutyl]Fe=CHCH_3\}^+$, in more recent studies by Maurice Brookhart.[19] Related optically active complexes have also been examined by Thomas Flood[20] and Henri Brunner.[21]

Especially commonly studied have been derivatives of Pettit's compound in which other groups are seen in place of the hydrogen substituents of the methylene carbon. Many of these derivatives are more stable than the parent methylene complex because of the cation-stabilizing ability of the additional substituents. These compounds are too numerous to discuss in detail here, but some of them bearing carbon groups in place of hydrogen include Maurice Brookhart's benzylidene complexes,[22] Warren Giering's benzocyclobutenylidene complexes,[23] William M. Jones' cyclopropylidene and cycloheptatrienylidene complexes,[24] and Maurice Brookhart's cyclopropylmethylene complexes.[25] The hetero-substituted carbene complexes having either one or two alkoxy, amino, or alkylthio groups attached to the carbene center are generally much more stable than the simple hydrogen- or carbon-substituted carbene complexes. These compounds have been studied in detail by Robert Angelici,[26a-d] Anthony Barrett,[26e] Charles Casey,[27] Alan Cutler,[28] Stephen Davies,[29] Alan Davison,[30] Malcolm Green,[31] Véronique Guerchais,[17c] Russell Hughes,[32] Wolfgang Malisch,[33] D. F. Marten,[34] Myron Rosenblum,[35] Paul Treichel,[36] and Joachim Wachter.[37]

Other aspects of the chemistry of these and related iron carbene complexes have been reported by several other workers as cited in the references[38] and as discussed in later parts of this chapter. Furthermore, Maurice Brookhart has recently published a review of alkene cyclopropanation reactions that employ many of these as well as other types of carbene complexes.[39] Except for a few special cases, related complexes of metals besides iron will not be included in the present chapter.

II. OUR EARLY STUDIES: DEVELOPMENT OF A METHYLENE TRANSFER REAGENT

With this background information in mind, the author's research group entered this area. We recognized that most of the early work with carbene complexes of the η_5-$C_5H_5(CO)_2$Fe system was of a structural, mechanistic, or theoretical nature. However, being primarily synthetic organic chemists, we could not help but notice that very little had been done up to this point in terms of applications of these iron complexes. We therefore became very interested in the question of whether this organometallic chemistry could be used to develop some new reagents that would be of use in organic synthesis. In particular, we believed that some new reagents for the cyclopropanation of alkenes could arise from these efforts.

As an initial goal, we set out to prepare complexes of the general structure **7** having a potential leaving group Z bonded to the iron-bearing carbon atom (Scheme 2). We hoped that these compounds would be sufficiently stable to serve as useful, easily handled laboratory reagents but that the group Z could be lost in the presence of alkene substrates, which upon reaction with the resulting carbene complex would give cyclopropanes directly. Perhaps the group Z would be lost spontaneously in solution, or perhaps it would first need to be activated by another reagent.

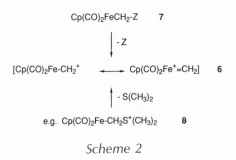

$$Cp(CO)_2FeCH_2\text{-}Z \quad \textbf{7}$$

$$\downarrow \; \text{-}Z$$

$$[Cp(CO)_2Fe\text{-}CH_2^+ \quad \longleftarrow \quad Cp(CO)_2Fe^+{=}CH_2] \quad \textbf{6}$$

$$\uparrow \; \text{-}S(CH_3)_2$$

$$\text{e.g.} \; Cp(CO)_2Fe\text{-}CH_2S^+(CH_3)_2 \quad \textbf{8}$$

Scheme 2

In principle, the choice of the leaving group Z may be based upon many heteroatomic functional groups. Of course, the ether derivative reported originally by Pettit[9] is one specific case, and several more recent investigators have studied a number of related ether complexes.[39] Alternatively, other possibilities that readily come to mind are functional groups containing nitrogen, sulfur, selenium, phosphorus, arsenic, etc.[40]

Our decision to concentrate upon sulfur derivatives (e.g. the sulfonium salt **8**) has its origin back in the period of the author's postdoctoral work with

Professor E. J. Corey at Harvard in 1973. Professor Corey had been responsible for much of the earlier development of sulfonium and oxysulfonium
ylides as reagents for the conversion of carbonyl compounds into epoxides
and of α,β-unsaturated carbonyl compounds into cyclopropyl derivatives.[41]
Professor Corey was interested in the possibility of extending the use of these
sulfur reagents to reactions with simple, unactivated alkenes to provide
cyclopropanes in general. This idea was based in part upon a proposed
biosynthetic pathway whereby the methyl carbon of the sulfonium salt,
S-adenosylmethionine, is incorporated into naturally occurring cyclopropanes.[42] Also, an indication of the possibility of effecting this kind of
reaction was given by Barry Trost in his studies of copper-catalyzed reactions
of a phenacylsulfonium ylide.[43a] In the Corey laboratory, attempts were made
to catalyze these reactions through use of various transition metal
compounds. Because of their ready availability, dimethylsulfonium ylides
were used in these studies. However, none of the reactions gave significant
amounts of cyclopropanes, although somewhat later, Theodore Cohen was
successful in employing a less readily available diphenylsulfonium ylide in
reactions catalyzed by copper(II) acetylacetonate.[43b] Much more recently,
Klaus-Richard Pörschke has shown that an oxysulfonium ylide complex of
nickel that contains ethylene ligands decomposes to give cyclopropane
among other products.[43c]

The lack of success in using dimethylsulfonium ylides in metal-catalyzed
reactions eventually led us to pursue a systematic study of possible stoichiometric pathways whereby a transition metal-bonded ylide would first be
generated as a discrete intermediate followed by a separate reaction with an
alkene. Ylide complexes of various metals were already known at the time,
and several more have subsequently been studied.[43,44] When faced with the
question of which metal system would be best for our studies, we were
attracted to the η_5-$C_5H_5(CO)_2$Fe system for three principal reasons. First of
all, the results of Pettit described above (Scheme 1)[9] provided a very good
basis for assuming that if the iron sulfonium derivative **8** (Scheme 2) were
prepared, it could serve as a precursor of the carbene complex **6** which, unlike
the complexes of many other metals, would then be capable of undergoing
the desired cyclopropanation reactions. A second reason for favoring η_5-
$C_5H_5(CO)_2$Fe complexes was that the preparative chemistry for σ-organo
derivatives of this system was already very well-developed as the result of
extensive work of organometallic chemists starting way back in the 1950s.
Finally, these derivatives were quite stable relative to many other organotransition metal compounds.[13]

Only preliminary investigations of this idea were pursued at Harvard.
Finally, in the late 1970s, after the author had been on the faculty at the State
University of New York at Stony Brook for several years, the decision was
made to pursue this work in much more detail. At that time, Steven Brandt

was an advanced Ph.D. student in the author's laboratory. Steve had been very successful with some other transition metal-based projects in organic synthesis, and he decided to take on the iron carbene area as yet another part of his dissertation research. Very quickly, Steve also achieved major success with this new project.[45] Our desired reagent was prepared by a straightforward route (Scheme 3). In analogy with Pettit's preparation of his ether derivative **5** (Scheme 1), the commercially available dinuclear complex **3** was treated with sodium amalgam, and the intermediate ferrate was allowed to react *in situ* with chloromethyl methyl sulfide to give the sulfide derivative **9** as a yellow–brown oil which could be obtained in pure form in 93% yield after low-temperature recrystallization. Basically the same procedure had been reported much earlier by Bruce King, although the yield of **9** was not as high in this previous work.[46] In turn, reaction with trimethyloxonium tetrafluoroborate[47] gave our reagent **8** as a yellow solid. Although the reaction of complexes of the type $Cp(CO)_2FeX$ with sulfonium ylides would appear to be a more direct means of obtaining **8** and related complexes,[44] our attempts to employ this reaction have never been successful in the case of the parent methylide or more highly substituted derivatives.

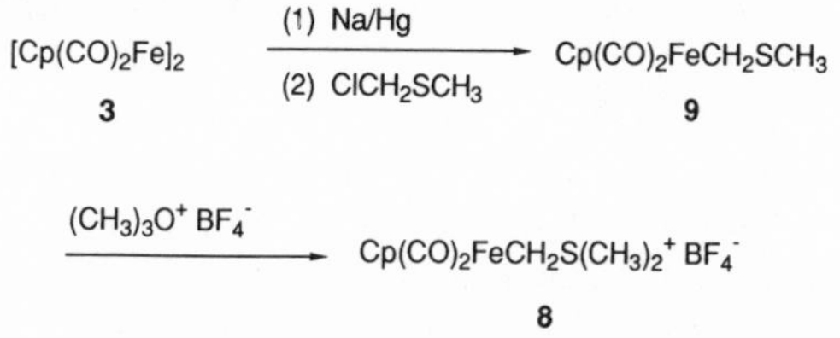
[Cp(CO)2Fe]2
3
(1) Na/Hg
(2) ClCH2SCH3
Cp(CO)2FeCH2SCH3
9
(CH3)3O+ BF4-
Cp(CO)2FeCH2S(CH3)2+ BF4-
8

Scheme 3

A brief survey of the reaction of this sulfonium salt with alkenes indicated that it did indeed have good potential as a useful methylene transfer reagent; a number of alkenes were converted directly into the corresponding cyclopropanes (Eq. 4). Our preliminary results were reported in communication form in 1979.[48]

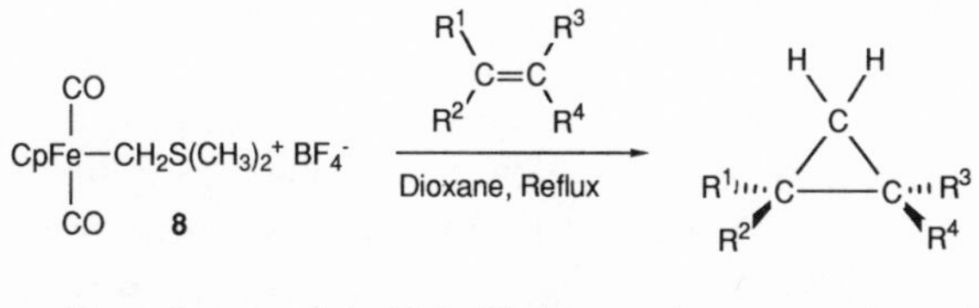
CO
CpFe—CH2S(CH3)2+ BF4-
CO
8
R1 R3 C=C R2 R4
Dioxane, Reflux
H H C
R1 C C R3
R2 R4

(4)

After Steve Brandt completed his Ph.D. work, very systematic and detailed studies of our reagent were taken up by Edward O'Connor, another very successful doctoral student in our laboratory at Stony Brook.[49] Ed first concentrated on improving and streamlining the preparation of **8**. He soon found that there is no need to isolate and purify the sulfide **9**. Instead, the solution containing the crude compound is simply filtered to remove inorganic salts and is then treated directly with the methylating agent to give the sulfonium salt **8**.

Ed also found that any of several alkylating agents may be used. These reagents include trimethyloxonium tetrafluoroborate as already mentioned but also methyl fluorosulfonate ("Magic Methyl"; CAUTION: this alkylating agent is very volatile and highly toxic), dimethoxycarbenium tetrafluoroborate, and methyl iodide. For the sulfonium salt to have good reactivity as a cyclopropanation reagent, however, the counterion must be relatively non-nucleophilic (*vide infra*). The tetrafluoroborate salt has shown the best reactivity to date. Ed found that for obtaining this salt, an especially convenient methylating agent among those listed above is the dimethoxycarbenium reagent, $(CH_3O)_2CH^+ \ BF_4^-$. It is very easily prepared by reaction of trimethyl orthoformate with boron trifluoride etherate according to the procedure of Richard Borch,[50] and there is no need to isolate the salt before use. This procedure is much simpler to perform than the preparation of trimethyloxonium tetrafluoroborate.[47] Trimethyloxonium salts are commercially available, but they are somewhat expensive, and their levels of purity vary from order to order.

There is no question that methyl iodide would be one of the most attractive methylating agents to use. In order to take advantage of the convenience of using this reagent, and in order to meet the requirement of having a non-nucleophilic counterion in our reagent **8**, Ed studied various anion exchange procedures. He found that upon obtaining the iodide salt from reaction of **9** with methyl iodide, the desired anion exchange occurs with aqueous sodium tetrafluoroborate. Also, the iodide and fluorosulfonate salts of **8** undergo exchange with aqueous sodium tetraphenylborate.

In order to provide even more convenient access to our reagent, we have sought to improve our procedure further. In particular, we set out to avoid the use of sodium amalgam because of the inconvenience and hazards of using large amounts of mercury. Also, we wished to have the preparation of **8** approach a true "one-pot" procedure as closely as possible. Matthew Mattson, a talented doctoral student presently in our laboratory, now at the University of Notre Dame, has tackled this problem. We employ the procedure of Daniel Reger[51] whereby the sodium ferrate **4** is generated by heating the dinuclear iron complex **3** and sodium dispersion in refluxing tetrahydrofuran. The solution is cooled to 0°C, and the chloromethyl methyl sulfide is added. After the alkylation reaction has occurred to produce a solution of the sulfide derivative **9**, methyl iodide is added to the same reaction flask. The mixture containing the iodide salt of the sulfonium complex **8** is concentrated by rotary evaporation, the residue is taken up in hot water, the mixture is filtered, and aqueous sodium tetrafluoroborate is added to the filtrate. The desired tetrafluoroborate salt of **8** precipitates from the reaction mixture. This procedure may readily be scaled up for the preparation of several tens of grams of our reagent at a time.

Before the reactions and other details of this reagent are discussed, a few comments will be made about its physical properties. It is initially obtained from the above procedures as a yellow powder or microcrystalline solid that is of sufficient purity for use in the reactions described below. However, recrystallization may be accomplished very easily from a variety of solvents (e.g. nitromethane) to give the reagent in very pure form as large, beautiful, amber-colored crystals. Especially noteworthy is that the salt **8** is an organometallic reagent of unique stability. Not only is it stable to the air, but it is also remarkably stable to water. This point is made especially clear by the fact that the anion exchanges described above are done in water at 90°C! The resulting salt is stored as a normal laboratory reagent exposed to the air in ordinary reagent bottles. A control sample was stored in this manner and was checked occasionally by ¹H-NMR and by using it in cyclopropanation reactions; no noticeable deterioration occurred after several years! In addition, the cyclopropanation reactions of **8** may be carried out in ordinary flasks open to the air.

The various salts **8** have been studied by the usual spectroscopic techniques, including IR, ¹H-NMR, ¹³C-NMR, and MS. These data are, for the most part, unexceptional and are fully consistent with their formulation as σ-bonded alkyl derivatives of the η^5-$C_5H_5(CO)_2$Fe system. Confirmation of this assignment was provided when the structure of the fluorosulfonate salt was determined by single-crystal X-ray diffraction (Figure 2).[52] The methylene carbon (C3) exhibits tetrahedral-like geometry, although the Fe–C3–S1 angle is opened up to 115.5(2)°. Also, the Fe–C3 bond length of 2.036(3) Å may be compared to values of 2.06–2.11 Å for a range of other alkyl derivatives in the η^5-$C_5H_5(CO)_2$Fe series.[52] These differences in parameters for **8** are probably too minor to be attributed to any special property of our sulfonium salt derivatives and do not give any clear indication of metal carbene-like character. Isolated carbene complexes on the other hand have iron-carbon bond lengths of approximately 1.8–1.9 Å (*vide infra*). One in-

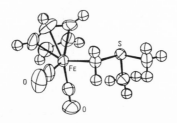

Figure 2. X-ray structure of $Cp(CO)_2FeCH_2S(CH_3)_2{}^+$ $FSO_3{}^-$ (**8**)

triguing observation, however, is the appearance of peaks for both the cation of **8** and the methylene complex **6** in the FAB MS spectrum of these salts. Beauchamp reported analogous findings in his ion cyclotron resonance investigation of Pettit's original system (Scheme 1).[14]

Before studying the scope of the reactions of **8** with alkenes, we determined the effects of various reaction parameters on the efficiency of the cyclopropanation. We first varied the counterion X^- in $Cp(CO)_2FeCH_2S^+(CH_3)_2$ X^- under a standard set of conditions employing equimolar amounts of *cis*-cyclooctene and the salt in 1,4-dioxane at reflux. As shown by the summary of our data in Table 1, the tetrafluoroborate salt was the most efficient reagent, although the fluorosulfonate was also reasonably good. The significantly lower efficiency of the iodide may be attributed to the much higher nucleophilicity of this counterion and the possibility that it may intercept intermediates that normally lead to the cyclopropane products. We have no satisfactory explanation for the lack of cyclopropanation ability of the tetraphenylborate.

We next studied the effect of solvent (Table 2) in the case of the fluorosulfonate rather than the tetrafluoroborate salt. Presumably, the former salt, being somewhat less reactive, may be expected to give a broader range of yields in a variety of solvents. Because 1,4-dioxane was found in our early studies to be a rather good solvent for these reactions, it was the most commonly used solvent in our subsequent investigations of the scope of the cyclopropanation reactions (*vide infra*). Nitromethane was found to be a superior solvent only much later in our work. This solvent gave somewhat higher yields, and the reaction times were considerably shorter. Of particular interest is that a protic solvent such as ethanol may be used effectively in these reactions. This observation is consistent with the previously mentioned stability of our reagents to hot water.

Modest yields (40–50%) of cyclopropanes are also obtained when the flasks containing the reaction mixtures are placed in a Rayonet Srinivasan–Griffin photochemical reactor (350–nm lamps) for 12 h or in a laboratory

Table 1. Counterion Effect in Reaction of $Cp(CO)_2FeCH_2S^+(CH_3)$ X^- (**8**) with *cis*-Cyclooctene in 1,4-Dioxane[a]

X	Yield of Bicyclo[6.1.0]octane (%)[b]
BF_4	95
FSO_3	72
I	38
BPh_4	0

[a] 1 mmol of sulfonium salt **8** and 1 mmol of alkene in 0.5 ml of 1,4-dioxane at reflux for 12 h. [b] Yield determined by GLPC.

Table 2. Solvent Effect in Reaction of $Cp(CO)_2FeCH_2S^+(CH_3)_2$ FSO_3^- (**8**) with *cis*-Cyclooctene[a]

Solvent	Temperature (°C)	Time (h)	Yield of Bicyclo[6.1.0]octane (%)[b]
Acetone	56	12	0
THF	65	24	0
CH_2Cl_2	40	12	2
$(Me_2N)_3PO$	95	12	2
DME	85	12	5
DMSO	120	12	15
$ClCH_2CH_2Cl$	83	12	35
DMF	120	12	70
1,4-Dioxane	102	12	75
EtOH	78	72	75
CH_3NO_2	104	3	95

[a] 1 mmol of sulfonium salt **8** and 1 mmol of alkene in 0.5 mL of solvent. [b] Yield determined by GLPC.

ultrasonic cleaning bath for 22 h. We have not attempted to distinguish possible photochemical or sonication effects from simple thermal effects. Based upon these results, a standard set of conditions was chosen to study the reactions of the tetrafluoroborate salt **8** with a variety of alkenes (Table 3).[58] We employed one molar-equivalent of alkene and two molar-equivalents of the salt in dioxane at reflux for a typical reaction time of 12 h. An excess of **8** was employed in order to increase the conversion of alkenes to cyclo-propanes. When equimolar amounts of **8** and alkene are used, substantial amounts of unreacted alkene are recovered in many cases, but the conversion increases when excess **8** is employed.

Our results show that the reagent **8** can be used to obtain modest to high yields of cyclopropanes from a wide range of alkenes. The best substrates are generally monoalkyl, dialkyl-, and aryl-substituted alkenes (see entries 1–10 of Table 3). The cyclopropanations are highly stereospecific in that the configurations of the alkene starting materials are retained in the resulting cyclopropanes (see entries 9 and 10).

There are various limitations on the utility of this reaction, however. The sensitivity of the reagent **8** to steric hindrance is indicated by the low yields in the cases of trisubstituted alkenes (e.g. entry 11), by the lower yields obtained from trans-disubstituted alkenes compared to their cis isomers (entries 9 and 10), and by the failure of norbornene to react with **8**. Poor results are usually obtained with cyclohexenes. For examples, cyclohexene itself and 1-methylcyclohexene are both nearly completely consumed, but complex mixtures of products are obtained. In each of these two cases, products are obtained that have molecular weights (MS) indicating the transfer of either one or two methylene groups. Two of the products that

Table 3. Cyclopropanation of Alkenes with $Cp(CO)_2FeCH_2S^+(CH_3)_2BF_4^-$ (**8**, Eq. 4)[a]

Entry	Alkene	Cyclopropane(s)	Conversion of Alkene (%)[b]	Corrected yield of cyclopropane (%)[c]
1	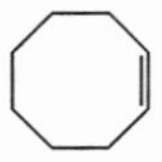	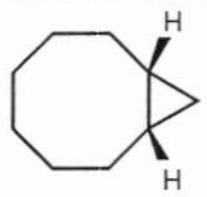	100	92, 76[d]
2	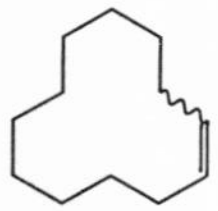	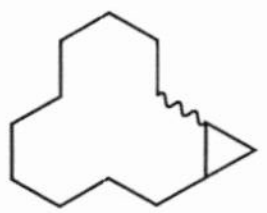	100	90[e]
3	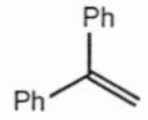	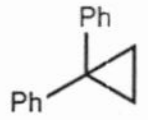	100	99, 88[d]
4	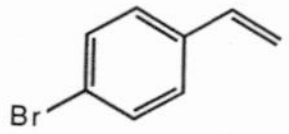	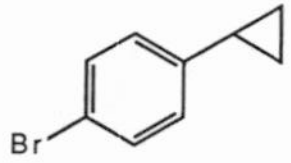	100	> 90[d,f,g]
5	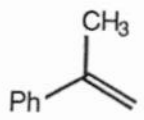	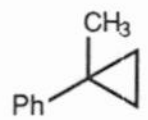	100	64
6	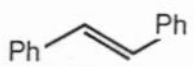		62	64[h]

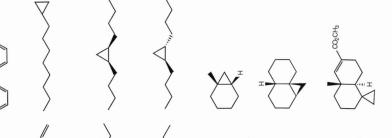

7		38	82
8		81	70[d]
9		59	87
10		49	67
11		90	26
12		—	—[i]
13		100	86[d,j]

(continued over)

157

Table 3 (Continued)

Entry	Alkene	Cyclopropane(s)	Conversion of Alkene (%)[b]	Corrected yield of cyclopropane (%)[c]
14	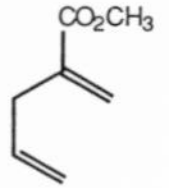 (structure, CO_2CH_3)	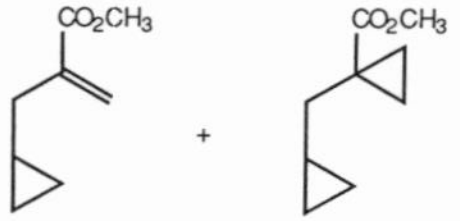 (structures, CO_2CH_3) +	99	62 + 5
15	 (structure, PhS)	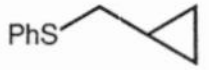 (structure, PhS)	100	60[k]
16	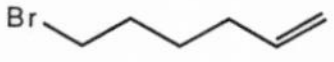 (structure, Br)	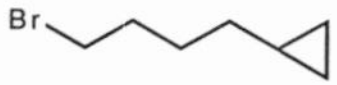 (structure, Br)	100	55
17	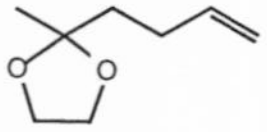 (dioxolane structure)	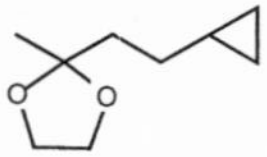 (structure)	100	22
18	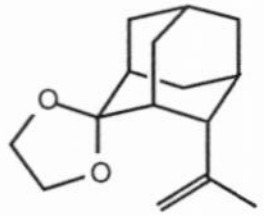 (adamantane dioxolane structure)	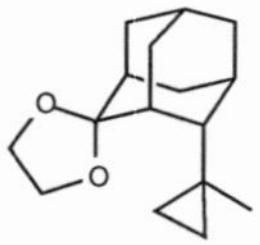 (structure)	< 100	—[l]

[a]Unless otherwise indicated, these reactions were run under a standard set of conditions employing 2 molar equivalents of **8a** and a 2 M solution of alkene in dioxane at reflux for 12 h. [b]Conversion of alkene indicates the amount of alkene consumed in the reaction as determined by measuring the amount of unreacted alkene by GLPC using an internal standard. [c]Unless otherwise specified, the yields of cyclopropanes were determined by GLPC using an internal standard and a sample of pure, isolated product for calibration and are corrected for the amount of unreacted alkene. [d]Yield of pure, isolated product. [e]Both the starting cyclododecene and the product were ca. 1:1 mixtures of cis and trans isomers. [f]The reaction solvent was nitromethane. [g]Ref. 53. [h]The stereochemistry of the product was determined by ^{1}H-NMR as reported in Ref. 54. [i]Ref. 55. This reaction was performed primarily for the purpose of a structure proof; the yield of the product was not determined. [j]Ref. 56. These workers reported the use of 3 molar equivalents of **8a**. [k]PhSCH₃ comprised the remaining 40% of the product mixture. [l]Ref. 57. These workers simply reported incomplete conversion.

158

have been identified in the case of 1-methylcyclohexene as the substrate (entry 1) are the expected 1-methylbicyclo[4.1.0]heptane and the unexpected 1,6-dimethylbicyclo[4.1.0]heptane. Complex mixtures are also obtained from substrates containing exocyclic alkenes (e.g. 6-methylenenorbornene and methylenecycloheptane). In the case of 4-vinylcyclohexene, a mixture of at least four cyclopropane-containing products is obtained due to reaction of both double bonds. A comparison can be made in this case with our ethylidenation reagent, $Cp(CO)_2FeCH(SPH)CH_3$ (**18**, *vide infra*), which reacts selectively with the double bond of the vinyl group. Another complication that apparently arises in some cases is rearrangement of the initially obtained product. One case is the use of phenanthrene (entry 7) which leads to a mixture of the corresponding cyclopropane and approximately a 10% yield of 9-methylphenanthrene; this minor product has previously been reported as a rearrangement product of the cyclopropane.[59] We have not ruled out the possibility, however, that the methylated product arises from an electrophilic aromatic substitution pathway not involving the cyclopropane as an intermediate.

The reagent **8** is compatible with some functional groups such as alkyl halides (entries 4 and 16), esters (entries 13 and 14), and thioethers (entry 15). The last of these groups undergoes partial cleavage to give thioanisole as a side product, and in the case of 3-(phenylthio)cyclohexene, this reaction occurs to the exclusion of the cyclopropanation. Ketal-containing substrates (entries 17 and 18) give low yields of products, and the reaction fails to give detectable quantities of cyclopropanes in the cases of 3-bromocyclohexene, various unsaturated alcohols, and 5-hexen-2-one. Based upon **8** being an electrophilic cyclopropanation reagent, we were not surprised to find that it is not useful for reactions with electron-deficient alkenes. An α,β-unsaturated ester such as ethyl crotonate gives only a 10–15% yield of the corresponding cyclopropane, and α,β-unsaturated ketones fail entirely. As a result, the cyclopropanation of simple alkene units can be effected chemoselectively in the presence of conjugated, electron-deficient alkenes (entries 13 and 14). At the other extreme, very electron-rich alkenes such as enamines, enol ethers, and vinyl sulfides show very high reactivity toward the reagent **8**, but except for a 10% yield obtained from (phenylthio)ethylene, we have not been able to isolate the corresponding cyclopropanes in these cases. Conceivably, the cyclopropanes are produced, but being hetero-substituted, they are very susceptible to electrophilic cleavage[3d] by the cationic byproducts (e.g. $Cp(CO_2)(L)Fe^+$) of these reactions.

For the purpose of comparison of our reagent with a better-known, more commonly used method of alkene cyclopropanation,[3] a very appropriate choice is the Simmons-Smith reaction. This method employs 1,1-dihaloalkanes and zinc–copper couple, diethylzinc, or other metal reagents, or even photochemical conditions in place of the metal species.[3a,60] Many of the simpler types of alkenes shown in Table 3 give similar results with the Simmons–Smith reaction and our reagent. The Simmons–Smith reaction is clearly superior with cyclohexenes and certain functionalized alkenes such as unsaturated alcohols, some α,β-unsaturated carbonyl compounds, enol derivatives, and some enamines. A major advantage of the Simmons–Smith reaction is the stereochemical directing effect observed with many unsaturated alcohols. However, compared to the use of 8, the Simmons–Smith reaction sometimes gives lower yields for some aryl-substituted alkenes (see entries 3–7 of Table 3), perhaps due to Lewis acid-promoted alkene polymerization as a competing reaction. Also, the Simmons–Smith reaction fails in cases of some sulfur-containing substrates such as vinylic and allylic sulfides; instead of undergoing cyclopropanation, rearrangement of allylic sulfides to homoallylic derivatives has been noted. The reaction has also failed when simple organic sulfides have been present with alkenes that normally react successfully with Simmons–Smith reagents.[61] From our own experience, the Simmons–Smith reaction failed with the substrate shown in entry 12 of Table 3.[55] We should also comment again upon the use of two molar-equivalents of 8 in our reactions by mentioning that Simmons–Smith reactions are frequently performed with several-fold excesses of reagents. The same is often true of another commonly used method of cyclopropanation, namely, the use of metal-catalyzed reactions of diazo compounds.[62] In defense of the Simmons–Smith reaction, the recent development of trialkylaluminum reagents for this reaction by Hisashi Yamamoto is especially promising and may provide means for overcoming some of the problems that have been reported in the past.[60d]

Nevertheless, a clear advantage of using 8 is that once this reagent is obtained, it is extremely convenient to use because of its unusual stability. Unlike some other reagents, there is no need to activate it before use. The use of 8 also avoids the handling of hazardous or sensitive materials such as diethylzinc or triethylaluminum as in some modifications of the Simmons–Smith reaction or diazomethane as used in some other metal-catalyzed cyclopropanations.

We have also studied various modifications of our reagent. First of all, we determined the effect of changing the nature of the sulfonium group. The phenylthiomethyliron complex 10 was prepared from chloromethyl phenyl sulfide in analogy with the route shown in Scheme 3. This complex is a golden-yellow, crystalline solid that may be handled in the air for at least an hour and that is stable indefinitely upon storage at 25°C under nitrogen. Reaction with trimethyloxonium tetrafluoroborate gives the sulfonium salt 11 as a solid that is stable for at least several hours at 25°C. This compound is more reactive than the original dimethylsulfonium salt 8 in that 11 reacts with cis-cyclooctene in methylene chloride at 25°C to give the corresponding cyclopropane in 20–30% yield within 12 h; however, higher yields comparable to those from **8** are not obtained. Kent Barefield has prepared the diphenylsulfonium derivative **12**, and as expected, it is an even more reactive complex that converts *cis*-cyclooctene to the cyclopropane in 85% yield in only three hours at 22°C.[38j]

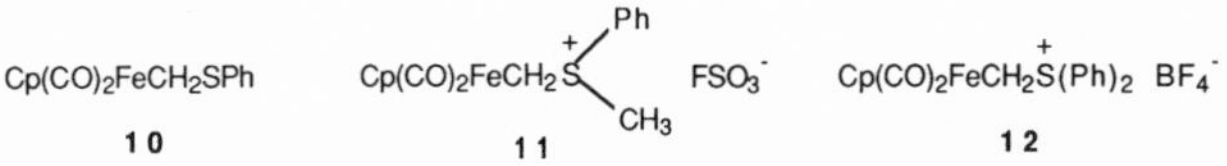

Cp(CO)₂FeCH₂SPh   Cp(CO)₂FeCH₂S⁺(Ph)/CH₃ FSO₃⁻   Cp(CO)₂FeCH₂S⁺(Ph)₂ BF₄⁻

    10   **11**   **12**

We have also been interested in the use of other metals. There is no inherent reason that the types of cyclopropanations discussed above should be limited to complexes of iron. Based upon another result from the pioneering work of Pettit,[9] molybdenum is another possible choice of a metal for these systems. We have prepared the methylthiomethylmolydenum complex **13** and subsequently the dimethylsulfonium salt **14** (Scheme 4). Unfortunately, **14** is much less stable than the iron reagent **8**, and it gives only very low yields of cyclopropanes from *cis*-cyclooctene and 1,1-diphenylethylene. Maurice Brookhart[63] and Alan Cutler[28c] have also studied related moybdenum complexes, and Kent Barefield has examined nickel analogues,[44s] but the iron reagents remain superior at this time.

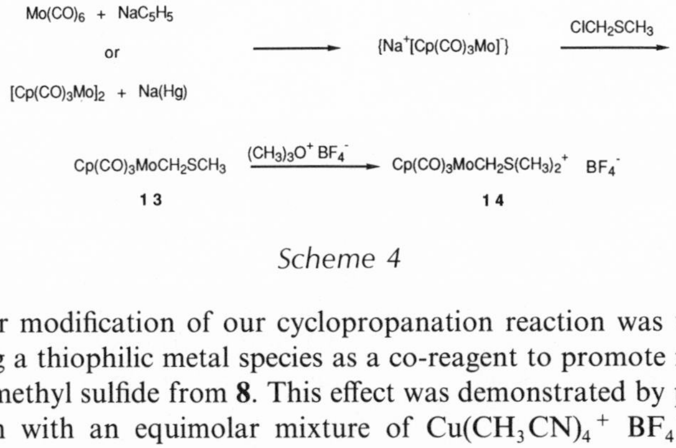

Mo(CO)₆ + NaC₅H₅

    or   ⟶   {Na⁺[Cp(CO)₃Mo]⁻} $\xrightarrow{\text{ClCH}_2\text{SCH}_3}$

[Cp(CO)₃Mo]₂ + Na(Hg)

Cp(CO)₃MoCH₂SCH₃ $\xrightarrow{\text{(CH}_3)_3\text{O}^+ \text{BF}_4^-}$ Cp(CO)₃MoCH₂S(CH₃)₂⁺ BF₄⁻

    13   **14**

Scheme 4

Another modification of our cyclopropanation reaction was the idea of employing a thiophilic metal species as a co-reagent to promote more facile loss of dimethyl sulfide from **8**. This effect was demonstrated by performing a reaction with an equimolar mixture of Cu(CH₃CN)₄⁺ BF₄⁻, the iron reagent **8**, and 1,1-diphenylethylene in nitromethane. A 50% conversion of the alkene to the cyclopropane was observed within only 2 h at 25°C, although no further increase of product was seen after several more hours.

III. An Ethylidene Transfer Reagent

We certainly did not want to be restricted to cyclopropanation reactions that transfer only the simple methylene group to alkenes. Rather, we wished to explore the generality of our types of iron reagents for *alkylidene* transfer in

general. Admittedly, we were concerned that this further work with sulfonium salts **15** could be frustrated by the possibility of a competing process
whereby the resulting substituted carbene complexes **16** would undergo a
1,2-hydrogen shift leading to π-alkene complexes **17** (Scheme 5) instead of the
desired cyclopropanation reactions. This type of rearrangement was
precedented by earlier work of Myron Rosenblum.[64] The alkene complexes
17 have many useful applications,[13] but they would most likely be of no use
to us in the context of our cyclopropanation studies.

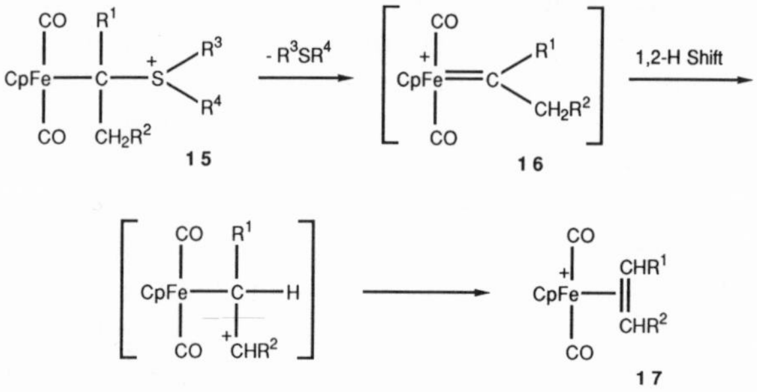

Scheme 5

The first logical step in addressing this issue was to determine whether we
could develop an ethylidene transfer reagent that would permit the synthesis
of methyl-substituted cyclopropanes. This phase of our work was taken on
by Kenneth Kremer, another Ph.D. student at Stony Brook. Ken proved very
quickly to be a very successful member of our research team. Within a month
of joining the group, he obtained the 1-(phenylthio)ethyliron complex **18** as
a possible precursor of the ethylidene complex **16** ($R^1 = R^2 = H$) and found
that it was indeed a very useful ethylidene transfer reagent (Scheme 6).[65] In
this case, the subsequently generated sulfonium salt is not sufficiently stable
to be isolated. Therefore, we use the neutral sulfide complex 18 as our
storable laboratory reagent for ethylidenations. It can be purified by a

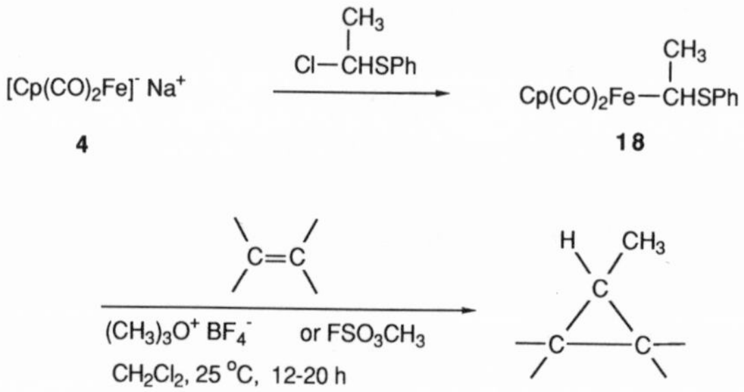

Scheme 6

modified[66] flash chromatography procedure[67] and by recrystallization to give yellow crystals that may be weighed and transferred in the air for short periods of time. For long-term storage, the reagent is placed in nitrogen-filled containers in a freezer. Samples of **18** have been kept in this manner for at least a few years.

For the purpose of performing cyclopropanation reactions, **18** may be converted to the corresponding sulfonium salt *in situ*. Among the alkylating agents that result in good yields of cyclopropanes are trimethyl- or triethyloxonium tetrafluoroborate[47,68] and methyl fluorosulfonate ("Magic Methyl"; CAUTION: this alkylating agent is very volatile and highly toxic), although the trimethyloxonium salt again gives somewhat superior yields. Methyl iodide and trimethyloxonium hexachloroantimonate are not useful at all. Unlike the methylene transfer reactions using **8**, the reactions of **18** must be conducted under an inert atmosphere. Somewhat higher conversions of alkenes to cyclopropanes are again seen when the reagent **18** is used in excess, but the effect is smaller than in the case of **8**. Therefore, the use of nearly stoichiometric amounts of reagents and alkenes are sufficient for our ethylidene transfer reactions. The sulfonium salt appears to form at temperatures as low as $-80°C$, but the rate of reaction with olefins becomes significant only above $0°C$. Of the several solvents that have been used in this reaction, the best is simply methylene chloride. Based upon our optimization studies, the standard conditions for these reactions are to use 1 M solutions of the alkenes in methylene chloride, one molar-equivalent of **18**, and 1.25 molar-equivalents of trimethyloxonium tetrafluoroborate or methyl fluorosulfonate at $25°C$ for 12–20 hours. Our results for several alkenes as substrates are summarized in Table 4.

The range of alkenes that undergo efficent ethylidenation with our reagent is similar to that seen for our methylene transfer reaction. A peculiar observation is that the results obtained with *trans*-1,2-disubstituted alkenes are highly dependent upon the choice of alkylating agent for conversion of the reagent **18** to the sulfonium salt; whereas the use of methyl fluorosulfonate leads to no cyclopropane, trimethyloxonium tetrafluoroborate gives at least a very modest yield of the desired product (entry 4). Also to be noted is the selectivity for simple, monosubstituted alkenes in the presence of a cyclohexene double bond (entry 10); in contrast, the Simmons–Smith reaction occurs at both sites.[71] Remote halide, acetal, and ketone groups (entries 12–14) are at least somewhat tolerated. In addition to the electron-deficient alkene, methyl crotonate (entry 15), the reaction also fails with 3-methylcyclohexene, norbornene, phenanthrene (compare with entry 7 of Table 3 for reagent **8**), 3-bromocyclohexene, 4-(2'-hydroxyethyl)cyclohexene, 4-penten-1-ol, allyl phenyl sulfide (compare with entry 15 of Table 3), and *cis*-1-methoxycyclooctene.

The cyclopropanations using **18** show excellent stereospecificity for reten-

Table 4. Ethylidene Transfer to Olefins with $Cp(CO)_2FeCH(CH_3)SPh$ (**18**)

Entry	Olefin	Cyclopropane	Conversion of olefin (%)[a]		Corrected Yield (%)[a]	
			MM[b]	TMO[c]	MM[b]	TMO[c]
1	 (cyclooctene)	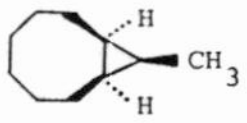 (cyclopropane, H, CH₃, H)	85	100	70[d,e]	73
2	 (cyclononene)	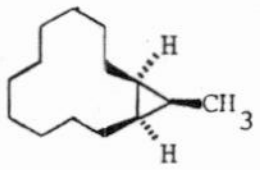 (cyclopropane, H, CH₃, H)	49		70[h]	
3	 (alkene)	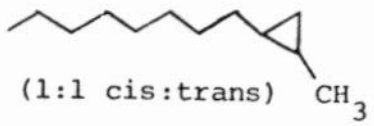 (cyclopropane, CH₃) (1:1 cis:trans)	55	54	70	87
4	 (alkene)	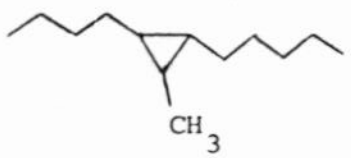 (cyclopropane, CH₃)	—	29	0	48[e]
5	 (alkene)	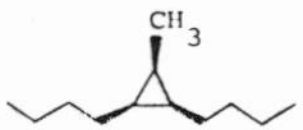 (cyclopropane, CH₃)	62	65	44	66[e]
6	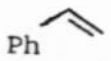 (Ph alkene)	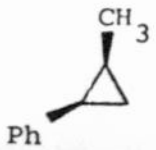 (cyclopropane, CH₃, Ph)	81	100	67[e,f]	67

164

7	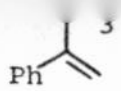 (structure: Ph, CH₃ vinyl)	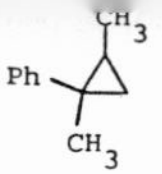 (structure: cyclopropane Ph, CH₃, CH₃)	100	100	58[g]	65
8	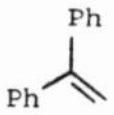 (structure: Ph vinyl, Ph)	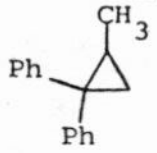 (structure: Ph, CH₃, CH₃, Ph) (41:17 Z:E)	59		81	
9	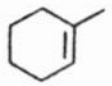 (structure: methylcyclohexene)	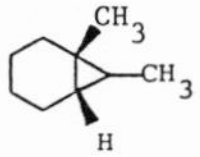 (structure: CH₃, CH₃, H) (>5.6:1 endo:exo)		34		66
10	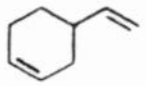 (structure: vinylcyclohexene)	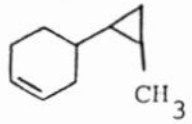 (structure: CH₃) (1:1 cis:trans)		50[i]		80[i]
11	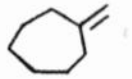 (structure)	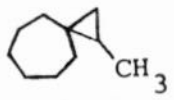 (structure: CH₃)		100		
12	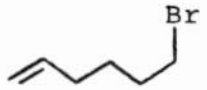 (structure: Br)	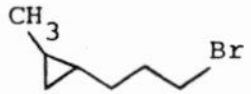 (structure: CH₃, Br) (.9:1 cis:trans)		59		84
13	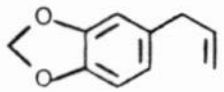 (structure: methylenedioxyphenyl allyl)	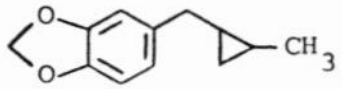 (structure: CH₃) (1.3:1 trans:cis)		81		46

(continued over)

Table 4 (Continued)

Entry	Olefin	Cyclopropane	Conversion of olefin (%)[a]		Corrected Yield (%)[a]	
			MM[b]	TMO[c]	MM[b]	TMO[c]
14	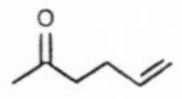	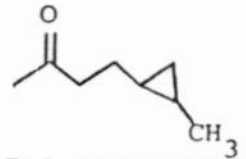 (1.7:1 trans:cis)		77		27
15	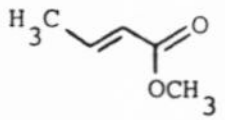	—		—		0

[a] These values were determined as described in Table 3 (notes b and c). [b] MM: Magic Methyl (methyl fluorosulfonate) was used as the alkylation agent. [c] TMO: trimethyloxonium tetrafluoroborate was used as the alkylating agent. [d] Stereochemistry was determined by ^{1}H-NMR as in Ref. 60a. [e] This product was the only cyclopropane isomer detected. [f] Stereochemistry was determined by ^{1}H-NMR as in Refs. 7b and 69; the CH$_3$ resonance appeared at δ 0.79. [g] The C-2 CH$_3$ group appeared in the ^{1}H-NMR spectrum at δ 0.76 for the Z-isomer and at δ 1.16 for the E-isomer. [h] When a mixture of cis- and trans- cyclododecene is used as in our preliminary communication (see notes 65a and 70), the expected mixture of cyclopropane is obtained. [i] These values were estimated from GLPC recorder tracings.

166

tion of the configuration of the original alkene configuration (e.g. entries 4 and 5 of Table 4) and often excellent stereoselectivity for formation of the *endo*-methylcyclopropanes (entries 1, 2, 5, 6, 7, and 9). Poor stereoselectivity is seen for most monosubstituted alkenes (entries 3, 10, and 12–14), except for styrene which shows very high stereoselectivity (entry 6). Of interest is that when this stereoselectivity is seen, the favored product is the more sterically encumbered *endo-* or *syn*-isomer rather than the more stable *exo-* or *anti*-product. This interesting phenomenon has also been reported for some other methods of cyclopropanation.[7b,69,72]

Concurrent with our studies of **18** as an ethylidene transfer reagent, Maurice Brookhart developed the 1-methoxyethyl complex, $Cp(CO)_2FeCH(OCH_3)CH_3$, for the same purpose.[73] This homolog of Pettit's complex **5** undergoes silylative cleavage with trimethylsilyl trifluoromethanesulfonate in the presence of alkenes to give methyl-substituted cyclopropanes in yields quite comparable to ours. Brookhart's stereochemical results were also basically the same as ours except that when *p*-methoxystyrene was used as a substrate, a slight predominance of the *trans*-disubstituted cyclopropane (*cis:trans* = 1.1:1.0) was observed due to the operation of a special isomerization pathway in this particular case of a styrene bearing a very good electron-donating substituent.[73c] Isotopic labeling studies also revealed that the configuration of the double bond of this styrene derivative was not left intact. These findings have important mechanistic implications for these cyclopropanation reactions.

In neither our work nor Brookhart's work with ethylidene transfer reagents did the originally anticipated problem of 1,2-hydrogen rearrangements (Scheme 5) present itself to any significant or detectable extent. However, this general type of problem did finally arise when we and Brookhart attempted to perform n-propylidene transfer reactions. For example, the 1-(phenylthio)propyl complex **19** was prepared by our usual route, but when it was subjected to methylation in the presence of alkenes, no cyclopropanes were obtained. Instead, the alkenyl sulfide **20** was isolated (Eq. 5), which is indicative of a type of β-hydride elimination of an alkylmetal complex. However, it is not the product expected from the 1,2-hydrogen shift of a carbene complex (Scheme 5), which should have produced the propene complex of iron in this particular case. Brookhart's results are more consistent with the expected rearrangement (Eq. 6).[73] The n-propylidene (R = H) and the isobutylidene (R = CH_3) precursors (**21**, L = CO or PPh_3) both give the corresponding alkene complexes **22**; very small amounts of cyclopropanes are produced when L = CO and R = H, but at least the potential exists for obtaining better yields when L = PPh_3. Triphenylphosphine, as a better donor than the carbonyl ligand, stabilizes the intermediate alkylidene complexes to the extent that they can be observed by low-temperature NMR, but whether these consequently somewhat less reactive species will be of general use for the cyclopropanation of a wide range of alkenes has not yet been established. The differences in behavior between our sulfur-containing precursors and Brookhart's ether derivatives may suggest subtle differences in the mechanisms by which they effect alkylidene transfer reactions.

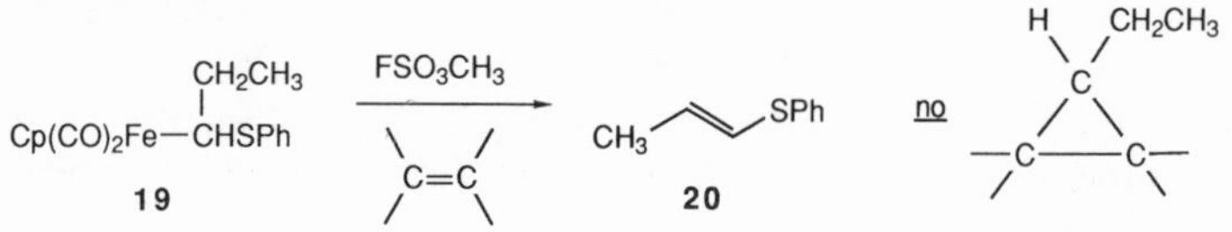

(5)

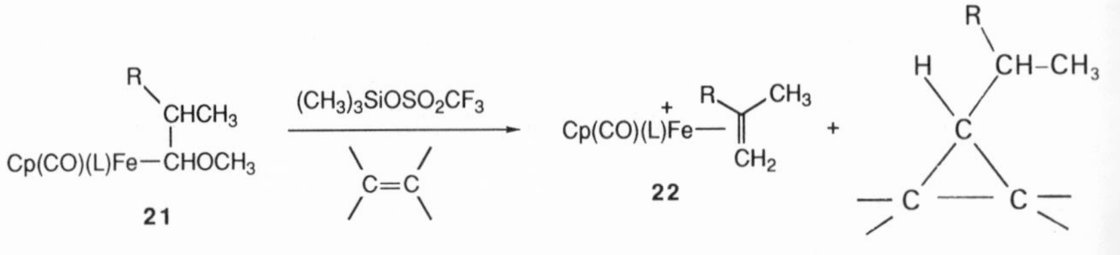

(6)

IV. AN ALTERNATIVE APPROACH TO ALKYLIDENE COMPLEXES: PROTONATION OF ALKENYL COMPLEXES

Keeping in mind our goal of extending the use of our reagents to the transfer of alkylidene groups in general, we next wished to examine 1,1-disubstituted carbene complexes. We recognized that the dimethylcarbene, or isopropylidene, complex 23 could potentially be one of the most important species of this type to study; reaction with alkenes would supposedly provide *gem*-dimethylcyclopropanes (Eq. 7), a structural feature of a very large number of naturally occurring compounds (see compilation of representative structures in Section V). Although isopropylidene addition to the requisite alkenes would in principle provide a general approach for the synthesis of many of these natural products, relatively few methods were available for effecting this transformation.[74]

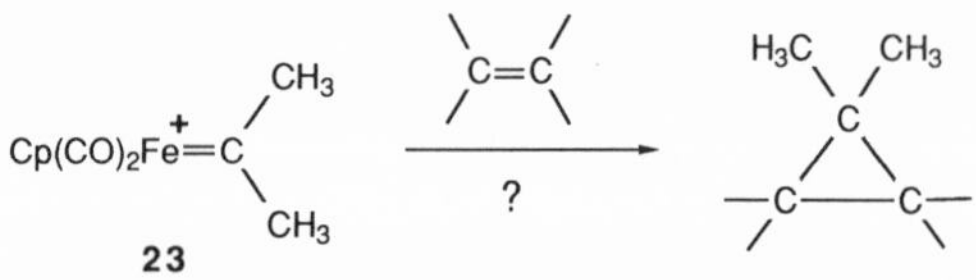

(7)

If we were to continue to use sulfur-containing precursors as for the simpler systems, we recognized that we would face a severe problem in the case of the isopropylidene system. In particular, the preparation of the precursor 25 would require the reaction of the sodium ferrate with the highly substituted chlorosulfide 24 (Eq. 8). We expected this reaction to be difficult, and indeed, it failed when it was attempted.

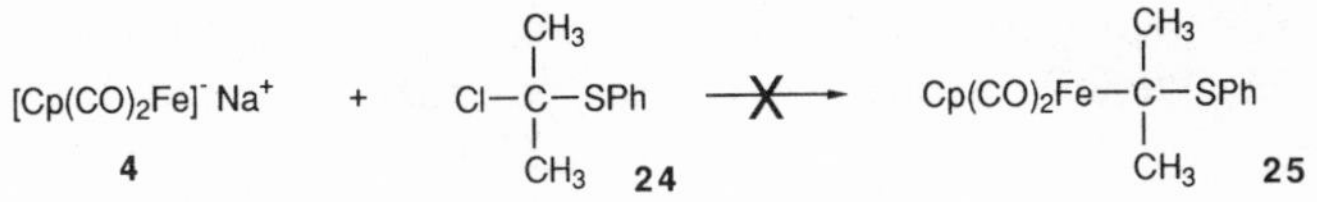

$$[Cp(CO)_2Fe]^- \ Na^+ \quad + \quad \underset{24}{Cl-\overset{\overset{\displaystyle CH_3}{|}}{\underset{\underset{\displaystyle CH_3}{|}}{C}}-SPh} \quad \xrightarrow{\quad \times \quad} \quad \underset{25}{Cp(CO)_2Fe-\overset{\overset{\displaystyle CH_3}{|}}{\underset{\underset{\displaystyle CH_3}{|}}{C}}-SPh} \qquad (8)$$

We quickly realized that an alternative approach would be required for the generation of the more highly substituted carbene complexes. Our basic idea was to employ simple alkenyl complexes **26** that could conceivably undergo protonation with acid at the β-carbon to give the desired substituted carbene complexes **16** (Scheme 7). However, we were also concerned that the protonation may instead occur at the α-carbon to give a cation that would need to undergo only a small change of geometry to give a stable alkene complex **17**. Even if the desired carbene complex **16** were formed, once again we had to be concerned with a possible 1,2-hydrogen shift as another means of ultimately forming the alkene complex **17**. Please recall that these concerns also arose in our earlier work with the ethylidene and attempted n-propylidene transfer reactions (see previous section). Nevertheless, we were able to proceed with some confidence, based upon reports of the addition of acid and other types of electrophiles to alkenyl and alkynyl complexes of various metals.[28b,30,33,34,35b,38a,c,d,75]

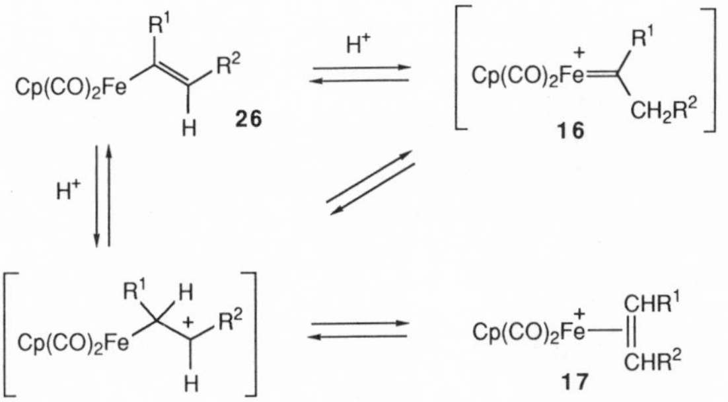

Scheme 7

To pursue this approach, our group joined forces with Robert Kerber, a faculty colleague at Stony Brook who had had an interest in this type of reaction for many years. The student who played the most important role in the actual experimental work was Gee-Hong Kuo, another Ph.D. candidate who had many successes in our laboratory. Ken Kremer and Ed O'Connor also made significant contributions to this phase of the project.

Our starting point was first to investigate the generation of the ethylidene complex **16** ($R^1 = R^2 = H$) by protonation of the parent vinyliron complex **27**. Based upon the much earlier work of Green,[76] the vinyl complex was made very simply by the reaction of the iron iodide derivative with vinylmagnesium bromide (Scheme 8). The protonation needed to be performed in the presence of the alkene substrate in order to obtain the corresponding cyclopropanes; even at a temperature at low as $-82°C$, the ethylidene complex cannot be detected by NMR, apparently due to very rapid decomposition. The best conditions that we found were to add fluoroboric acid to a solution of the alkene in methylene chloride in a temperature range of approximately $-45-0°C$. However, the yields of cyclopropanes were never acceptably high but were typically between 10 and 20%. Because of these poor results, the study of this reaction was restricted to *cis*-cyclooctene and α-methylstyrene as substrates. Interestingly, the same *endo*-selectivity seen for our earlier ethylidene transfer reagent **18** as well as for Brookhart's reagent[73] was again seen when the vinyliron complex was employed.[65a,77]

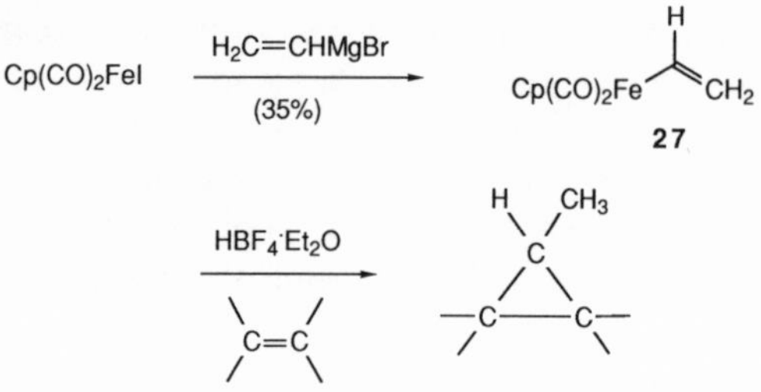

Scheme 8

Concurrent with our work, Alan Cutler had also obtained evidence for the formation of the ethylidene complex by this protonation approach.[28b] Also, he reported that the principal product of decomposition of the ethylidene complex is the dinuclear bridged species **28**, which we subquently found in our reaction mixtures as well. It is apparently formed by reaction of the ethylidene complex with the vinyliron starting material.

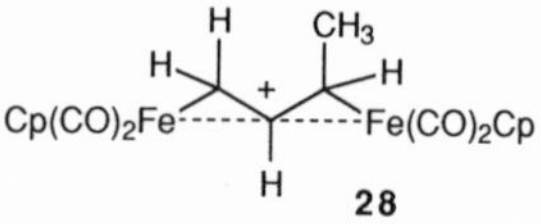

The protonation reaction was extended to the isopropenyl complex **30** in order to provide access to the isopropylidene complex **23** which was not available by our sulfur-based approach (see Eq. 8 above). Unknown to us until very late in our studies, an essentially identical investigation was being performed simultaneously by Charles Casey at the University of Wisconsin.[78] The isopropenyl complex could be obtained in one step in yields up to 60% yield from the iron iodide and isopropenylmagnesium bromide in analogy to the preparation of the parent vinyl complex **27**. Overall yields as high as 90% could be obtained by an alternative two-step route proceeding by acylation of the sodium ferrate with methacryloyl chloride followed by photochemical decarbonylation (Scheme 9).[79] The isopropenyl complex **30** is a reasonably stable, yellow, crystalline solid. Upon addition of fluoroboric acid to a diethyl ether solution of **30** at $-30°C$ in the absence of an alkene, a yellow precipitate is formed quantitatively. Low-temperature NMR showed it to be the desired isopropylidene complex **23**, and thus, it was the first instance in our work of the formation of a carbene complex that could actually be isolated and characterized.[80]

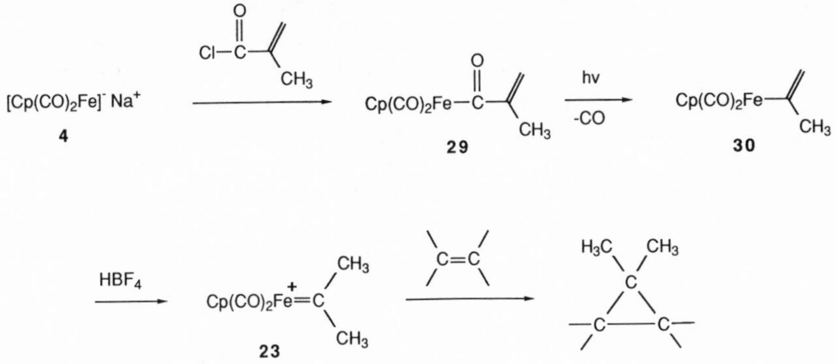

Scheme 9

The two methyl groups appear together as one singlet in the ^{1}H-NMR spectrum down to $-110°C$, the lower operating limit of our apparatus. This observation can be interpreted in terms of either a low barrier to rotation about the iron–carbon "double" bond or adoption of a "cross-wise" conformation **23A**. In either case, the two methyl groups would be equivalent. More will be said later about the conformations of carbene complexes when we present results from the study of hetero-substituted derivatives (Section V).

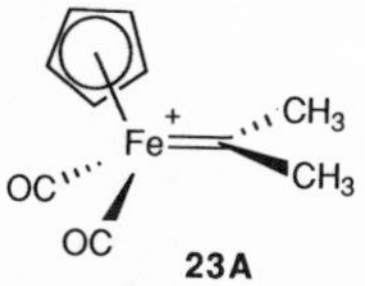

We explored various aspects of the chemistry of the isopropylidene complex. In the absence of solvent, **23** is stable for at least several minutes at 25°C. Under these conditions, it melts to give a dark, orange–brown oil that resolidifies at low temperature; re-examination by NMR shows that **23** survives this treatment. However, in solution above $-10°C$, it rapidly undergoes the anticipated 1,2-hydrogen shift rearrangement to give the propene complex **31** (Eq. 9). When alkenes are added to solutions of **23** in methylene chloride at $-78°C$ and the mixtures are then warmed to 25°C, *gem-*

dimethylcyclopropanes are produced (Scheme 9), but the yields are not acceptable for synthetic applications. With styrene, α-methylstyrene, 1,1-diphenylethylene, 1-octene, and isobutylene as substrates, the yields of cyclopropanes ranged from 5 to 20%, although the yields could be raised to as high as 40% by using large excesses of alkenes (10–20 molar-equivalents). Several alkenes, including *cis*-5-decene, 4-methylcyclohexene, *cis*-cyclooctene, and *cis*-cyclododecene, failed to give detectable amounts of cyclopropanes, whereas methylenecycloheptane interestingly gave isobutenylcycloheptane (**32**, Eq. 10) in 35% yield, possibly by ring-opening of the cyclopropane. The electrophilic character of **23** was also demonstrated by its reaction with heteroatomic nucleophiles such as triphenylphosphine and sodium methylmercaptide (Eqs. 11 and 12). The adduct of the latter reaction could be converted into a sulfonium salt derivative of the type that we have discussed in detail above, but the same species was not obtained upon treatment of the isopropylidene complex **23** with dimethyl sulfide.

$$\text{Cp(CO)}_2\overset{+}{\text{Fe}}=\!\!\!\begin{array}{c}\text{CH}_3\\ \\ \text{CH}_3\end{array} \quad\xrightarrow{\;>-10\,^\circ\text{C}\;}\quad \text{Cp(CO)}_2\overset{+}{\text{Fe}}\!\!-\!\!\begin{array}{c}\text{CHCH}_3\\ \| \\ \text{CH}_2\end{array} \qquad (9)$$
$$\mathbf{23} \qquad\qquad\qquad\qquad\qquad \mathbf{31}$$

$$\qquad\qquad\qquad\qquad\qquad\qquad\qquad\qquad\qquad\qquad\qquad\qquad (10)$$
$$\mathbf{32}$$

$$\mathbf{23} \quad\xrightarrow[\;(86\%)\;]{\text{PPh}_3}\quad \text{Cp(CO)}_2\text{Fe}\!-\!\overset{\overset{+}{\text{PPh}_3}}{\underset{\text{CH}_3}{\text{C}}}\!-\!\text{CH}_3 \qquad (11)$$

$$\mathbf{23} \quad\xrightarrow{\text{NaSCH}_3}\quad \text{Cp(CO)}_2\text{Fe}\!-\!\overset{\overset{\text{SCH}_3}{}}{\underset{\text{CH}_3}{\text{C}}}\!-\!\text{CH}_3 \quad\xrightarrow{(\text{CH}_3)_3\text{O}^+\,\text{BF}_4^-}\quad \text{Cp(CO)}_2\text{Fe}\!-\!\overset{\overset{+}{\text{S(CH}_3)_2}}{\underset{\text{CH}_3}{\text{C}}}\!-\!\text{CH}_3 \quad (12)$$

In order to obtain an isopropylidene complex of greater stability, we wished to prepare a complex having a good donor ligand in place of one of the carbonyl ligands. We proceeded by irradiation of the methacryloyl complex **29** in the presence of trimethylphosphite; the resulting isopropenyl complex **33** containing the phosphite ligand reacts with fluoroboric acid to give the modified isopropylidene complex **34** as a bright yellow solid (Scheme 10). This complex is considerably more stable than the originally obtained **23**. Samples of **34** have proven to be stable not only as a solid but also in solution for several hours at room temperature. It would appear to be a good candidate for X-ray crystal structure determination. As has been seen for other iron carbene complexes bearing good donor ligands, their stability is increased to the point that the complexes can often be isolated and characterized, but at the same time, they are of reduced reactivity in cyclopropanation reactions.

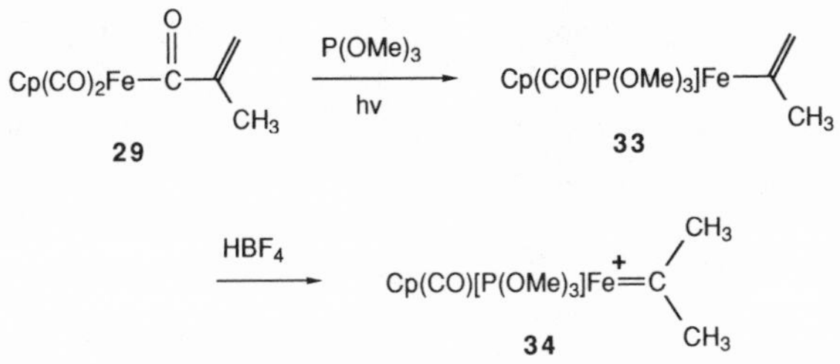

Scheme 10

We have already noted that concurrent studies of the isopropylidene system were performed by Casey.[78] The results obtained by our two groups paralleled one another almost exactly. These findings together with the results obtained in the study of other related reagents still leave much to be desired with respect to the development of truly useful isopropylidene transfer reagents for the efficient, direct synthesis of *gem*-dimethylcyclopropanes.[81]

We have extended the alkenyliron protonation reactions to additional systems. We had assumed that the α-styryl complex **35** would undergo conversion to a relatively stable α-methylbenzylidene complex **36**, especially since Maurice Brookhart had succeeded in obtaining the parent benzylidene complex.[22] However, we instead isolated the styrene complex **37** (Scheme 11). We do not know whether this product results from 1,2-hydrogen shift rearrangement of the desired benzylidene complex **36** or from initial protonation at the α-carbon of the styryl complex **35**, a pathway that we had foreseen earlier as a possible competing reaction (see Scheme 7). When the protonation of **35** was performed in the presence of alkenes, no cyclopropanes were produced. Therefore, even if the benzylidene complex is produced, it undergoes rearrangement faster than it reacts with alkenes.

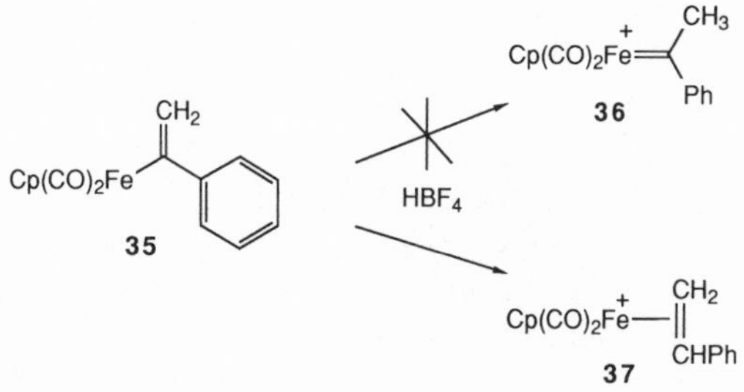

Scheme 11

We had considerably greater success with the dienyliron complexes **39**. These compounds were prepared by the previously described acylation/ photo-decarbonylation route, starting with dienoyl chlorides[82] and proceeding via the doubly unsaturated acyliron derivatives **38** (Scheme 12). Depending upon whether trimethyl phosphite was present during the photolysis step, either the parent dicarbonyl **39a,c** or the monocarbonyl monophosphite dienyl complexes **39b,d** were obtained. The dienyl complexes were found to undergo reaction with trifluoromethanesulfonic acid at low temperature to give the desired allylidene, or vinylcarbene, complexes **40**.[83,84]

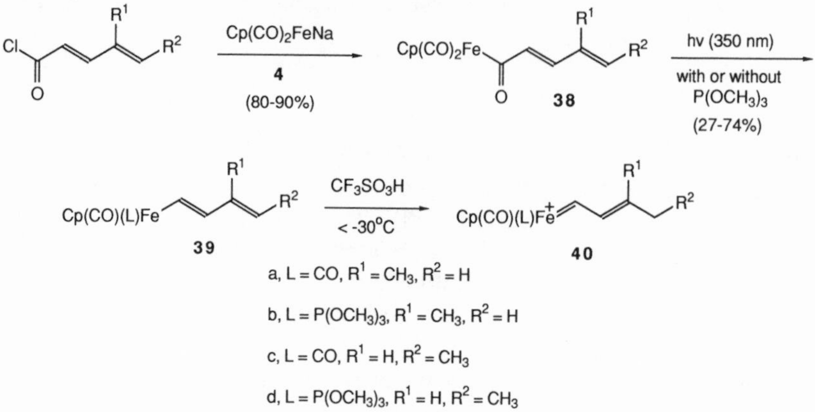

a, L = CO, R^1 = CH_3, R^2 = H

b, L = P(OCH$_3$)$_3$, R^1 = CH_3, R^2 = H

c, L = CO, R^1 = H, R^2 = CH_3

d, L = P(OCH$_3$)$_3$, R^1 = H, R^2 = CH_3

Scheme 12

Concurrent with our studies, Charles Casey also developed a route to an allylidene complex. However, his route, in closer analogy with Pettit's original generation of the methylene complex, was based upon treatment of the alcohol derivative **41** with acid (Eq. 13).[78b,85]

(13)

The allylidene complexes were characterized by low-temperature NMR. Typical of carbene complexes in general,[22a,c,73b,78,80,86] the hydrogen substituent on the α-carbon of each of the complexes **40** appeared at low field in the ^{1}H-NMR spectrum, with the range of values being δ 15.88–17.16. For complex **40b**, the ^{13}C-NMR spectrum was also obtained and showed the typically very far downfield position of δ 314 for the α-carbon. Evidence was also given for significant delocalization of positive charge in the allylidene ligand in that the γ-carbon was seen to have a chemical shift of δ 162.

None of the four allylidene complexes were sufficiently stable to be isolated at room temperature, but of these compounds, **40b** was the most stable. We were able to isolate it at $-25°C$ as a yellow–orange solid in a yield of at least 60%. Casey similarly isolated **40a**.[78b,85] Both of these isolated complexes were found to undergo reaction with alkenes to give the vinylcyclopropanes **42** (Scheme 13). With a large excess of styrene as the substrate with the phosphite-containing complex **40b**, we obtained a 5:1 mixture of the *cis*- and *trans*-disubstituted cyclopropanes, but the yield was only 15% based upon the iron reagent. This stereochemical outcome is consistent with the previously discussed *syn*- or *endo*-selectivity seen for simpler carbene complexes of this iron series. Casey had also prepared a triphenylphosphine-substituted analogue of **40b** (L = PPh$_3$) and found that it gave no cyclopropane upon attempted reaction with styrene. On the other hand, he found that the dicarbonyl complex **40a** gives 37–56% yields of cyclopropanes upon reaction with excess isobutylene, *cis*-cyclooctene, and styrene. Thus, the reduced cyclopropanation reactivity of iron carbene complexes containing donor ligands such as phosphites and phosphines is again apparent. A further, important observation by Casey was that although *cis*-cyclooctene again gave the *endo*-product, styrene gave a 2:1 mixture of the *trans*- and *cis*-disubstituted cyclopropanes due to iron-promoted isomerization of the initially formed *cis*-isomer.[78b,85]

In the course of studying the allylidene complexes, we remained alert for the possible occurrence of another conceivable pathway for reaction with alkenes leading to cyclopentenes, or their iron complexes (Scheme 13). We felt that if the alkene were to add initially to the γ-position of the allylidene complex, then a subsequent cationic ring closure may lead to the cyclopentenes. We have yet to see any evidence for this pathway, although we certainly realize that the vinylcyclopropanes that we have obtained could be used as substrates for the well-known thermal rearrangement to produce the desired cyclopentenes by an overall two-step sequence.[87]

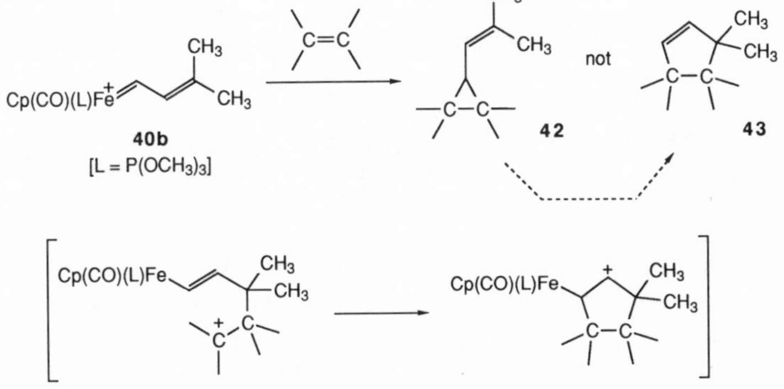

Scheme 13

The allylidene complexes exhibited other expected types of reactivity. For example, when **40c** was warmed to room temperature, the 1,3-pentadiene complex **44** was obtained as a result of the 1,2-hydrogen shift rearrangement. Subsequent treatment with sodium iodide released the free diene (Scheme 14). Allylidene complex **40b** could be trapped by triphenylphosphine and sodium methylmercaptide as nucleophiles to give the adducts **45** and **46**, respectively (Scheme 15). The latter product, at least, was obtained as a 4:1 mixture of diastereomers (note the chirality at the iron center). Taking a cue from our earlier work with sulfonium derivatives, we treated the methylthio derivative **46** with trimethyloxonium tetrafluoroborate in the presence of styrene, and again the corresponding cyclopropane was produced, albeit in low yield.

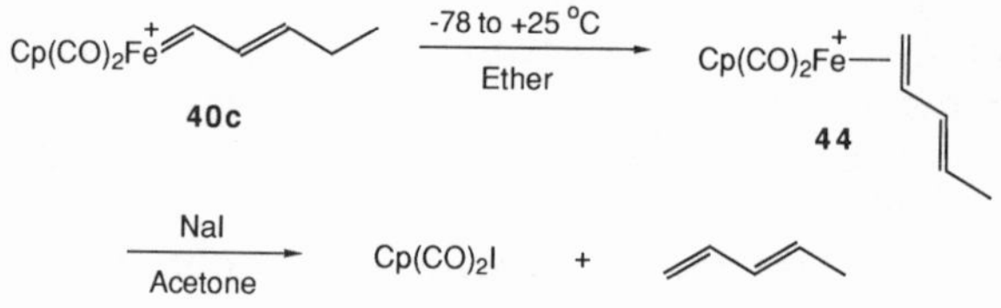

Scheme 14

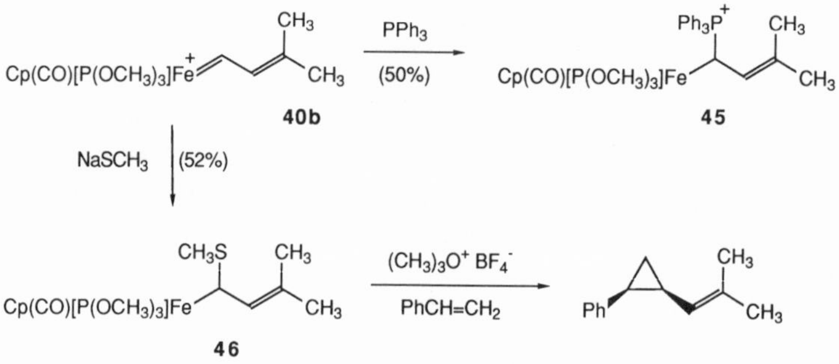

Scheme 15

V. INTRAMOLECULAR CYCLOPROPANATION REACTIONS AND THE DEVELOPMENT OF APPROPRIATE PRECURSORS

Intramolecular cycloaddition reactions have proven to be very valuable methods in synthetic organic chemistry. The advantages of these reactions over their intermolecular counterparts are that they form more than one ring at a time, they often exhibit high degrees of stereochemical control, and they often proceed under milder conditions because of more favorable entropy factors than bimolecular reactions. As a subclass of these internal additions, intramolecular cyclopropanations have also proven to be valuable.[87-89] Therefore, an obvious extension of our earlier work was to do some preliminary experiments to determine whether the cyclopropanation reactions of our types of iron carbene complexes could be applied intramolecularly.

As a very simple starting point, we studied the formation of norcarane (Scheme 16).[55] This system was also used as a basis of comparison of our alkenyliron and sulfonium salt approaches. The alkenyliron reaction gave only a 4% yield of norcarane whereas the sulfur-based approach gave an overall yield of 25% (from the corresponding α-chlorosulfide); no attempts were made to optimize the yields of these reactions, and only crude organometallic intermediates were employed in the cyclizations.

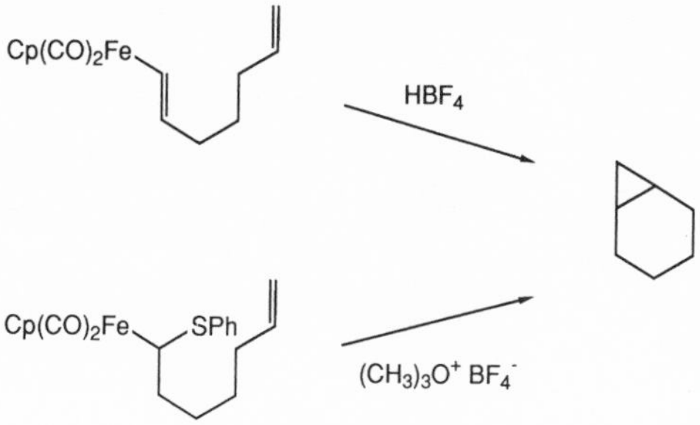

Scheme 16

Based upon these observations and upon the generally much higher yields of the sulfonium salt reactions in our intermolecular reactions as well, we chose to emphasize the use of sulfur-containing precursors in our further studies of intramolecular reactions. A key illustrative example is shown in greater detail (Scheme 17).[55]

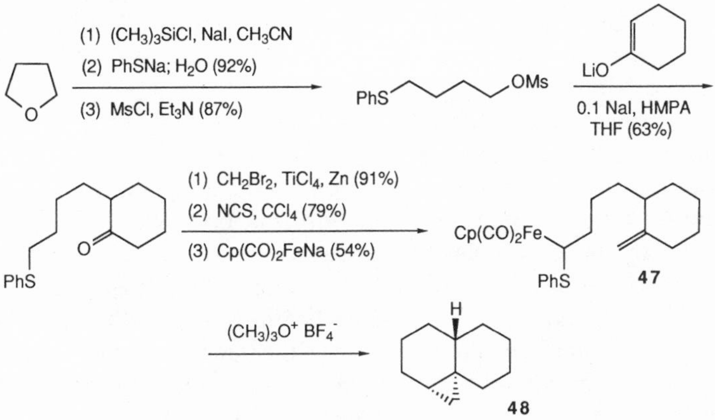

Scheme 17

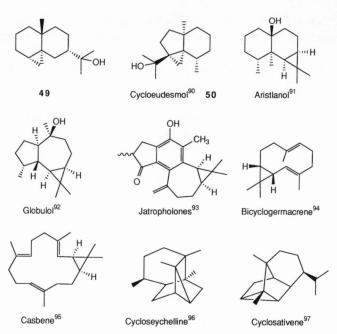

49   Cycloeudesmol[90] 50   Aristlanol[91]

Globulol[92]   Jatropholones[93]   Bicyclogermacrene[94]

Casbene[95]   Cycloseychelline[96]   Cyclosativene[97]

Figure 3.   Representative examples of cyclopropane-containing polycyclic natural products.

One important point about this example is that it demonstrates the potential usefulness of the intramolecular cyclopropanation reactions of our organoiron reagents for the synthesis of cyclopropane-containing, polycyclic compounds in general. The importance of this point becomes clearer upon consideration of the structures of some representative, naturally occurring compounds that contain these polycyclic ring systems (Figure 3). Of special interest is that our cyclization product **48** is very similar to structure **49** (or other stereoisomers) that at one time had been assigned to cycloeudesmol, but which has since been revised to **50**.[90]

A second point that Scheme 18 illustrates is that if our intramolecular cyclopropanations are to be used in any practical sense, more direct methods are needed for the preparation of the iron alkylidene precursors. The route shown for the preparation of the precursor **47** is unattractively long. Instead of first preparing the organic sulfide, converting it into the corresponding α-chlorosulfide (unpleasant!), and then using it to alkylate the ferrate, followed by *S*-alkylation and finally cyclization, we would prefer to have methods that permit direct incorporation of the intact iron alkylidene precursor into our cyclization substrates. The need for special reagents for this purpose is outlined in Scheme 18 in which we depict the alkylidene complex **47A**; this species is the intermediate that supposedly results from loss of sulfide from **47** and which undergoes cyclization to give the observed tricyclic product **48**. In a formal sense, we can imagine very direct construction of the basic skeleton of **47A** by formation of the indicated carbon-carbon bonds 1, 2, or 3. These constructions could be accomplished if we had access to the appropriate iron-containing 1-carbon, 2-carbon, or 3-carbon reagents, respectively, that could be coupled with the corresponding organic fragments. In these highly schematic representations, the asterisks simply denote centers of nucleophilic, electrophilic, or radical reactivity as required for effecting the desired coupling reactions. Thus, these reagents would permit not only the direct introduction of the intact iron carbene units (or their precursors) but also simultaneous introduction of varying numbers of carbon atoms of the rest of the skeleton of the intramolecular cyclopropanation substrates.

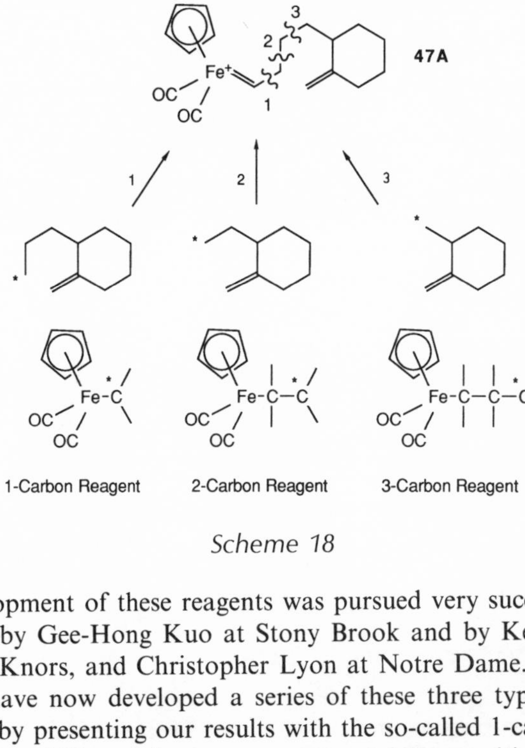

Scheme 18

The development of these reagents was pursued very successfully in our laboratories by Gee-Hong Kuo at Stony Brook and by Kerry Brinkman, Christopher Knors, and Christopher Lyon at Notre Dame. Through their efforts, we have now developed a series of these three types of reagents.

We begin by presenting our results with the so-called 1-carbon reagents. Because of the ability of the iron group to stabilize positive charge at an adjacent carbon atom, these reagents would most logically have electrophilic reactivity (* = + in the 1-carbon reagent in Scheme 18). Our simplest example is obtained by starting with the readily available phenylthiomethyl complex **10** that we have mentioned previously in the context of simple methylene transfer reagents. Hydride abstraction with trityl hexafluorophosphate gives the desired phenylthiocarbene complex **51** (Eq. 14).[98] This compound is obtained as a golden yellow, crystalline solid that may be handled at room temperature in air for at least short periods of time and which may be stored indefinitely under nitrogen. Hydride abstraction reactions have been used for many related systems.[17c,23,24,28a,75f,99,100b] Also, Robert Angelici had previously reported other methods for obtaining alkyl-thiocarbene complexes very similar to **51**,[26b,d,100] but our preparation of **51** is especially straightforward. In addition, alkylthiocarbene complexes have been obtained for other metals besides iron.[44l,u,v,101]

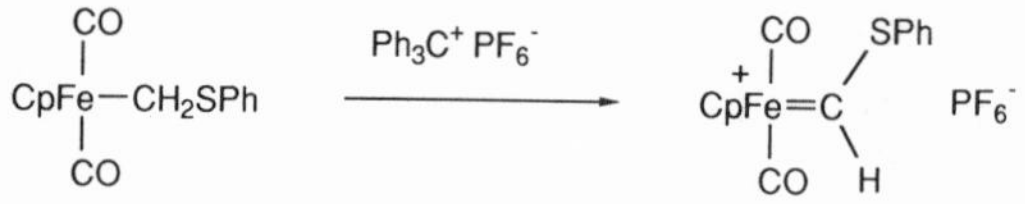

$$
\underset{\mathbf{10}}{\overset{\displaystyle\overset{CO}{|}}{\underset{\displaystyle\underset{CO}{|}}{CpFe-CH_2SPh}}}
\quad\xrightarrow{Ph_3C^+\,PF_6^-}\quad
\underset{\mathbf{51}}{\overset{\displaystyle\overset{CO}{\underset{+}{|}}}{\underset{\displaystyle\underset{CO\quad H}{|}}{CpFe{=}C{\overset{SPh}{\underset{\diagdown}{\diagup}}}}}}\;\; PF_6^-
\qquad(14)
$$

Through use of a procedure that had been developed by Russell Hughes for phosphine-coordinated derivatives,[32] we have been able to convert the readily available acetyliron complex **52**[79a] into the more highly substituted alkylthiocarbene complexes **54**. This conversion apparently occurs via the vinylidene complex **53**[102] for which we have low-temperature ^{1}H-NMR evidence (Scheme 19). The initially obtained trifluoromethanesulfonate salts were subjected to anion exchange to give the more stable hexafluorophosphate salts of **54** that may be exposed to air for several days at room temperature.[77,98a]

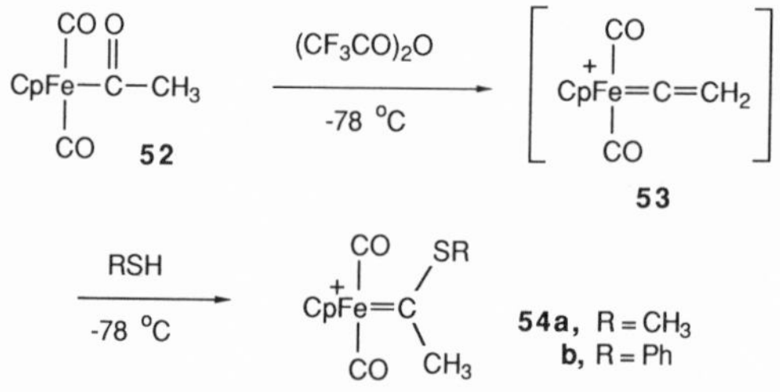

Scheme 19

The alkylthiocarbene complexes **51** and **54** were characterized by IR, ^{1}H-NMR, and ^{13}C-NMR. In addition, the structures of **51** and **54a** were determined by single-crystal X-ray diffraction (Figures 4 and 5).[98a] Although disorder problems limited the quality of our data, some important structural features are clearly evident. First of all, the iron–carbon bond distances for the alkylidene moieties are 1.88(1) and 1.94(2) Å, respectively, for these two complexes. These values are in reasonable agreement with previous literature values for other iron–carbene complexes.[24b,52,103]

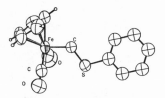

Figure 4. X-ray crystal structure of **51** (major conformer).

A second point of interest is that whereas the less substituted alkylthiocarbene complex **51** adopts a so-called "upright" conformation in which the plane of the carbene ligand nearly bisects the OC–Fe–CO bond angle, the more highly substituted complex **54a** adopts a "cross-wise" conformation in which the plane of the carbene ligand is nearly orthogonal to that of **51** (Figure 6). In this regard, Roald Hoffmann had predicted on the basis of his calculations that the "upright" conformation **6A** should be favored for the parent methylene complex **6** (Figure 6; see also **6A** and **6B**, Section I).[15] However, the calculated barrier to rotation about the iron–carbon bond of **6** and the rotational barriers that have been measured (dynamic NMR studies) for various derivatives are relatively low (ca. 6–10 kcal/mol).[22c,24a,c,78b,104] Consequently, the "cross-wise" conformation of **54a** may readily be rationalized on the basis of steric and electronic effects associated with the more highly substituted carbene center, thus overriding the usual conformational preferences of these systems. Also, we cannot rule out solid-state, crystal-packing effects. On the other hand, we see no significant changes in the NMR spectra of **51** and **54a** down to − 100°C. To be noted is that Jones has also reported "cross-wise" conformations for cycloheptrtrienylidene complexes of this series,[24b] and Casey has considered the possibility of this conformation for the isopropylidene complex.[78b]

The reason that we developed these complexes was to employ them as reagents for introducing potentially useful iron carbene moieties into organic systems (Scheme 18). Therefore, we have studied the addition of various carbon-based nucleophiles to **51** and **54**. Among the species that undergo addition are Grignard reagents, alkyllithiums, organocuprates, and enolates

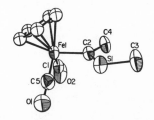

Figure 5. X-ray crystal structure of **54a**

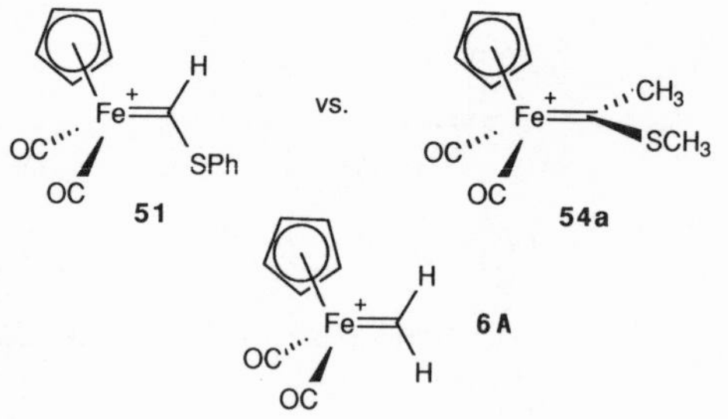

Figure 6. Comparison of conformations of iron carbene complexes

(Eq. 15–17). Of these reactions, we believe that the addition of enolates (Eq. 17) may prove to be especially useful for applications in organic synthesis.

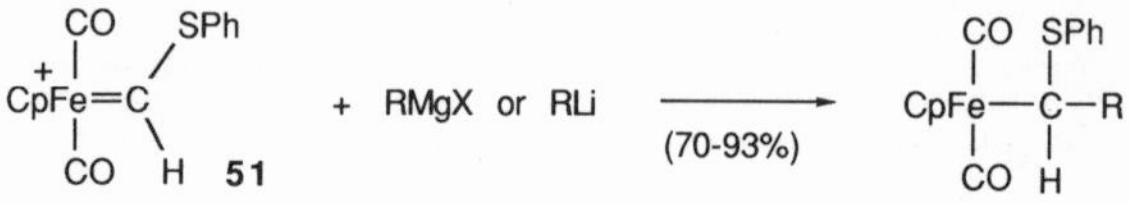

$$\qquad\qquad\qquad\qquad\qquad\qquad\qquad\qquad\qquad\qquad\qquad\qquad (15)$$

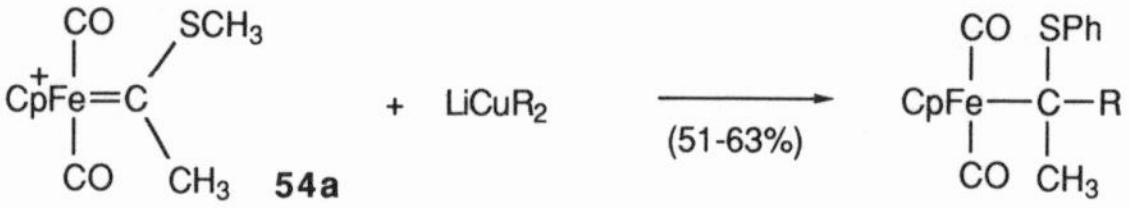

$$\qquad\qquad\qquad\qquad\qquad\qquad\qquad\qquad\qquad\qquad\qquad\qquad (16)$$

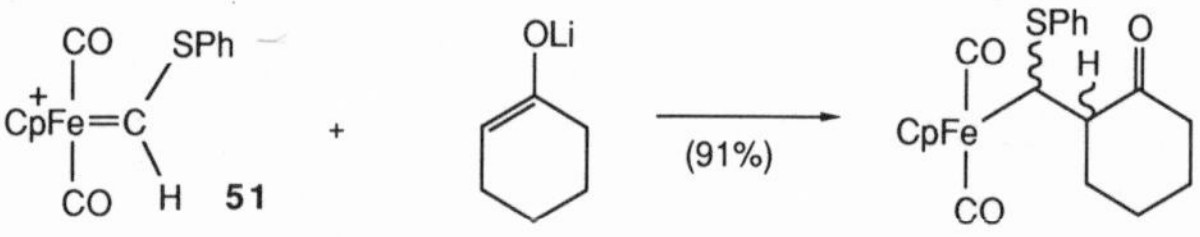

$$\qquad\qquad\qquad\qquad\qquad\qquad\qquad\qquad\qquad\qquad\qquad\qquad (17)$$

With reference back to Scheme 18, we next needed to develop an appropriate 2-carbon reagent. The use of enolates formed from acyliron complexes appeared to be especially well-suited for our purposes in that these enolates would function as readily available nucleophilic 2-carbon reagents (* = – in Scheme 18). These species have been thoroughly developed by Stephen Davies, especially for applications in asymmetric synthesis,[105] and many other workers have also contributed to this area.[106] In our laboratory, Kerry Brinkman found that the simple acetyl complex 52[79a] in the parent cyclopentadienyldicarbonyliron system could be converted efficiently into the corresponding lithium enolate 55, which then reacts with several electrophilic reagents to give the chain-extended acyl complexes 56 (Eq. 18, Table 5).[107] Similarly, the propionyl complex was also found to undergo efficient enolate formation and subsequent alkylation (Eq. 19). The use of lithium

Table 5. Reaction of Cp(CO)$_2$Fe—C=CH$_2$ (**55**) with Electrophiles, where C bears an OLi substituent.

Electrophile (E$^+$)[a]	Product (**56**)[b]	Yield (%)[c]
CH$_3$I	FpCOCH$_2$CH$_3$	84
CH$_3$CH$_2$CH$_2$CH$_2$OTf	FpCOCH$_2$CH$_2$CH$_2$CH$_2$CH$_3$	85
(CH$_3$)$_2$CHOTf	FpCOCH$_2$CH(CH$_3$)$_2$	60
H$_2$C=CHCH$_2$Br	FpCOCH$_2$CH$_2$CH=CH$_2$	54
H$_2$C=C(CH$_3$)CH$_2$I	FpCOCH$_2$CH$_2$C(CH$_3$)=CH$_2$	92
C$_6$H$_5$CH$_2$Br	FpCOCH$_2$CH$_2$C$_6$H$_5$	56
4—CH$_3$OC$_6$H$_4$CH$_2$I	FpCOCH$_2$CH$_2$C$_6$H$_4$—4—OCH$_3$	96
(epoxide structure)	FpCOCH$_2$CH$_2$CHOHCH$_2$CH$_3^d$	68
CH$_3$CH$_2$CHO	FpCOCH$_2$CHOHCH$_2$CH$_3^d$	62

[a]The molar ratios of electrophile: **3** were 1.2:1.0. [b]Fp = Cp(CO)$_2$Fe. [c]All yields are for products that have been purified by silica gel chromatography. [d]One molar equivalent of BF$_3$·Et$_2$O was employed.

hexamethyldisilazide as the base for enolate formation is crucial; with other bases (e.g. lithium diisopropylamide), a competing deprotonation reaction of the cyclopentadienyl ligand predominates.[106d,e,k]

$$\text{Cp(CO)}_2\text{Fe}\overset{O}{\underset{}{\|}}\text{CH}_3 \xrightarrow{\text{LiN(SiMe}_3)_2} \left[\text{Cp(CO)}_2\text{Fe}\overset{OLi}{\underset{}{}}\text{CH}_2\right] \xrightarrow{E^+} \text{Cp(CO)}_2\text{Fe}\overset{O}{\underset{}{\|}}\text{E}$$

$$\textbf{52} \qquad\qquad \textbf{55} \qquad\qquad \textbf{56}$$

$$(18)$$

$$\text{Cp(CO)}_2\text{Fe}\overset{O}{\underset{}{\|}}\text{CH}_2\text{CH}_3 \xrightarrow[\text{(90\%)}]{\text{LiN(SiMe}_3)_2 \quad \text{CH}_3\text{I}} \text{Cp(CO)}_2\text{Fe}\overset{O}{\underset{\text{CH}_3}{\|}}\text{CH}_3 \qquad (19)$$

The importance of these chain-extension reactions in the context of our work is that olefinic groups, as sites for later intramolecular cyclopropanation reactions, can be introduced as part of the electrophilic reagents, and the acyliron groups can be converted by several means into iron carbene units.[19,27,29,32,73,98]

Finally, we needed to develop an appropriate 3-carbon reagent. We have most straightforwardly been able to generate a species having electrophilic reactivity (* = + in Scheme 18). As we had done in many previous cases, we began by performing an acylation of the cyclopentadienyldicarbonylferrate, but this time with acryloyl chloride. The resulting acryloyl complex **57** was then subjected to O-methylation to give the methoxyallylidene complex **58** (Scheme 20).[98b,108]

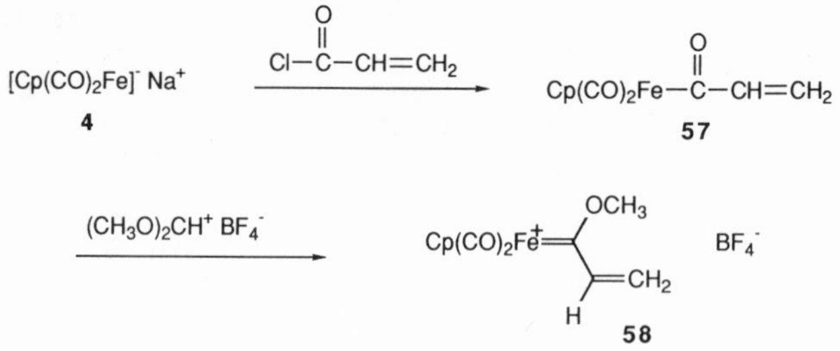

Scheme 20

The 3-carbon reagent **58** undergoes efficient addition of various carbon-based nucleophilic reagents, including Grignard reagents, organocuprates, and enolates. The use of cyclohexanone enolate is representative of these addition reactions (Scheme 21). The initially formed adduct is a methoxyvinyliron, or enol ether complex **59**. This compound may be isolated, or it may be hydrolyzed directly to give the ketoacyl complex **60** in high yield.

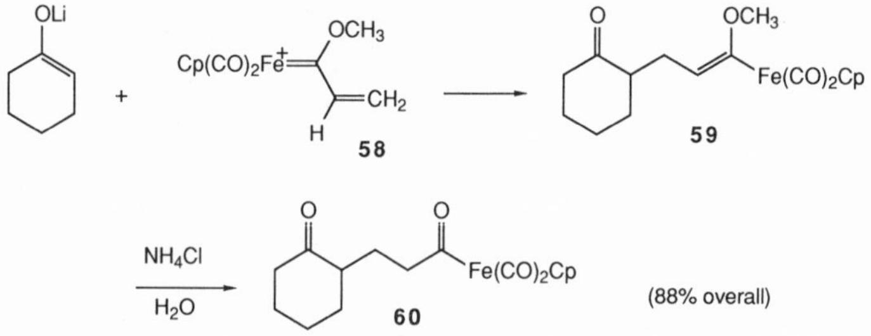

Scheme 21

Michael-type addition reactions to α,β-unsaturated acylmetal complexes such as the acryloyl complex **57** and related compounds have been developed extensively by Stephen Davies,[105d] Lanny Liebeskind,[106j,n] and Iwao Ojima,[106h] but the methoxyallylidene derivative **58** appears to be especially versatile in these addition reactions. Also, James Herndon[84o,r] and N. T. Allison[84q] have recently reported other application of complexes related to **58**.

Now that we have met our goal of finding 1-, 2-, and 3-carbon reagents for the introduction of iron alkylidene precursors into organic frameworks, we have begun to use these reagents to construct actual substrates for intramolecular cyclopropanations. Very illustrative of our ongoing work are some reaction sequences employing our phenylthiocarbene complex **51** (Schemes 22 and 23).[98b] The attractiveness of these routes lies in their directness compared to the multi-step approach that was used previously for the preparation of the cyclization substrates (e.g. Scheme 17).

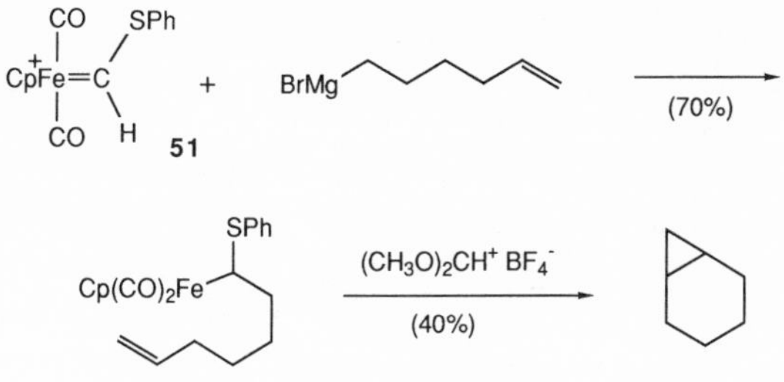

Scheme 22

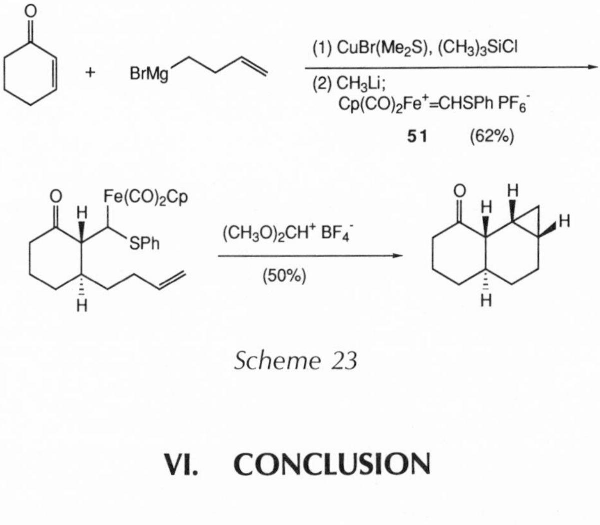

Scheme 23

VI.   CONCLUSION

Our work with iron carbene complexes began several years ago with the development of a sulfonium ylide complex of iron that serves as a very useful reagent for the methylenation of alkenes. The basic chemistry of this relatively simple reagent has since been extended to several other complexes that permit the transfer of different alkylidene groups other than just the methylene group, thus leading to cyclopropanes of various substitution patterns. These reactions often occur with excellent stereochemical control. Most recently, we have been emphasizing the use of these cyclopropanation reactions in an intramolecular fashion. For this purpose, we have developed a series of reagents that greatly facilitate the preparation of the substrates for these cyclization reactions. Clearly, much work remains to be done in this area. For one thing, we need to optimize the yields of our intramolecular cyclopropanation reactions, and we need to develop more fully the scope of this potentially valuable approach to polycyclic compounds. The effects of ligands need to be explored more fully in order to achieve a good balance between the stability and reactivity of the intermediate carbene complexes. This fine-tuning will be especially important if a truly useful isopropylidene complex is to be developed in this iron series and if generally applicable reagents for asymmetric synthesis are to arise from this work.[18-21]

We have said little about mechanism. Among several other workers in this area, Maurice Brookhart has been particularly instrumental in shedding light on the pathways by which these iron carbene complexes undergo reactions,[19,39,73b,c] but needless to say, there is a need for much more mechanistic work before we can hope to have a reasonably complete understanding of these reactions.

As the preceding aspects of the chemistry of iron carbene complexes are addressed in further studies, several attractive applications of these complexes in natural products synthesis can also be investigated. We hope that not only our group but that several others will also be encouraged to make good use of these reagents in these future applications.

ACKNOWLEDGEMENTS

The author wishes to express his sincere appreciation to many people, without whom this research would not have been possible. First among them is Professor E. J. Corey who provided the initial guidance in entering this area. Many of the excellent contributions of Steve Brandt, Ken Kremer, Ed O'Connor, Gee-Hong Kuo, and Ramnath Iyer at Stony Brook and of Chris Knors, Kerry Brinkman, Chris Lyon, Shi-Kai Zhao, Pascale Seutet, and Matthew Mattson at Notre Dame have been cited throughout the text of this chapter. The author wishes to thank this group of fine co-workers for their hard work and the experimental expertise that they have provided over the years. Maurice Brookhart of the University of North Carolina has been and continues to be a most thoughtful and helpful colleague in this area through his very valuable discussions, suggestions, and sharing of data with us. The author also wishes to extend his thanks to several other colleagues, including Joseph Lauher and Robert Kerber at Stony Brook, Charles Eigenbrot, Robert Scheidt, and Marvin Miller at Notre Dame, Björn Åkermark at the Royal Institute of Technology in Stockholm, and Charles Casey and William Miles at the University of Wisconsin. Finally, the author thanks the National Science Foundation, the National Institutes of Health, the Petroleum Research Fund administered by the American Chemical Society, the Natural Science Research Council (Sweden), the State University of New York at Stony Brook, and the University of Notre Dame for very generous funding of this research.

REFERENCES AND NOTES

1. In this chapter, no distinction will be made between the terms 'carbene' and 'alkylidene' complexes, although some other workers have preferred to use the two terms to distinguish complexes having different types of reactivity.

2. For some recent reviews of carbenes, see: (a) Wentrup, C. and Wulfman, D. S.; Poling, B. In 'Reactive Intermediates'; Abramovitch, R. A., Ed.; Plenum Press: New York, 1980; Vol. 1, Chapters 4 and 5. (b) Wentrup, C. 'Reactive Molecules: The Neutral Reactive Intermediates in Organic Chemistry'; Wiley: New York, 1984. (c) "Reactive Intermediates;" Jones, M.; Moss, R. A., Eds.; Wiley: New York, 1985; Vol. 3, and the earlier volumes of this series. (d) Moody, C. J. In "Organic Reaction Mechanisms. 1985"; Knipe, A. C.; Watts, W. E.; Eds.; Wiley: Chichester, 1987; Chapter 6, pp.. 223–243, and the earlier volumes of this series.

3. For some reviews concerned with cyclopropanes see: (a) Simmons, H. E., Cairns, T. L., Vladuchick, S. A., Hoiness, C.M. *Org. React.* 1973, *20*, 1. (b) Ferguson, L. N. "Highlights of Alicyclic Chemistry"; Franklin: Palisade, New Jersey, 1973; pp. 21–271. Ferguson, L. N.; Paulson, D. R. *Ibid.* 1977, Part 2, pp. 89–109. (c) Burke, S. D.; Grieco, P. A. *Org. React.* 1979, *26*, 361. (d) Wenkert, E. *Accounts Chem. Res.* 1980, *13*, 27. (e) Moss, R. A. *Accounts Chem. Res.* 1980, *13*, 58. (f) Freidlina, R. Kh.; Kamyshova, A. A.; Chukovskaya, E. Ts. *Russ. Chem. Rev.* 1982, *51*, 368. (g) McQuillin, F. J.; Baird, M. S. "Alicyclic Chemistry", 2nd edn.; Cambridge University Press: Cambridge, 1983; pp. 92-112. (h) Hudlicky, T.; Kutchan, T. M.; Naqvi, S. M. *Org. React.* 1985, *33*, 247. (i) Larock, R. C. "Organomercury Compounds in Organic Synthesis"; Springer-Verlag: Berlin, 1985; pp. 327–380. (j) Aratani, T. *Pure Appl. Chem.* 1985, *57*, 1839. (k) Doyle, M. P. *Chem. Rev.* 1986, *86*, 919. A summary of more recent methods for the synthesis of cyclopropanes may be found in a detailed, full paper from our laboratory. See Ref. 58.

4. Emerson, G. F.; Watts, L.; Pettit, R. *J. Am. Chem. Soc.* 1965, *87*, 131.

5. For some recent reviews of carbene complexes see: (a) Brown, F. J. *Prog. Inorg. Chem.* 1980, *27*, 1. (b) Dötz, K. H.; Fischer, H.; Hofmann, P.; Kreissl, F. R.; Schubert, U.; Weiss, K. "Transition Metal Carbene Complexes"; Verlag Chemie: Weinheim, 1983. (c) Dötz, K. H. *Angew. Chem., Int. Ed. Engl.* 1984, *23*, 587. (d) Schubert, U. *Coord Chem. Rev.* 1984, *55*, 261. (e) Hahn, J. E. *Prog. Inorg. Chem.* 1984, *31*, 205. (f) Pourreau, D. B.; Geoffroy, G. L. *Adv. Organomet. Chem.* 1985, *24*, 249. (g) Casey, C. P. In "Reactive Intermediates"; Jones, M.; Moss, R. A., Eds.; Wiley: New York, 1985; Vol. 3 and the earlier volumes of this series. (h) Winter, M. J. *Organometal. Chem.* 1986, *14*, 225. Smith, A. K. *Ibid.* 1986, *14*, 278 and the earlier volumes of this series. See also Refs 3j,k and 39.

6. For very early work by Fischer in this area see: (a) Fischer, E. O.; Maasböl, A. *Angew. Chem., Int. Ed. Engl.* 1964, *3*, 580. For brief reviews of Fischer's early work see: (b) Fischer, E. O. *Pure Appl. Chem.* 1970, *24*, 407; 1972, *30*, 353.

7. For a few examples of Fischer carbene complexes undergoing cyclopropanation reactions, see ref 6b. For some other examples of carbene complexes in general undergoing these reactions, see: (a) Tebbe, F. N.; Parshall, G. W.; Reddy, G. S. *J. Am. Chem. Soc.* 1978, *100*, 3611. (b) Casey, C. P.; Polichnowski, S. W.; Shusterman, A. J.; Jones, C. R. *Ibid.* 1979, *101*, 7282. (c) Dötz, K. H.; Pruskil, I. *Chem. Ber.* 1981, *114*, 1980. (d) Casey, C. P.; Shusterman, A. J.; Vollendorf, N.W.; Haller, K. J. *J. Am. Chem. Soc.* 1982, *104*, 2417. (e) Casey, C. P.; Cesa, M. C. *Organometallics* 1982, *1*, 87. (f) Weiss, K.; Hoffmann, K.; *J. Organometal. Chem.* 1983, *255*, C24. (g) Toledano, C. A.; Rudler, H.; Daran, J.-C.; Jeannin, Y. *J. Chem. Soc., Chem. Commun.* 1984, 574. (h) Casey, C. P.; Vollendorf, M. W.; Haller, K. J. *J. Am. Chem. Soc.* 1984, *106*, 3754. (i) Doyle, M. P.; Griffin, J. H.; Chinn, M. S.; van Leusen, D. *J. Org. Chem.* 1984, *49*, 1917. (j) Doyle, M. P.; Griffin, J. H.; Bagheri, V.; Dorow, R. L. *Organometallics*, 1984, *3*, 53. (k) Mackenzie, P. B.; Ott, K. C.; Grubbs, R. H. *Pure Appl. Chem.* 1984, *56*, 59.(l) Jacobson, D. B.; Freiser, B. S. *J. Am.*

Chem. Soc. 1985, *107*, 2605. (m) Doyle, M. P.; Griffin, J. H.; da Conceicao, J. *J. Chem. Soc., Chem. Commun.* 1985, 328. (n) Doyle, M. P.; Dorow, R. L.; Terpstra, J. W.; Rodenhouse, R. A. *J. Org. Chem.* 1985, *50*, 1663. (o) Parlier, A.; Rudler, H.; Platzer, N.; Fontanille, M.; Soum, A. *J. Organomet. Chem.* 1985, *287*, C8. (p) Casey, C. P.; Shusterman, A. J. *Organometallics* 1985, *4*, 736. (q) Casey, C. P.; Hornung, N. L.; Kosar, W.P. *J. Am. Chem. Soc.* 1987, *109*, 4908. (r) Parlier, A.; Rudler, H.; Platzer, N.; Fontanille, M.; Soum, A. *J. Chem. Soc., Dalton Trans.* 1987, 1041. See also the reviews cited in Ref 5.

8. For some recent examples of other applications of Fischer carbene complexes, see:(a) Semmelhack, M. F.; Bozell, J.; Keller, L.; Sato, T.; Spiess, E. J.; Wulff, W. D.; Zask, A. *Tetrahedron* 1985, *41*, 5803. (b) Aumann, R.; Heinen, H. *Chem. Ber.* 1986, *119*, 253. (c) Cambie, R. C.; Rutledge, P. S.; Tercel, M.; Woodgate, P. D. *J. Organomet. Chem.* 1986, *315*, 171. (d) Yamashita, A. *Tetrahedron Lett.* 1986, *27*, 5915. (e) Borel, C.; Hegedus, L. S.; Krebs, J.; Satoh, Y. *J. Am. Chem. Soc.* 1987, *109*, 1101. (f) Xu, Y.-C.; Wulff, W. D. *J. Org. Chem.* 1987, *52*, 3263. (g) Dötz, K. H.; Sturm, W.; Alt, H. G. *Organometallics* 1987, *6*, 1424. See also the reviews cited in Ref. 5.

9. (a) Jolly, P. W.; Pettit, R. *J. Am. Chem. Soc.* 1966, *88*, 5044. (b) Riley, P. E.; Capshew, C. E.; Pettit, R.; Davis, R. E. *Inorg. Chem.* 1978, *17*, 408.

10. [Cp(CO)₂Fe]₂ is commercially available (Alfa; Aldrich), but we usually prepare it very straightforwardly from iron pentacarbonyl and cyclopentadiene dimer according to the procedure described in: King, R. B.; Stone, F. G. A. *Inorg. Synth.* 1963, *7*, 110. Rather than the 38% yield reported therein, we routinely obtain yields of 80–90%.

11. King, R. B. *Accounts Chem. Res.* 1970, *3*, 417.

12. (a) Green, M. L. H.; Ishaq, M.; Whiteley, R. N. *J. Chem. Soc. A* 1967, 1508. (b) Green, M. L. H.; Mitchard, L. C.; Swanwick, M. G. *J. Chem. Soc. A* 1971, 794.

13. (a) Rosenblum, M.; Bucheister, A.; Chang, T. C. T.; Cohen, M.; Marsi, M.; Samuels, S. B.; Scheck, D.; Sofen, N.; Watkins, J. C. *Pure Appl. Chem.* 1984, *56*, 129. (b) Pearson, A. J. In "Comprehensive Organometallic Chemistry"; Wilkinson, G.; Stone, F. G. A.; Abel, E. W., Eds.; Pergamon: Oxford, 1982; Vol. 8, pp. 939–1011.

14. Stevens, A. E.; Beauchamp, J. L. *J. Am. Chem. Soc.* 1978, *100*, 2584.

15. Schilling, B. E. R.; Hoffmann, R.; Lichtenberger, D. L. *J. Am. Chem. Soc.* 1979, *101*, 585.

16. Brookhart, M.; Tucker, J. R.; Flood, T. C.; Jensen, J. *J. Am. Chem. Soc.* 1980, *102*, 1203.

17. (a) Guerchais, V.; Astruc, D. *J. Chem. Soc., Chem. Commun.* 1985, 835. (b) Guerchais, V.; Lapinte, C. *J. Chem. Soc., Chem. Commun.* 1986, 663. (c) Guerchais, V.; Lapinte, C.; Thépot, J.-Y. *Organometallics* 1988, *7*, 604.

18. Davison, A.; Krusell, W. C.; Michaelson, R. C. *J. Organomet. Chem.* 1974, *72*, C7.

19. Brookhart, M.; Timmers, D.; Tucker, J. R.; Williams, G. D.; Husk, G. R.; Brunner, H.; Hammer, B. *J. Am. Chem. Soc.* 1983, *105*, 6721.

20. (a) Flood, T. C.; DiSanti, F. J.; Miles, D. L. *J. Chem. Soc., Chem. Commun.* 1975, *336*. (b) Flood, T. C.; DiSanti, F. J.; Miles, D. L. *Inorg. Chem.* 1976, *8*, 1910.

21. Brunner, H.; Leblanc, J.-C. *Z. Naturforsch.* 1980, *35b*, 1491.

22. (a) Brookhart, M.; Nelson, G. O. *J. Am. Chem. Soc.* 1977, *99*, 6099. (b) Brookhart, M.; Broom, B. H.; Kratzer, H. J.; Nelson, G. O. *Ibid.* 1980, *102*, 7802. (c) Brookhart, M.; Tucker, J. R.; Husk, G. R. *J. Organomet. Chem.* 1980, *193*, C23.

23. (a) Sanders, A.; Cohen, L.; Giering, W. P.; Kenedy, D.; Magatti, C. V. *J. Am. Chem. Soc.* 1973, *95*, 5430. (b) Sanders, A.; Bauch, T.; Magatti, C. V.; Lorenc, C.; Giering, W. P. *J. Organomet. Chem.* 1976, *107*, 359.

24. (a) Allison, N. T.; Kawqda, Y.; Jones, W. M. *J. Am. Chem. Soc.* 1978, *100*, 5224. (b) Riley, P. E.; Davis, R. E.; Allison, N. T.; Jones, W. M. *J. Am. Chem. Soc.* 1980, *102*, 2458. (c) Manganiello, F. J.; Radcliffe, M. D.; Jones, W. M. *J. Organomet. Chem.* 1982, *228*, 273. (d) Radcliffe, M. D.; Jones, W. M. *Organometallics* 1983, *2*, 1053. (e) Lisko, J. R.; Jones, W. M. *Organometallics* 1985, *4*, 612.

25. (a) Brookhart, M.; Studabaker, W. B.; Husk, G. R. *Organometallics*, 1985, *4*, 943. (b) Brookhart, M.; Studabaker, W. B.; Husk, G. R. *Ibid.* 1987, *6*, 1142.

26. (a) McCormick, F. B.; Angelici, R. J. *Inorg. Chem.* 1979, *18*, 1231. (b) McCormick, F. B.; Angelici, R. J. *Inorg. Chem.* 1981, *20*, 1111, 1118. (c) McCormick, F. B.; Angelici, R. J. *J. Organomet. Chem.* 1981, *205*, 79. (d) Yu, Y. S.; Angelici, R.J. *J. Organometallics* 1983, *2*, 1018, 1583. (e) Barrett, A. G. M.; Carpenter, N. E. *Ibid.* 1987, *6*, 2249.

27. Casey, C. P.; Miles, W. H. *J. Organomet. Chem.* 1983, *254*, 333.

28. (a) Cutler, A. R. *J. Am. Chem. Soc.* 1979, *101*, 604. (b) Bodnar, T.; Cutler, A. R. *J. Organomet. Chem.* 1981, *213*, C31. (c) Bodnar, T. W.; Cutler, A. R. *Organometallics* 1985, *4*, 1558.

29. (a) Curtis, P. J.; Davies, S. G.; *J. Chem. Soc., Chem. Commun.* 1984, 747. (b) Baird, G. J.; Davies, S. G.; Maberly, T. R. *Organometallics* 1984, 3, 1764. (c) Ayscough, A. P.; Davies, S. G. *J. Chem. Soc. Chem. Commun.* 1986, 1648

30. Davison, A.; Reger, D. L. *J. Am. Chem. Soc.* 1972, *94*, 9237.

31. (a) Green, M. L. H.; Hurley, C. R. *J. Organomet. Chem.* 1967, *10*, 188. (b) Green, M. L. H.; Mitchard, L. C.; Swanwick, M. G. *J. Chem. Soc. (A)* 1971, 794.

32. (a) Boland, B. E.; Fam, S. A.; Hughes, R. P. *J. Organomet. Chem.* 1979, *172*, C29. (b) Boland-Lussier, B. E.; Hughes, R. P. *Organometallics* 1982, *1*, 635.

33. Grötsch, G.; Malisch, W. *J. Organomet. Chem.* 1984, *262*, C38.

34. (a) Marten, D. F. *J. Chem. Soc. Chem. Commun.* 1980, 341. (b) Marten, D. F. *J. Org. Chem.* 1981, *46*, 5422.

35. (a) Priester, W.; Rosenblum, M. *J. Chem. Soc., Chem. Commun.* 1978, 26. (b) Bates, D. J.; Rosenblum, M.; Samuels, S. B. *J. Organomet. Chem.* 1981, *209*, C55.

36. Treichel, P. M.; Firsich, D. W.; Lemmen, T. H. *J. Organomet. Chem.* 1980, *202*, C77.

37. Brunner, H.; Kerkien, G.; Wachter, J. *J. Organomet. Chem.* 1982, *224*, 295.

38. (a) Davison, A.; Selegue, J. P. *J. Am. Chem. Soc.* 1978, *100*, 7763. (b) Nesmeyanov, A. N.; Petrovskaya, E. A.; Rybin, L. V.; Rybinskaya, M. I. *Bull. Acad. Sci. USSR, Div. Chem. Sci.* 1979, *28*, 1887. (c) Adams, R. D.; Davison, A.; Selegue, J. P. *J. Am. Chem. Soc.* 1979, *101*, 7232. (d) Selegue, J. P. *Ibid.* 1982, *104*, 119. (e) Bodnar, T. W.; Cutler, A. R. *Ibid.* 1983, *105*, 5926. (f) Richmond, T. G.; Shriver, D. F. *Organometallics* 1984, *3*, 305. (g) Reger, D. L.; Swift, C. A. *Ibid.* 1984, *3*, 876. (h) Bly, R. S.; Hossain, M. M.; Lebioda, L. *J. Am. Chem. Soc.* 1985, *107*, 5549. (i) Lisko, J. R.; Jones, W. M. *Organometallics* 1985, *4*, 944. (j) Barefield, E. K.; McCarten, P.; Hillhouse, M. C. *Ibid.* 1985, *4*, 1682. (k) Bly, R. S.; Bly, R. K. *J. Chem. Soc. Chem. Commun.* 1986, 1046.

39. Brookhart, M.; Studabaker, W. B. *Chem. Rev.* 1987, *87*, 411.

40. For a thorough compilation of references to complexes of this type see footnote 1 in Ref. 38j. See also: Barefield, E. K.; Sepelak, D. J. *J. Am. Chem. Soc.* 1979, *101*, 6542.

41. For an early report and for reviews of sulfonium ylide chemistry see: (a) Corey, E. J.; Chaykovsky, M. *J. Am. Chem. Soc.* 1965, *87*, 1353. (b) Trost, B. M.; Melvin, Jr., L. S. "Sulfur Ylides"; Academic Press: New York, 1975. (c) Block, E. "Reactions of Organo-sulfur Compounds"; Academic Press: New York, 1978; Chapter 3. (d) Johnson, C. R. In "Comprehensive Organic Chemistry"; Barton, D.; Ollis, W. D., Eds.; Pergamon Press: Oxford, 1979; Vol. 3, pp. 247–260.

42. Law, J. H. *Accounts Chem. Res.* 1971, *4*, 199.

43. (a) Trost, B. M. *J. Am. Chem. Soc.* 1967, *89*, 138. (b) Cohen, T.; Herman, G.; Chapman, T. M.; Kuhn, D. *Ibid.* 1974, *96*, 5627. (c) Pörschke, K.-R. *Chem. Ber.* 1987, *120*, 425.

44. For some other examples of sulfonium ylide complexes see: (a) Kilbourn, B. T.; Felix, D. *J. Chem. Soc. A* 1969, *163*. (b) Sato, T.; Higuchi, J. *Tetrahedron Lett.* 1972, 407. (c) Schmidbaur, H. *Accounts Chem. Res.* 1975, *8*, 62. (d) Weleski, E. T.; Silver, J. L.; Jansson, M. D.; Burmeister, J. L. *J. Organomet. Chem.* 1975, *102*, 365. (e) Koezuka, H.; Mat-subayashi, G.; Tanaka, T. *Inorg. Chem.* 1976, *15*, 417. (f) Bravo, P.; Fronza, G.; Ticozzi, C. *J. Organomet. Chem.,* 1976, *118*, C78. (g) Seno, M.; Tsuchiya, S. *J. Chem. Soc., Dalton Trans.* 1977, 751. (h) Yoshida, G.; Kurosawa, H.; Okawara, R. *J. Organomet. Chem.* 1977, *131*, 309. (i) Matsubayashi, G.; Kondo, Y.; Tanaka, T. Nishigaki, S.; Nakatsu, K. *Chem. Lett.* 1979, *375*. (j) Nishiyama, H. *J. Organomet. Chem.* 1979, *165*, 407. (k) Stein,

J.; Fackler, J. P.; Paparizos, C.; Chen, H.-W. *J. Am. Chem. Soc.* 1981, *103*, 2192. (l) McCormick, F. B.; Gladysz, J. A. *J. Organomet. Chem.* 1981, *218*, C57. (m) Puddephatt, R. J. In "Comprehensive Organometallic Chemistry"; Wilkinson, G.; Stone, F. G. A.; Abel, E. W., Eds.; Pergamon Press: Oxford, 1982, Vol. 2, pp. 797–798. (n) Maitlis, P. M.: Espinet, P.; Russell, M. J. H. *Ibid.*; Vol. 6, pp. 299–303. (o) Hartley, F. R. *Ibid.*; Vol. 6, pp. 511–513. (p) Weber, L. *Angew Chem. Int. Ed. Engl.* 1983, *22*, 516. (q) Fischer, H.; Weber, L. *Chem. Ber.* 1984, *117*, 3340. (r) Touchard, D.; Dixneuf, P. H.; Adams, R. D.; Segmüller, B. E. *Organometallics* 1984, *3*, 640. (s) Davidson, J. G.; Barefield, E. K.; Van Derveer, D. G. *Ibid.* 1985, *4*, 1178. (t) Weber, L.; Wewers, D.; Lücke, E. *Z. Naturforsch.* 1985, *40b*, 968. (u) Roper, W. R. *J. Organomet Chem.* 1986, *300*, 167. (v) McCormick, F. B.; Gleason, W. B.; Zhao, X.; Heah, P. C.; Gladysz, J. A. *Organometallics* 1986, *5*, 1778. (w) Weber, L.; Lücke, E. *Organometallics* 1986, *5*, 2114. (x) Fischer, H.; Schmid, J.; Zeuner, S. *Chem. Ber.* 1987, *120*, 583.

45. Brandt, S. Ph.D. Dissertation, State University of New York at Stony Brook (1979).
46. King, R. B.; Bisnette, M. B. *Inorg. Chem.* 1965, *4*, 486.
47. (a) Meerwein, H. *Org. Synth. Coll. Vol. V.* 1973, 1096. (b) Curphey, T. J. *Org. Synth.* 1971, *51*, 142.
48. Brandt, S.; Helquist, P. *J. Am. Chem. Soc.* 1979, *101*, 6473.
49. O'Connor, E. J. Ph.D. Dissertation, State University of New York at Stony Brook (1984).
50. Borch, R. F. *J. Org. Chem.* 1969, *34*, 627.
51. Reger, D. L.; Fauth, D. J.; Dukes, M. D. *Syn. React. Inorg. Metal-Org. Chem.* 1977, *7*, 151.
52. O'Connor, E. J.; Helquist, P. *J. Am. Chem. Soc.* 1982, *104*, 1869.
53. Creary, X.; Mehrsheikh-Mohammadi, M. (University of Notre Dame), personal communication (1986).
54. Curtin, D. Y.; Gruen, H.; Shoulders, B. A. *Chem. Ind. (London)* 1958, 1205.
55. (a) Iyer, R. S.; Kuo, G.-H.; Helquist, P. *J. Org. Chem.* 1985, *50*, 5898. (b) Iyer, R. S. Ph.D. Dissertation, State University of New York at Stony Brook, 1985.
56. Wender, P. A.; Eck, S. L. *Tetrahedron Lett.* 1982, *23*, 1871.
57. le Noble, W. J.; Srivastava, S.; Cheung, C. K. *J. Org. Chem.* 1983, *48*, 1099.
58. O'Connor, E. J.; Brandt, S.; Helquist, P. *J. Am. Chem. Soc.* 1987, *109*, 3739.
59. Müller, E.; Kessler, H.; Suhr, H. *Tetrahedron Lett.* 1965, 423.
60. (a) Kawabata, N.; Nakagawa, T.; Nakao, T.; Yamashita, S. *J. Org. Chem.* 1977, *42*, 3031. (b) Kropp, P. J. *Accounts Chem. Res.* 1984, *17*, 131. (c) Arai, I.; Mori, A.; Yamamoto, H. *J. Am. Chem. Soc.* 1985, *107*, 8254. (d) Maruoka, K.; Fukutani, Y.; Yamamoto, H. *J. Org. Chem.* 1985, *50*, 4412. (e) Friedrich, E. C.; Domek, J. M.; Pong, R. Y. *Ibid.* 1985, *50*, 4640. (f) Nelson, K. A.; Mash, E. A. *Ibid.* 1986, *51*, 2721.
61. (a) Kosarych, Z.; Cohen, T. *Tetrahedron Lett.* 1982, *23*, 3019. (b) Liebowitz, S. M.; Johnson, H. J. *Synth. Commun.* 1986, *16*, 1255.
62. For a discussion of this problem and its solution based upon slow addition of diazo compounds see: Doyle, M. P.; van Leusen, D.; Tamblyn, W. H. *Synthesis* 1981, 787. See also Refs. 3k and 7m.
63. Kegley, S. E.; Brookhart, M.; Husk, G. R. *Organometallics* 1982, *1*, 760.
64. Cutler, A.; Fish, R. W.; Giering, W. P.; Rosenblum, M. *J. Am. Chem. Soc.* 1972, *94*, 4354.
65. (a) Kremer, K. A. M.; Helquist P.; Kerber, R. C. *J. Am. Chem. Soc.* 1981, *103*, 1862. (b) Kremer, K. A. M.; Helquist, P. *J. Organomet. Chem.* 1985, *285*, 231. (c) Kremer, K. A. M. Ph.D. Dissertation, State University of New York at Stony Brook (1982).
66. Kremer, K. A. M.; Helquist, P. *Organometallics* 1984, *3*, 1743.
67. Still, W. C.; Kahn, M.; Mitra, A. *J. Org. Chem.* 1978, *43*, 2923.
68. Meerwein, H. *Org. Synth. Coll. Vol. V.* 1973, 1080.
69. (a) Closs, G. L.; Moss, R. A. *J. Am. Chem. Soc.* 1964, *86*, 4042. (b) Mathias, R.; Weyerstahl, P. *Chem. Ber.* 1979, *112*, 3041

70. In our preliminary communication (Ref. 65a), we had reported the formation of two isomeric cyclopropanes from cyclododecene, but further work showed that in fact pure *cis*-cyclododecene gives the *endo* isomer as the only detectable product. The earlier result was due to our original sample of cyclododecene being a 1:1 mixture of *cis* and *trans* isomers.

71. Burger, U.; Huisgen, R. *Tetrahedron Lett.* 1970, 3057.

72. (a) Kirmse, W.; "Carbene Chemistry"; 2nd edn.; Academic Press: New York, 1971; pp. 286–293. (b) "Carbenes"; Jones, M.; Moss, R. A., Eds.; Wiley: New York, 1973; Vol. 1, pp. 126, 263–264, 285, 294, and 297. (c) Wulfman, D. S.; Poling, B. In "Reactive Intermediates"; Abramovitch, R. A., Ed.; Plenum: New York, 1980; Vol. 1, pp. 365–366.

73. (a) Brookhart, M.; Tucker, J. R.; Husk, G. R. *J. Am. Chem. Soc.* 1981, *103*, 979. (b) Brookhart, M.; Tucker, J. R.; Husk, G. R. *Ibid.* 1983, *105*, 258. (c) Brookhart, M.; Kegley, S. E.; Husk, G. R. *Organometallics,* 1984, *3*, 650.

74. (a) Cocker, W.; Geraghty, N. W. A.; Grayson, D. H. *J. Chem. Soc., Perkin Trans. I* 1978, 1370. (b) Bury, A.; Johnson, M. D.; Stewart, M. J. *J. Chem. Soc. Chem. Commun.* 1980, *622*. (c) Arlt, D.; Jautelat, M.; Lantzsch, R. *Angew. Chem., Int. Ed. Engl.* 1981, *20*, 703. (d) Franck-Neumann, M.; Sedrati, M.; Vigneron, J.-P.; Bloy, V. *Angew. Chem., Int. Ed. Engl.* 1985, *24*, 996. (e) Chaboteaux, G.; Krief, A. *Bull. Soc. Chim. Belg.* 1985, *94*, 495. (f) Harayama, T.; Fukushi, H.; Ogawa, K.; Yoneda, F. *Chem. Pharm. Bull.* 1985, *33*, 3564. (g) Saha, M.; Muchmore, S.; van der Helm, D.; Nicholas, K. M. *J. Org. Chem.* 1986, *51*, 1960. (h) Harayama, T.; Fukushi, H.; Ogawa, K.; Aratani, T.; Sonehara, S.; Yoneda, F. *Chem. Pharm. Bull.* 1987, *35*, 4977. (i) Wender, P. A.; Keenan, R. M.; Lee, H. Y. *J. Am. Chem. Soc.* 1987, *109*, 4390.

75. (a) Bellerby, J. M.; Mays, M. J. *J. Organomet. Chem.* 1976, *117*, C21. (b) Bell, R. A.; Chisholm, M. H. *Inorg. Chem.* 1977, *16*, 698. (c) Bruce, M. I.; Wallis, R. C. *Austral. J. Chem.* 1980, *33*, 1471. (d) Bruce, M. I.; Swincer, A. G. *Ibid.* 1980, *33*, 1471. (e) Reger, D. L.; Swift, C. A. *Organometallics* 1984, *3*, 876. (f) Bodner, G. S. Smith, D. E.; Hatton, W. G.; Heah, P. C.; Georgiou, S.; Rheingold, A. L.; Geib, S. J.; Hutchinson, J. P.; Gladysz, J. A. *J. Am. Chem. Soc.* 1987, *109*, 7688.

76. Green, M. L. H.; Ishaq, M.; Mole, T. *Z. Naturforsch.* 1965, *20b*, 598.

77. Kuo, G.-H. Ph.D. Dissertation, State University of New York at Stony Brook (1985).

78. (a) Casey, C. P.; Miles, W. H.; Tukada, H.; O'Connor, J. M. *J. Am. Chem. Soc.* 1982, *104*, 3761. (b) Casey, C. P.; Miles, W. H.; Tukada, H. *J. Am. Chem. Soc.* 1985, *107*, 2924.

79. (a) King, R. B.; Bisnette, M. G. *J. Organomet. Chem.* 1964, *2*, 15. (b) Nesmeyanov, A. N.; Rybin, L. V.; Rybinskaya, M. I.; Kaganovich, V. S.; Petrovskii, P. V. *Bull. Acad. Sci. USSR, Div. Chem. Sci.* 1971, *20*, 2592. (c) Quinn, S.; Shaver, A. *Inorg. Chim. Acta*, 1980, *38*, 243.

80. Kremer, K. A. M.; Kuo, G.-H.; O'Connor, E. J.; Helquist, P.; Kerber, R. C. *Ibid.* 1982, *104*, 6119.

81. (a) Fischer, E. O.; Clough, R. L.; Besl, G.; Kreissl, F. R. *Angew. Chem., Int. Ed. Engl.* 1976, *15*, 543. (b) Fischer, E. O.; Clough, R. L.; Stuckler, P. *J. Organomet. Chem.* 1976, *120*, C6. (c) Levisalles, J.; Rudler, H.; Dahan, F.; Jeannin, Y. *J. Organomet. Chem.* 1980, *188*, 193. (d) Dyke, A. F.; Knox, S. A. R.; Naish, P. J. *Ibid.* 1980, *199,* C47. (e) Dyke, A. F.; Knox, S. A. R.; Mead, K. A.; Woodward, P. *J. Chem. Soc. Chem. Commun.* 1981, 861. (f) Cooke, M.; Davies, D. L.; Guerchais, J. E.; Knox, S. A. R.; Mead, K. A.; Roué, J.; Woodward, P. *Ibid.* 1981, 862.

82. (a) Lee, V. J.; Branfamn, A. R.; Herrin, T. R.; Rinehart, K. L. *J. Am. Chem. Soc.* 1978, *100*, 4225. (b) MacMillan, J. H.; Washburne, S. S. *J. Org. Chem.* 1973, *38*, 2982. (c) Sundberg, R. J.; Bukowick, P. A.; Holcombe, F. O. *Ibid.* 1967, *32*, 2938. (d) Widmer, U.; Heimgartner, H.; Schmid, H. *Helv. Chim. Acta* 1978, *61*, 815.

83. Kuo, G.-H.; Helquist, P.; Kerber, R. C. *Organometallics* 1984, *3*, 806.

84. For several earlier examples of allylidene, or vinylcarbene, complexes, see the papers cited

in footnote 16 of Ref. 83. For some more recent examples, see: (a) Green, M.; Orpen, A. G.; Schaverien, C. J.; Williams, I. D. *J. Chem. Soc. Chem. Commun.* 1983, *583*. (b) Mitsudo, T.; Ogino, Y.; Komiya, Y.; Watanabe, H.; Watanabe, Y. *Organometallics* 1983, 2, 1202. (c) Wulff, W. D.; Gilbertson, S. R. *J. Am. Chem. Soc.* 1985, 107, 503. (d) Morrow, J. R.; Tonker, T. L.; Templeton, J. L. *Ibid.* 1985, 107, 5004. (e) Manganiello, F. J.; Oon, S. M.; Radcliffe, M. D.; Jones, W. M. *Organometallics* 1985, *4*, 1069. (f) Aumann, R.; Heinen, H. *Chem. Ber.* 1986, 119, 3801. (g) Chan, K. S.; Wulff, W. D. *J. Am. Chem. Soc.* 1986, 108, 5229. (h) Mitsudo, T.; Ishihara, A.; Kadokura, M.; Watanabe, Y. *Organometallics* 1986, *5*, 238. (i) Parlier, A.; Rudler, H.; Platzer, N.; Fontanille, M.; Soum A. *J. Chem. Soc. Dalton Trans.* 1987, 1041. (j) Aumann, R.; Heinen, H. *Chem. Ber.* 1987, 120, 537. (k) Casey, C. P.; Woo, L. K.; Fagan, P. J.; Palermo, R. E.; Adams. B. R. *Organometallics* 1987, *6*, 447. (l) Dötz, K. H.; Sturm, W.; Alt, H. G. *Ibid.* 1987, *6*, 1424. (m) Raubenheimer, H. G.; Kruger, G. J.; Viljoen, H. W. *J. Organomet. Chem.* 1987, *319*, 361. (n) Mayr, A.; Asaro, M. F.; Glines, T. J. *J. Am. Chem. Soc.* 1987, *109*, 2215. (o) Herndon, J. W. *Ibid.* 1987, *109*, 3165. (p) Macomber, D. W.; Liang, M.; Rogers, R. D. *Organometallics* 1988, *7*, 416. (q) Landrum, B. E.; Lay, J. O.; Allison, N. T. *Ibid.* 1988, *7*, 787. (r) Herndon, J. W. *J. Org. Chem.* 1986, *51*, 2853.

85. Casey, C. P.; Miles, W. H. *Organometallics* 1984, *3*, 808.

86. For a general treatment of the NMR chemical shifts of carbene complexes, see Ref. 78b and Fenske, R. F. In "Organometallic Compounds: Synthesis, Structure and Theory"; Shapiro, B. L., Ed.; Texas A & M University Press: College Station, Texas, 1983.

87. Hudlicky, T.; Kutchan, T. M.; Naqvi, S. M. *Org. React.* 1985, *33*, 247.

88. For reviews and other leading references pertaining to intramolecular cyclopropanations, see: (a) Büchi, G.; White, J. D. *J. Am. Chem. Soc.* 1964, 86, 2884. (b) Kirmse, W.; "Carbene Chemistry"; 2nd edn.; Academic Press: New York, 1971; pp. 328–342. (c) Mori, K.; Ohki, M.; Kobayashi, A.; Matsui, M. *Tetrahedron* 1970, *26*, 2815. (d) McMurry, J. E.; Blaszczak, L. C. *J. Org. Chem.* 1974, 39, 2217. (e) Branca, S. J.; Lock, R. L.; Smith, A. B. *Ibid.* 1977, *42*, 3165. (f) Trost, B. M.; Vladuchick, W. C. *Ibid.* 1979, 44, 148. (g) Burke, S. D.; Grieco, P. A. *Org. React.* 1979, *26*, 361. (h) White, J. D.; Ruppert, J. F.; Avery, M. A.; Torii, S.; Nokami, J. *J. Am. Chem. Soc.* 1981, 103, 1813. (i) Ghatak, U. R.; Roy, S. C. *J. Chem. Res. (M)* 1981, 159. (j) Desimoni. G.; Tacconi, G.; Barco, A.; Pollini, G. P. "Natural Products Synthesis Through Pericyclic Reactions"; American Chemical Society: Washington, D.C., 1983; pp. 108–110. (k) Mandai, T.; Hara, K.; Kawada, M.; Nokami, J. *Tetrahedron Lett.* 1983, *24*, 1517. (l) Smith, A. B.; Toder, B. H.; Branca, S. J. *J. Am. Chem. Soc.* 1984, *106*, 3995. (m) Marshall, J. A.; Peterson, J. C.; Lebioda, L. *Ibid.* 1984, *106*, 6006. (n) Rousseau, G.; Slougui, N. *Ibid.* 1984, *106*, 7283. (o) Doyle, M. P.; Trudell, M. L. *J. Org. Chem.* 1984, *49*, 1196. (p) Callant, P.; D'Haenens, L.; Vandewalle, M. *Synth. Commun.* 1984, *14*, 155. (q) Corey, E. J.; Myers, A. G. *J. Am. Chem. Soc.* 1985, *107*, 5574. (r) Hudlicky, T.; Ranu, B. C.; Naqvi, S. M.; Srnak, A. *J. Org. Chem.* 1985, *50*, 123. (s) Remy, D. C.; King, S. W.; Cochran, D.; Springer, J. P.; Hirshfield, J. *Ibid.* 1985, *50*, 4120. (t) Imanishi, T.; Ninbari, F.; Yamashita, M.; Iwata, C. *Chem. Pharm. Bull.* 1986, *34*, 2268. (u) Ceccherelli, P.; Curini, M.; Marcotullio, M. C.; Wenkert, E. *J. Org. Chem.* 1986, *51*, 738. (v) Taber, D. F.; Amedio, J. C.; Sherrill, R. G. *Ibid.* 1986, *51*, 3382. (w) Adams, J.; Belley, M. *Ibid.* 1986, *51*, 3878. (x) Padwa, A.; Wisnieff, T. J.; Walsh, E. J. *Ibid.* 1986, *51*, 5036. (y) Schultz, A. G.; Eng, K. K.; Kullnig, R. K. *Tetrahedron Lett.* 1986, *27*, 2331. (z) Imanishi, T.; Matsui, M.; Yamashita, M.; Iwata, C. *Ibid.* 1986, *27*, 3161. (aa) Maas, G. *Topics Curr. Chem.* 1987, *137*, 75. (ab) Corey, E. J.; Reid, J. G.; Myers, A. G.; Hahl, R. W. *J. Am. Chem. Soc.* 1987, *109*, 918. (ac) Rigby, J. H.; Senanayake, C. *Ibid.* 1987, *109*, 3147. (ad) Majerski, Z.; Zuanic, M. *Ibid.* 1987, *109*, 3496. (ae) Adams, J.; Frenette, R.; Belley, M.; Chibante, F.; Springer, J. P. *Ibid.* 1987, *109*, 5432. (af) Singh, A. K.; Bakshi, R. K.; Corey, E. J. *Ibid.* 1987, *109*, 6187. (ag) Rigby, J. H.; Senanayake, C. *J. Org. Chem.* 1987, *52*, 4634. (ah) Hudlicky, T.; Natchus, M. G.; Sinai-Zingde, G. *Ibid.* 1987, *52*, 4641. (ai) Nakatani, K.; *Tetrahedron Lett.* 1987, *28*, 165.

89. For examples of carbene complexes undergoing intramolecular cyclopropanation reactions, see Refs 7d,e,g,h,o–r.
90. (a) Chen, E. Y. *J. Org. Chem.* 1984, *49*, 3245. (b) Ando, M.; Sayama, S.; Takase, K. *Ibid.* 1985, *50*, 251, and references cited therein.
91. Matsuo, A.; Ishii, O.; Suzuki, M.; Nakayama, M.; Hayashi, S. *Z. Naturforsch.* 1982, *37b*, 1636.
92. (a) Marshall, J. A.; Ruth, J. A. *J. Org. Chem.* 1974, *39*, 1971. (b) Papageorgiou, V. P.; Argyriadou, N. *Phytochemistry*, 1981, *20*, 2295, and references cited therein.
93. Smith, A. B.; Liverton, N. J.; Hrib, N. J.; Sivaramakrishnan, H.; Winzenberg, K. *J. Org. Chem.* 1985, *50*, 3239.
94. McMurry, J. E.; Bosch, G. K. *J. Org. Chem.* 1987, *52*, 4885.
95. Toma, K.; Miyazaki, E.; Murae, T.; Takahashi, T. *Chem. Lett.* 1982, 863.
96. Welch, S. C.; Gruber, J. M.; Morrison, P. A. *J. Org. Chem.* 1985, *50*, 2676.
97. Smedman, L.; Zavarin, E. *Tetrahedron Lett.* 1968, 3833.
98. (a) Knors, C.; Kuo, G.-H.; Lauher, J. W.; Eigenbrot, C.; Helquist, P. *Organometallics* 1987, *6*, 988. (b) Knors, C. Ph.D. Dissertation, University of Notre Dame (1987).
99. (a) Kiel, W. A.; Lin, G.-Y.; Bodner, G. S.; Gladysz, J. A. *J. Am. Chem. Soc.* 1983, *105*, 4958. (b) Patton, A. T.; Strouse, C. E.; Knobler, C. B.; Gladysz, J. A. *Ibid.* 1983, *105*, 5804.
100. (a) McCormick, F. B.; Angelici, R. J.; Pickering, R. A.; Wagner, R. E.; Jacobson, R. A. *Ibid.* 1981, *20*, 4108. (b) Matachek, J. R.; Angelici, R. J.; Schugart, K. A.; Haller, K. J.; Fenske, R. F. *Ibid.* 1984, *3*, 1038.
101. (a) Collins, T. J.; Roper, W. R.; *J. Organomet. Chem.* 1978, *159*, 73. (b) Pickering, R. A.; Angelici, R. J. *Inorg. Chem.* 1981, *20*, 2977. (c) Battioni, J.-P.; Chottard, J.-C.; Mansuy, D. *Ibid.* 1982, *21*, 2056. (d) Steinmetz, A. L.; Hershberger, S. A.; Angelici, R. J. *Organometallics* 1984, *3*, 461. (e) Kim, H. P.; Kim, S.; Jacobson, R. A.; Angelici, R. J. *Ibid.* 1984, *3*, 1124. (f) Raubenheimer, H. G.; Kruger, G. J.; van A. Lombard, A.; Linford, L.; Viljoen, J. C. *Ibid.* 1985, *4*, 275. See also Ref 5b, pp. 11, 15–16, and 23–24.
102. For some examples of related vinylidene complexes, see: (a) Wong, A.; Gladysz, J. A. *J. Am. Chem. Soc.* 1982, *104*, 4948. (b) Bruce, M. I.; Swincer, A. G. *Adv. Organomet. Chem.* 1983, *22*, 59. See also Refs 32, 34, 38c,d.
103. (a) Dötz, K. H.; Fischer, H.; Hofmann, P.; Kreissl, F. R.; Schubert, U.; Weiss, K. "Transition Metal Carbene Complexes"; Verlag Chemie: Weinheim, 1983; p. 97. (b) Crespi, A. M.; Shriver, D. F. *Organometallics* 1985, *4*, 1830. (c) Stenstrøm, Y.; Klauck, G.; Koziol, A.; Palenik, G. J.; Jones, W. M. *Ibid.* 1986, 5, 2155. (d) Iyer, R. S.; Selegue, J. P. *J. Am. Chem. Soc.* 1987, *109*, 910. (e) Stenstrøm, Y.; Koziol, A. E.; Palenik, G. J.; Jones, W. M. *Organometallics* 1987, *6*, 2079.
104. Consiglio, G.; Bangerter, F.; Darpin, C.; Morandini, F.; Lucchini, V. *Organometallics* 1984, *3*, 1446.
105. For leading papers from the Davies group concerned with enolates of acyliron complexes, see:(a) Aktogu, N.; Felkin, H.; Davies, S. G. *J. Chem. Soc. Chem. Commun.* 1982, 1303. (b) Aktogu, N.; Felkin, H.; Baird, G. J.; Davies, S. G.; Watts, O. *J. Organomet. Chem.* 1984, *262*, 49. (c) Davies, S. G.; Dordor-Hedgecock, I. M.; Warner, P.; Jones, R. H.; Prout, K. *Ibid.* 1985, *285*, 213. (d) Davies, S. G.; Walker, J. C. *J. Chem. Soc. Chem. Commun.* 1985, 209. (e) Davies, S. G.; Dordor-Hedgecock, I. M.; Sutton, K. H.; Walker, J. C.; Bourne, C.; Jones, R. H.; Prout, K. *Ibid.* 1986, 607. (f) Davies, S. G.; Easton, R. J. C.; Gonzalez, A.; Preston, S. C.; Sutton, K. H.; Walker, J. C. *Tetrahedron* 1986, *42*, 3987. (g) Brown, S. L.; Davies, S. G.; Foster, D. F.; Seeman, J. I.; Warner, P. *Tetrahedron Lett.* 1986, *27*, 623. (h) Davies, S. G.; Easton, R. J. C.; Sutton, K. H.; Walker, J. C.; Jones, R. H. *J. Chem. Soc., Perkin Trans. I* 1987, 489. (i) Bashiardes, G.; Davies, S. G. *Tetrahedron Lett.* 1987, *28*, 5563.
106. For other studies of the enolate chemistry of acyl complexes of iron and other metals, see: (a) Blau, H.; Malisch, W.; Voran, S.; Blank, K.; Krüger, C. *J. Organomet. Chem.* 1980, *202*, C33. (b) Theopold, K. H.; Becker, P. N.; Bergman, R. G. *J. Am. Chem. Soc.* 1982,

104, 5250. (c) Stasunik, A.; Malisch, W. *J. Organomet. Chem.* 1983, *247*, C47. (d) Liebeskind, L. S.; Welker, M. E. *Organometallics* 1983, *2*, 194. (e) Heah, P. C.; Gladysz, J. A. *J. Am. Chem. Soc.* 1984, *106*, 7636. (f) Liebeskind, L. S.; Welker, M. E. *Tetrahedron Lett.* 1984, *25*, 4341. (g) Wulff, W. D.; Gilbertson, S. R. *Ibid.* 1985, *107*, 503. (h) Ojima, I.; Kwon, H. B. *Chem. Lett.* 1985, 1327. (i) Cramer, R. E.; Higa, K. T.; Gilje, J. W. *Organometallics* 1985, *4*, 1140. (j) Liebeskind, L. S.; Welker, M. E. *Tetrahedron Lett.* 1985, *26*, 3079. (k) Heah, P. C.; Patton, A. T.; Gladysz, J. A. *J. Am. Chem. Soc.* 1986, *108*, 1185. (l) Rusik, C. A.; Tonker, T. L.; Templeton, J. L. *Ibid.* 1986, *108*, 4652. (m) Bassner, S. L.; Morrison, E. D.; Geoffroy, G. L. *Ibid.* 1986, *108*, 5358. (n) Liebeskind, L. S.; Welker, M. E.; Fengl, R. W. *Ibid.* 1986, *108*, 6328. (o) Weinstock, I.; Floriani, C.; Chiesi-Villa, A.; Guastini, C. *Ibid.* 1986, *108*, 8298. (p) Lukehart, C. M.; Myers, J. B.; Sweetman, B. J. *J. Organomet. Chem.* 1986, *316*, 319. (q) Akita, M.; Kondoh, A. *J. Organomet. Chem.* 1986, *299*, 369. (r) O'Connor, J. M.; Uhrhammer, R.; Rheingold, A. L. *Organometallics* 1987, *6*, 1987.

107.   Brinkman, K.; Helquist, P. *Tetrahedron Lett.* 1985, *26*, 2845.

108.   Brinkman, K.; Knors, C.; Lyon, C.; Helquist, P. unpublished results.

TRICARBONYL(η^6-ARENE)CHROMIUM COMPLEXES IN ORGANIC SYNTHESIS

Motokazu Uemura

OUTLINE

Advances in Metal-Organic Chemistry, Volume 2, pages 195–245
Copyright © 1991 JAI Press Ltd
All rights of reproduction in any form reserved
ISBN: 0-89232-948-3

I. INTRODUCTION

Although a wide range of η^6-arene transition metal compounds are known, only those of chromium have found significant applications in organic synthesis.[1] (η^6-Arene)Cr(CO)₃ complexes are readily prepared by several convenient methods.[2] Various solvents have been used, ranging from the arene itself to inert or polar solvents. The use or addition of donor solvents leads to significantly faster rate of complexation reactions.[3] Recent studies[4,5] indicated that the most suitable solvent medium for a wide range of high yield synthesis is an approximately 10/1 mixture of di-n-butyl ether and THF. This modification largely overcomes previous problems such as loss of Cr(CO)₆ via sublimation[6] and removal of high boiling solvents or excess arene from the products.

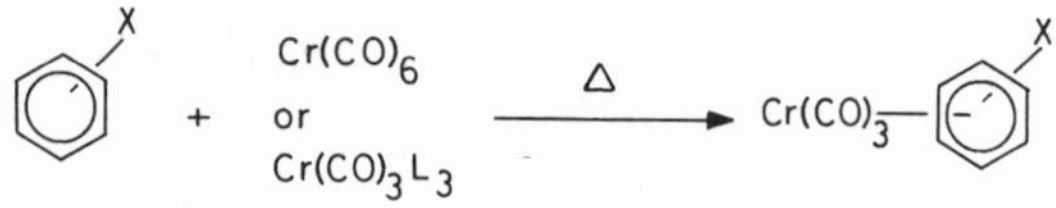

Scheme 1

In addition to their easy preparation, (η^6-arene)Cr(CO)₃ complexes are air-stable, crystalline solids, easily characterized by spectroscopic methods, and easily purified by chromatographic techniques or recrystallization. Among the most common substituted-arene complexes, those of benzaldehyde and benzoic acid are not accessible by these direct methods, but are stable when prepared by indirect routes via acetals or esters. Nitrobenzene is unknown as an (arene)chromium complex. At the end of synthetic sequences, the arene group is readily released from chromium complexes by mild oxidation of the metal by Ce(IV), I₂, or by exposure to sunlight.

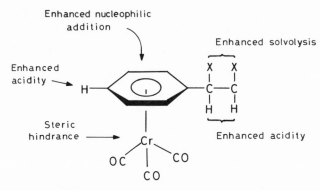

Figure 1. Effects of Cr(CO)$_3$ complexation to arene compounds.

The tricarbonyl(η^6-arene)chromium complexes have some significant properties due to their strong electron-withdrawing ability and the steric effect of the Cr(CO)$_3$ group. The chemical consequences of chromium complexation are summarized in Figure 1[7]: (1) activation of the aromatic ring to nucleophilic addition; (2) enhancement of the kinetic acidity of aromatic hydrogens; (3) steric hindrance of the arene face co-ordinated to the metal; (4) stabilization of carbanions at the benzylic position; and (5) stabilization of carbonium ions at the benzylic position.

The first of these effects, nucleophilic addition to the coordinated arene ring, is probably the most interesting and is now at a fairly advanced stage of exploitation, owing in large part to the efforts of Semmelhack and his group. Since a number of reviews dealing with nucleophilic addition to the chromium-complexed arene ring have appeared,[1,7,8] this article focuses on our investigations and the related works of (η^6-arene)chromium complexes in organic synthesis.

II. DIRECTED REGIOSELECTIVE LITHIATION OF (η_6-ARENE)CHROMIUM COMPLEXES

Co-ordination of the chromium tricarbonyl group to the arene enhances the kinetic acidity of the ring C–H bonds[9] due to the strong electron-withdrawing ability of Cr(CO)$_3$ group. Therefore, the directed lithiation of the arene ring is expected to occur more easily than the chromium-free parent arenes. These ring proton abstraction reactions by lithium bases to give chromium–aryllithium derivatives have been found to be particularly useful for the preparation of substituted (η^6-arene)Cr(CO)$_3$ complexes not easily obtained by the usual procedures. It is known, however, that the reaction of (η^6-arene)Cr-(CO)$_3$ complexes with various organo-lithium reagents can also lead to nucleophilic addition to the arene ring (ring alkylation[7,8,10]) or a carbonyl group of $Cr(CO)_3$ (formation of metal–carbene complexes[11]), depending on the reaction conditions and the nature of the lithium reagents (Scheme 2).

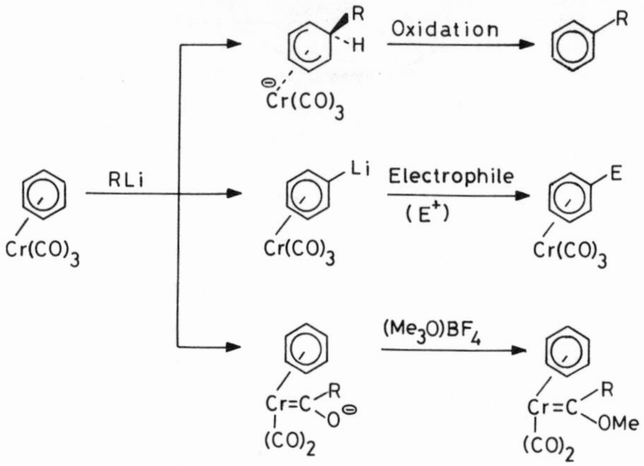

Scheme 2.   Reaction path of (arene)Cr(CO)$_3$ with RLi.

It is, therefore, of interest to develop a more effective route to (η^6-lithio-arene)Cr(CO)$_3$ complexes in order to extend the chemistry of arene chromium complexes. Generally, the reaction with n-BuLi at low temperature in the presence of N,N,N',N',-tetramethylethylenediamine (TMEDA) predominantly results in the directed lithiation of the arene ring.[12] For example, (benzene)Cr(CO)$_3$ gives (phenyllithium)Cr(CO)$_3$ (n-BuLi, 78°C), which can be substituted by quenching with various electrophiles. Similarly, the chromium complexes of anisole, fluoro- and chlorobenzene are lithiated at the *ortho* position of the hetero-atoms.[12,13] These results are particularly interesting, because benzene and chlorobenzene without Cr(CO)$_3$ complexation could not be directly lithiated at the ring.

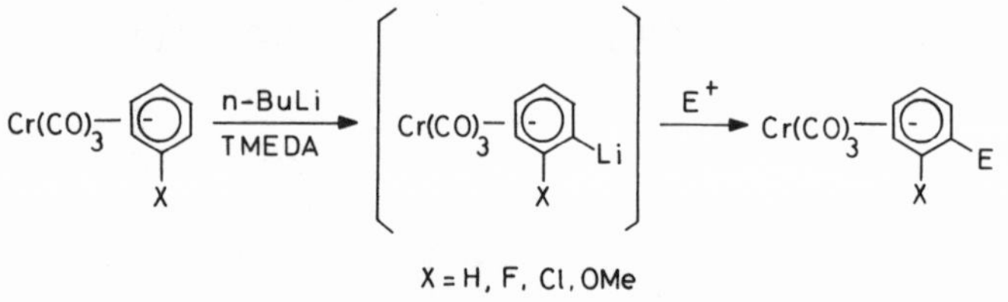

X = H, F, Cl, OMe

Scheme 3

Since the (arene)Cr(CO)$_3$ complexes easily undergo regioselective nucleophilic substitution via the addition of stabilized carbanions, *ortho* lithiation reaction of (arene)Cr(CO)$_3$, coupled with an intra- or intermolecular nucleophilic addition, represents some interesting synthetic schemes as follows[12a,14] (Scheme 4).

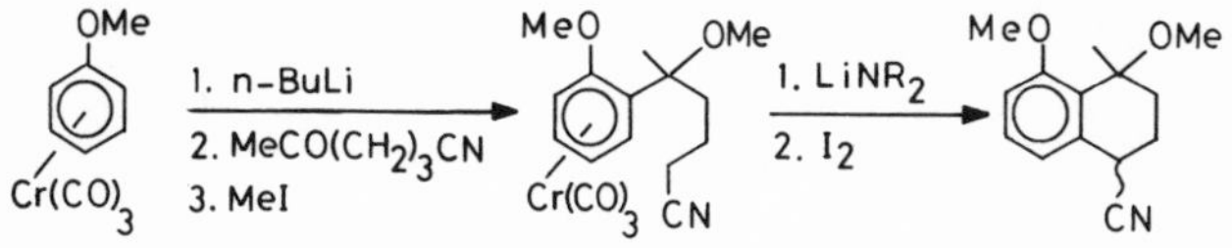

Scheme 4

Another attractive feature of such lithiations is that they are often highly selective with a different regioselectivity from that exhibited by chromium free arenes. The reaction of (*N,N*-dimethylaniline) Cr(CO)$_3$ (**6**) with n-BuLi gave a regioisomeric mixture of *N,N*-dimethyl toluidine (*ortho:meta:para* = 31:50:19) after quenching with MeI, followed by oxidative demetallation.[15] Since *N,N*-dimethylaniline itself was lithiated at the *ortho* position[16] via an intramolecular co-ordination of lithium to the nitrogen atom, the complexation by the Cr(CO)$_3$ group cleanly facilitated *meta* lithiation compared to the uncomplexed arene.

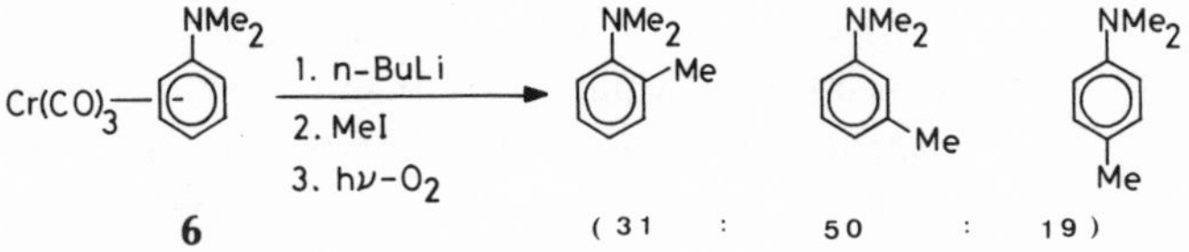

Scheme 5

In view of the fact that the introduction of various substituents at the *meta* position of an electron-donating group is difficult by usual aromatic substitution reactions, it is worthwhile developing the above *meta* lithiation of arene chromium complexes. The diversion of the lithiation site from *ortho* to *meta* could be realized by using sterically bulky protecting groups such as silyl derivatives developed by Widdowson's[17] and Oishi's[18] groups. Triisopropylsilyl and t-butyl-dimethylsilyl groups are very effective not only for the blocking of the *ortho* position but also for directing *meta* selectivity of chromium complexes of phenol and aniline (Scheme 6). Thus, the treatment of complex **7** with t-BuLi in THF followed by quenching with MeI gave *meta* and *para* methyl derivatives **8** and **9** in a ratio of 10:1 in 78% yield, accompanied with 3% yield of **10** derived from the *ortho* lithiated intermediate.[17a] Also, (*N*-t-butyl-dimethylsilyl-*N*-methylaniline)Cr(CO)$_3$ (**11**) afforded predominantly *meta*-lithiated product **12**.

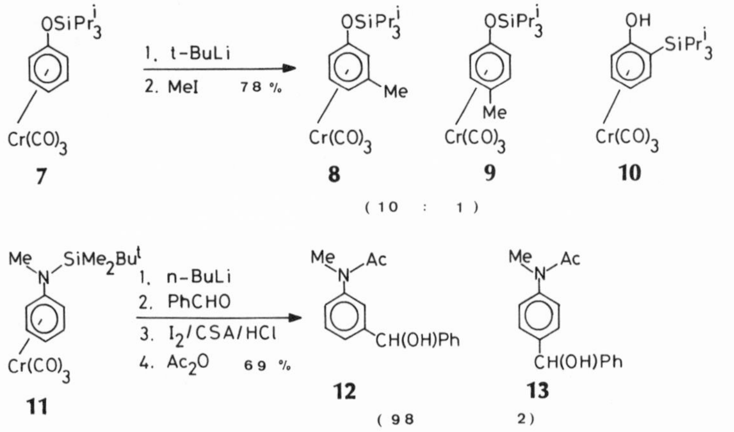

Scheme 6

The normally dominant inductive labilization of an *ortho* proton would be effectively negated without co-ordination of the lithium to hetero atoms owing to the steric bulkiness of protecting groups. Although the origin of *meta* lithiation is not known with certainty, the conformation of the Cr(CO)$_3$ group to the arene ring is an important factor. X-ray crystallography of the complex **11** shows a *syn*-eclipsed conformation **14** (Figure 2). For electron-donating groups such as alkyl or OMe, the *syn*-eclipsed conformer is normally found in the solid state, while for electron-withdrawing substituents (e.g. X = CO$_2$Me), the *anti*-eclipsed structure **15** is observed.[19] The preference for the non-sterically favored *syn*-eclipsed conformation in η^6-alkylbenzene chromium complexes has been attributed in simple terms to the greater

Figure 2. Syn- (**14**) and anti-eclipsed (**15**) conformers of tricarbonyl (η^6-arene) chromium complexes. In **14**, X = R or OMe, for example; in **15**, X = CO$_2$Me, for example.

π-electron density at the *ortho* and *para* positions to the alkyl groups.[20] The electronic factors behind these conformational preferences have been examined using extended Hückel-type MO calculations.[21] From ^{1}H-NMR data,[19b] conformer preferences in solution at low temperature follow a similar pattern to those established by X-ray crystallography.

In chromium complexes **7** and **11**, *meta* lithiation is assumed to proceed through the abstraction of a hydrogen eclipsed to the carbonyl ligand via an intramolecular co-ordination of the lithium to a carbonyl oxygen atom of Cr(CO)$_3$. An extension of current theories has also suggested[22] that an electron deficiency at the eclipsed carbon atom promotes either nucleophilic attack or lithiation, depending on the reagents employed. Similar regioselective lithiation reactions were found in the chromium complexes of *N*-protected indole derivatives.[23,24]

A. Lithiation of Tricarbonyl(3-methoxy benzylalcohol)chromium and Related Chromium Complexes

We have been interested in the regiochemistry of the nuclear lithiation in aromatic compounds where two *ortho* directing groups are located at the 1,3-positions. In such compounds, lithiation occurs predominantly or exclusively at the 2-position,[25] even if the *ortho* directing ability of these substituents is not so strong. For example, the lithiation[26] of 3-methoxy benzylalcohol (**16**) occurred at the 2-position with high selectivity via an intramolecular co-ordination intermediate **21**. The products were γ-lactone derivative **17** and hydroxy ester **18** in a ratio of 9:1 after quenching with carbon dioxide and treatment with diazomethane. Similarly, 7-methoxy-1-tetralol (**19**) with a rigid benzylic hydroxyl group afforded exclusively γ-lactone **20** under the same reaction conditions.

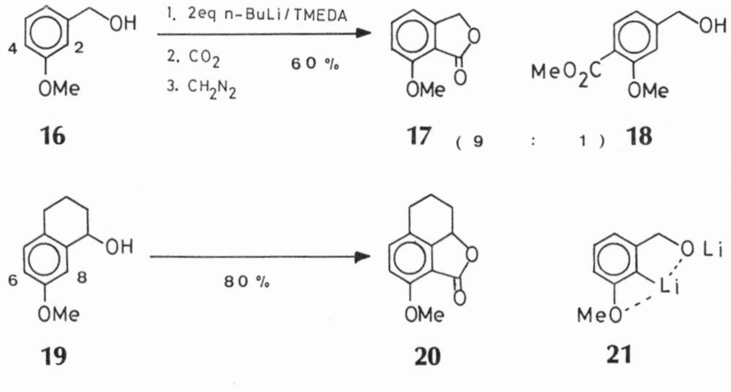

Scheme 7

On the other hand, (3-methoxy benzylalcohol)Cr(CO)$_3$ (22) was found to be lithiated with a different regioselectivity under mild conditions.[27] The complex 22 was treated with 2 equivalents of n-BuLi and TMEDA in THF at $-78°C$ and then reacted with CO_2 to give two products 17 and 18 in a ratio of 23:77 after demetallation and methylation with CH_2N_2 (Scheme 8).

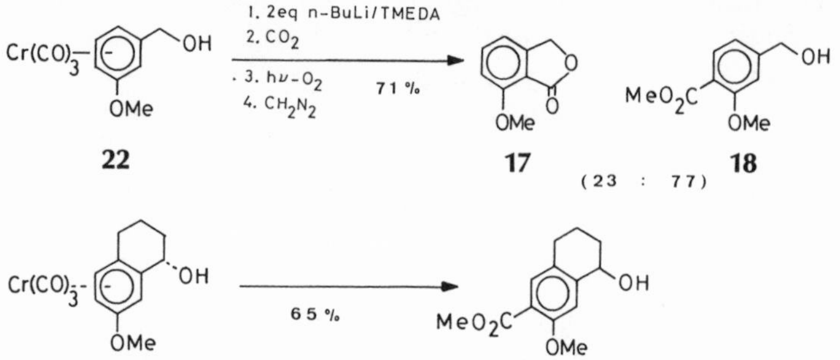

Scheme 8

The proportion of C-4 lithiated product of tricarbonyl(3-oxygenated benzylalcohol)chromium complexes increased with increasing steric bulkiness of the butyllithium reagents[28] (Table 1). However, the chromium complex 23 with a methoxymethyl ether group gave C-4 lithiated product exclusively even with n-BuLi, presumubly due to additional chelating ability

Table 1.   Lithiation of (3-Oxygenated benzylalcohol)Cr(CO)$_3$ Complexes

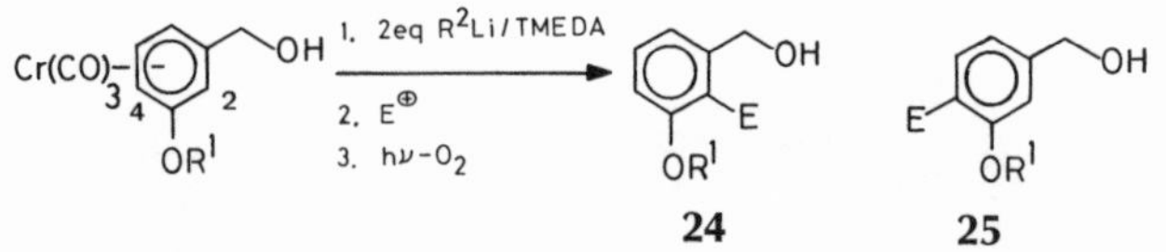

Entry	Complex	R^1	R^2	E$^+$	Ratio of 24:25	Yield (%)
1	22	Me	n-Bu	CO$_2$[a]	23:77	71
2	22	Me	s-Bu	CO$_2$[a]	15:85	55
3	22	Me	t-Bu	CO$_2$[a]	5:95	48
4	22	Me	t-Bu	Me$_2$SiCl	6:94	50
5	23	CH$_2$OMe	n-Bu	CO$_2$[a]	2:98	45
6	23	CH$_2$OMe	n-Bu	Me$_2$SiCl	2:98	67

[a] Isolated as methyl ester and γ-lactone after treatment with CH_2N_2.

of the methoxymethyl substituent. Surprisingly, (benzylalcohol)Cr(CO)$_3$ without a methoxyl on the aromatic ring was reacted with organolithium reagents in the presence or absence of TMEDA to give (cyclohexadiene)chromium complexes via a chelation-controlled nucleophilic addition of carbanions instead of a lithiation reaction.[29]

Since the effect of chromium co-ordination on the regiochemistry is expected to be manifested more clearly by conformational fixation of the benzylic hydroxyl group, we next studied the analogous lithiation reactions[27] with the chromium complexes of 7-methoxy-1-tetralol and its derivatives. Indeed, *endo*-(7-methoxy-1-tetralol)Cr(CO)$_3$ (**26**) gave exclusively 7-methoxy-6-methoxycarbonyl-1-tetralol (**27**) without C-8 lithiated product under the same reaction sequence (Scheme 8), in contrast to the result with Cr(CO)$_3$ free parent arene. Therefore, these two types of lithiation reactions (with or without Cr(CO)$_3$ complexation) supplement each other and provide useful tools in the organic synthesis of arene componds. Similarly, (2-substituted-7-methoxy-1-tetralol)Cr(CO)$_3$ afforded a single product lithiated at the 6-position in high yield (Table 2). The diastereoisomeric chromium complex with an *exo*-hydroxyl[30] still gave predominantly C-6 lithiated product (entry 10).

From X-ray crystallography,[28a] two carbonyl ligands are found to be located in proximity to the C-6 and C-8 positions of the arene ring (Figure 3). Lithiation of this type of chromium complex is presumably initiated by the

Table 2.   Lithiation of (7-Methoxytetralol)Cr(CO)$_3$

Entry	Complex	R^1	R^2	E$^+$	Yield (%)
1	**26**	H	H	CO$_2$[a]	65
2	**26**	H	H	Me$_3$SiCl	96
3	**26**	H	H	p-MeO-C$_6$H$_4$CHO	90
4	**26**	H	H	o-C$_6$H$_4$(CO$_2$Me)$_2$	91
5	**26**	H	H	MeCH=C(Me)COCl	82
6	**28**	Me	H	CO$_2$[a]	63
7	**28**	Me	H	DMF	90
8	**29**	Me	Me	DMF	92
9	**30**	CH(OH)Me	H	CO$_2$[a]	55
10	**31**[b]	H	H	CO$_2$[a]	52[c]

[a]Isolated as methyl ester. [b]Exo-hydroxyl complex. [c]Ratio at C-6 and C-8 is 86:14.

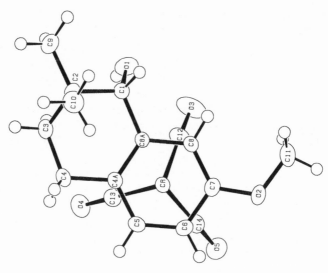

Figure 3.

co-ordination of lithium with both the methoxyl oxygen at C-7 and the oxygen atom of the carbonyl ligand, followed by abstraction of hydrogen at either C-6 or C-8 position. The C-8 position, however, is less susceptible to proton abstraction due to steric hindrance and/or electrostatic repulsion with the benzylic alkoxide.

Preparation of tricarbonyl(3-methoxy benzylalcohol)chromium (**22**). A mixture of 3-methoxy benzylalcohol (2.82 g, 20 mmol) and $Cr(CO)_6$ (3.3 g, 15 mmol) in butyl ether (150 mL), heptane (15 mL) and THF (15 mL) was refluxed in 300 mL flask equipped with long, wide Liebig condenser for 30 h under nitrogen atmosphere. After filtration and evaporation *in vacuo*, a crude product was purified by SiO_2 chromatography with ether/petroleum ether (1:4). Recrystalization with ether/hexane gave a complex **22** (3.5 g, 80%) as yellow crystals.

Lithiation of tricarbonyl(3-methoxy benzylalcohol)chromium (**22**). To a solution of **22** (274 mg, 1 mmol) and TMEDA (278 mg, 2.4 mmol) in dry ether (15 mL) at − 78°C was added 1.6 mL of *n*-BuLi (1.5 M in hexane, 2.4 mmol) under argon. After stirring for 4 h, the reaction mixture was poured into dry-ice in ether. After being allowed to stand overnight, the mixture was acidified with dilute HCl and extracted with ether. The ether solution was exposed to sunlight until the red solution changed to colorless. The precipitate was filtered off and then ether solution was treated with diazomethane. After evaporation of the solvent under reduced pressure, the crude product was purified by chromatography.

B. Lithiation of (η^6-Arene)chromium Complexes without Free Benzylic Hydroxyl group

As benzylic alcohol is a poor *ortho* directing group for the directed lithiation,[31] we next studied the lithiation reactions of the chromium complexes possessing a benzylic ether linkage. The chromium free arene compounds of this type, such as alkylbenzyl ethers and acetals of benzaldehydes, are generally not feasible for *ortho* lithiation because of the well known propensity for deprotonation at the benzylic position, followed by Wittig rearrangement[32] or acetal ring cleavage.[33] However, the increased acidity of the ring hydrogens by the chromium complexation allowed successful ring lithiation as follows. Tricarbonyl (3-methoxy benzylmethyl ether)chromium (**32**) was treated with 1 equivalent of n-BuLi to give C-2 lithiated product **33** and 2,4-bis-lithiated product **34** in a 83:15 ratio after quenching with methyl chloroformate and demetalation[28] (Scheme 9). The regioselectivity of this lithiation on complex **32** contrasts with those obtained with complex **22** which bears a free benzylic hydroxyl group as mentioned above.

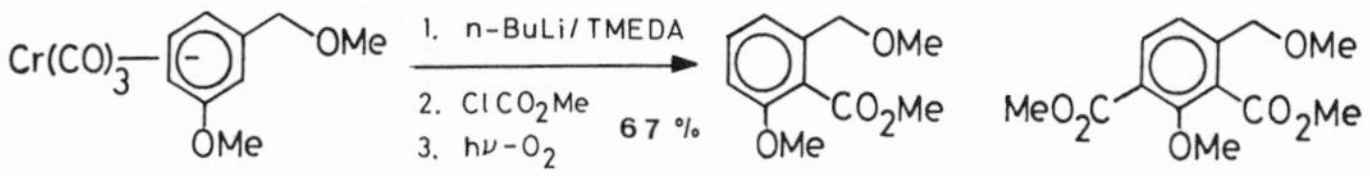

32 **33** (83 : 15) **34**

Scheme 9.

Similarly, (3-methoxybenzaldehyde ethylene acetal)$Cr(CO)_3$ (**35**) was lithiated predominantly at the 2-position with high selectivity, and the chromium complexes of other acetals were also lithiated at the position flanked by the

Table 3.

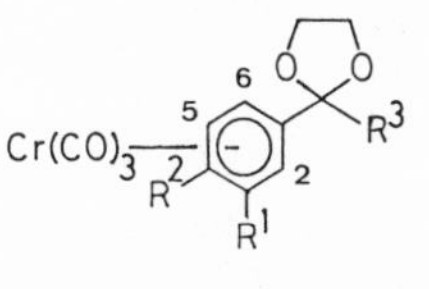

35 R^1=OMe R^2=R^3=H
36 R^1=R^2=OMe R^3=H
37 R^1, R^2= OCH_2O R^3=H
38 R^1=OMe R^2=H R^3= Me

Entry	Complex	Electrophile	Lithiated positions (ratio)	Yield (%)
1	**35**	$ClCO_2Me$	2/2,4 (93:7)	84
2	**35**	Me_3SiCl	2/2,4 (70:25)	90
3	**36**	$ClCO_2Me$	2/2,5 (80:15)	65
4	**37**	$ClCO_2Me$	2/2,5 (81:13)	92
5	**38**	$ClCO_2Me$	2/2,4 (59:38)	58

methoxyl and acetal groups (Table 3). Since the above mentioned electrostatic repulsion between the benzylic alkoxide anion and butyllithium in complex **22** is absent in these types of the chromium complexes, the directed lithiation is observed via a favorable co-ordination of the lithium with the proximal oxygen atoms of the two ether linkage.

However, this chelation effect is not a significant property of the lithiation reactions of the chromium complexes of cyclic compounds with a rigid benzylic ether. For example, (1-*endo*-methoxy-7-methoxytetralin) Cr(CO)₃ **(39)** gave no nuclear lithiated product under the usual conditions. The corresponding *exo*-complex[30] **40** afforded a single lithiated product **41** at the C-6 position, distinguished from the C-8 position via co-ordination effect. The failure of directed lithiation of the *endo*-methoxyl complex **39** could be attributed to deprotonation of an *exo*-hydrogen at the C-1 position because of the steric effect and enhancement of acidity[32b] of the benzylic hydrogen. However, even an *endo*-methoxyl complex, 2,2-dimethyl chromium complex **42**, was lithiated exclusively at the 6-position. Thus, we conclude that the steric bulkiness of the rigid benzylic methoxyl group in the cyclic compounds can override the co-ordination effect.

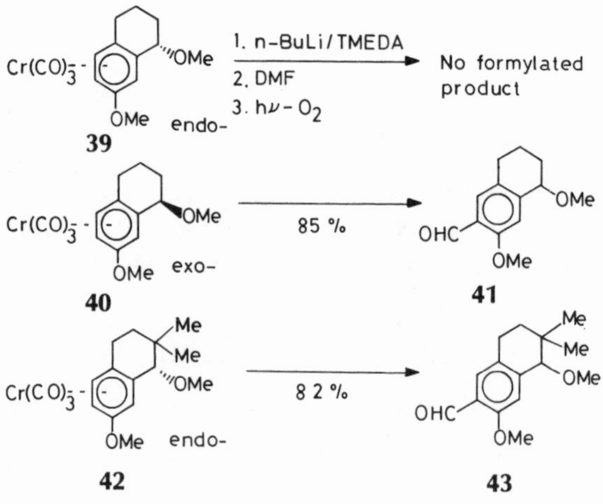

Scheme 10

The presence or absence of the benzylic oxygen atom in the tetralin and related cyclic compounds does not affect the regiochemistry of the directed lithiation of the chromium complexes as follows. The lithiation of chromium complexes **44** and **46**, followed by treatment with DMF, afforded single formylated complexes **45** and **47** at the less hindered *ortho* position of the methoxyl group[34] (Scheme 11). Similarly, the complex of protected 17β-estradiol was lithiated exclusively at the 2-position.[35]

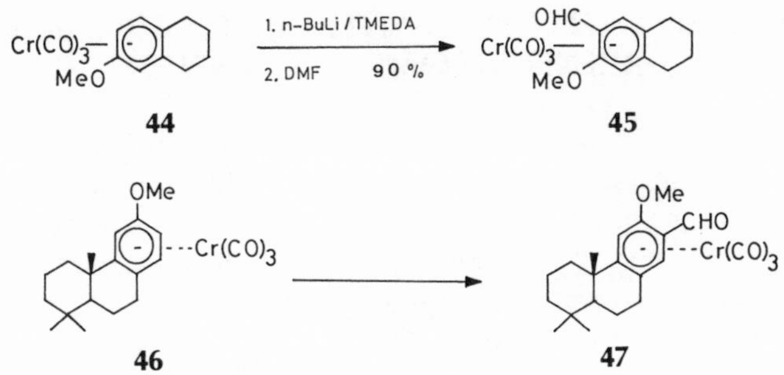

Scheme 11

In conclusion, the chromium complexes of methoxytetralines and their analogues are regioselectively lithiated at the less hindered *ortho* position.

C. Lithiation of (η^6-Arene)chromium Complexes Directed By Amide and Fluorine

Amides are one of the strongest *ortho* directing groups in the directed lithiation of the aromatic compounds.[25] The chromium complexes of *m*-methoxy-*N*-t-butylbenzamides **48** were lithiated at the *ortho* positions to the amide groups with little regioselectivity to give **49** and **50**. However, directed lithiation of the corresponding chromium free arenes gave only **49** but not **50**.

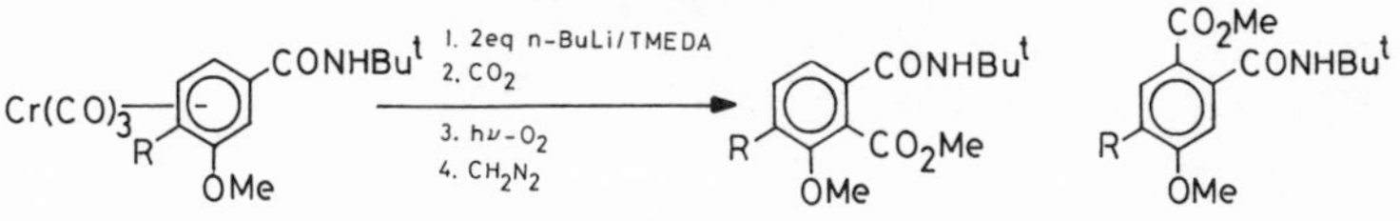

Scheme 12

The chromium complexes of fluorobenzene and anisole are metallated at the *ortho* position to the F and OMe groups, respectively. Methoxyl is more powerfully directing than fluorine on the lithiation reaction of chromium-free arenes. However, Widdowson *et al.* reported[14,36,37] interesting results in that the chromium complexes of fluoroanisoles, **51, 52** and **53**, were lithiated at the *ortho* position to the fluorine atom without any isomeric products.

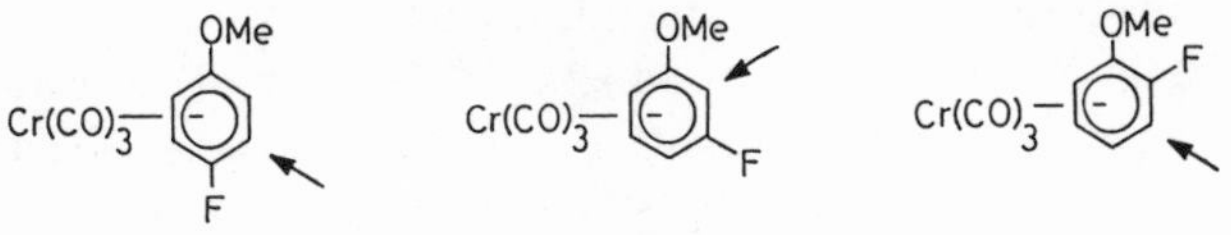

III. DIASTEREOSELECTIVE SYNTHESIS OF TRICARBONYL-(η_6-ARENE)CHROMIUM COMPLEXES

It is well documented that two possible diastereomeric chromium complexes, *exo*- and *endo*-isomers, are obtained as a mixture by direct complexation of substituted tetralin and indane derivatives with $Cr(CO)_6$ (Scheme 13), and the ratio of the diastereomers depends on the steric effect, the electronic nature of the substituents and reaction conditions. Similarly, direct complexation of *ortho*- and *meta*-substituted aromatic compounds possessing a chiral center at the side chain affords a diastereomeric mixture of chromium complexes. Stereoselective preparation of each diastereomeric chromium complex is important in organic synthesis, but there has been little previous investigation concerning this interesting problem. In this chapter, we discuss stereoselective synthesis of these types of complexes utilizing $Cr(CO)_3$-stabilized carbonium ions.

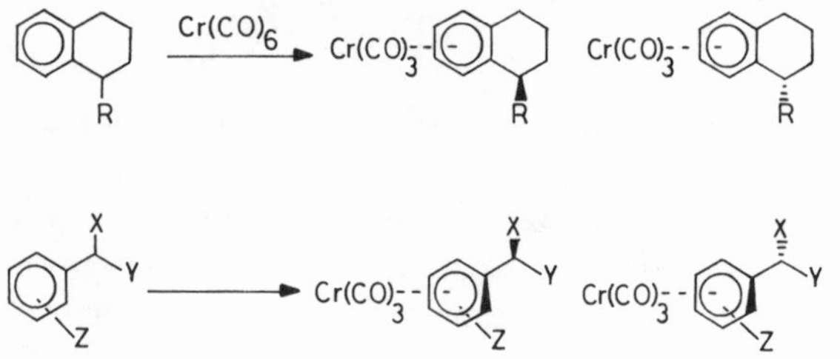

Scheme 13

A. Cyclic Series

The arene–$Cr(CO)_3$ unit can perform a dual purpose in stabilization of both carbanions and carbonium ions at the benzylic position (Figure 1). Of the two contrasting properties of $Cr(CO)_3$, i.e. acceptor and donor, with respect to an aromatic side chain, the inductive acceptor effect has been more widely explored with regard to its applications in organic synthesis. In particular, temporary complexation of the arene compounds favors proton abstraction from the side chain in basic media. Of a number of reviews[1] dealing with $Cr(CO)_3$-stabilized carbanions in organic synthesis, only few examples are illustrated here. The complexes **54** and **55** could be alkylated stereo- and regioselectively in this manner[38,39,40] (Scheme 14). The ability of the arene–$Cr(CO)_3$ system to stabilize such benzylic carbanions also leads to conjugate addition of nucleophiles to (η^6-styrene)$Cr(CO)_3$, *vide infra* (Scheme 46).

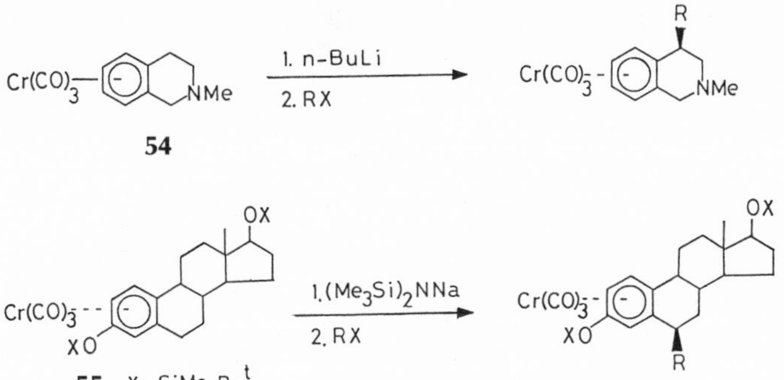

Scheme 14

In contrast to the above mentioned electron-withdrawing character of the Cr(CO)$_3$ group, it is well established that it can also act as a strong electron-donor for stabilizing carbonium ions at the benzylic position. This ability was first suspected from the rapid S$_N$1 solvolysis of (benzylchloride)Cr(CO)$_3$ complex.[41] There has been considerable debate concerning the mechanism by which the Cr(CO)$_3$ group stabilizes such carbonium ions. This extraordinary donating effect of the Cr(CO)$_3$ group has been interpreted as either a direct interaction between a filled chromium d orbital and a empty carbon p orbital, or as metal-ring σ–π delocalization in which the α-carbonium ion is established by π-conjugation.[42] However, the synthetic utility of Cr(CO)$_3$-stabilized α-carbonium ions has not been explored.

Jaouen *et al.* have reported that an α-carbonium ion **57** generated, *in situ*, from the *endo*-indanol complex **56** under adequate conditions could be trapped with O-, N- and S-nucleophiles such as H$_2$O, ROH, RNH$_2$ RCN and RSH to give *exo*-substituted chromium complexes with stereochemical inversion.[32b,43]

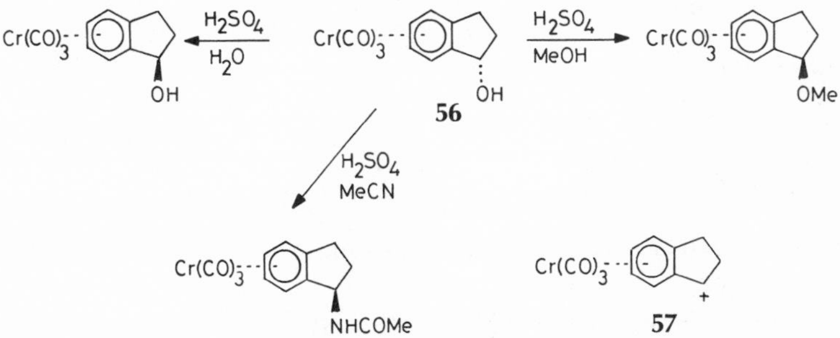

Scheme 15

Recently, Reetz reported the same α-carbonium ion could be trapped with enolsilyl ether as a carbon nucleophile to form a carbon–carbon coupling product[44] (Scheme 16).

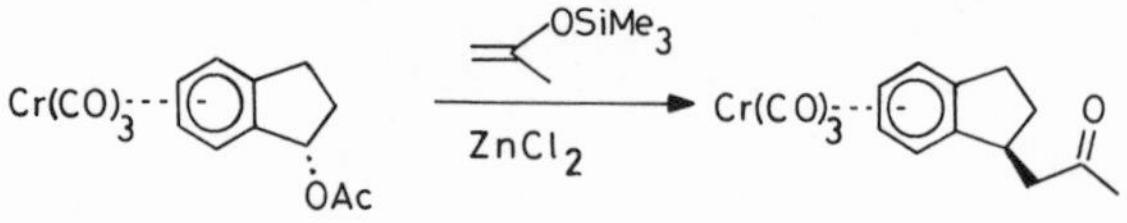

Scheme 16

We[45] have explored this interesting synthetic utility of the $Cr(CO)_3$-stabilized α-carbonium ions, particularly with respect to their function as benzyl electrophiles in carbon–carbon bond forming reactions, in order to synthesize stereoselectively both *exo*- and *endo*-diastereomeric products from a common intermediate. For example, (1-*exo*-methyl-1-*endo*-tetralol)-$Cr(CO)_3$ (59) obtained from (α-tetralone)$Cr(CO)_3$ (58) and MeLi, afforded (1-*endo*-methyltetralin)$Cr(CO)_3$ (60) by ionic hydrogenolysis[46] with an excess of triethylsilane and trifluoroacetic acid via a stereoselective *exo*-hydride displacement.[47] On the other hand, an *endo*-acetate complex 61, prepared by hydride reduction of 58 and subsequent acetylation, was converted into (1-*exo*-methyltetralin)$Cr(CO)_3$ (62) by treatment with trimethylaluminum (Scheme 17). The reaction results of some $Cr(CO)_3$-stabilized carbonium ions with carbon nucleophiles are summarized in Table 4.

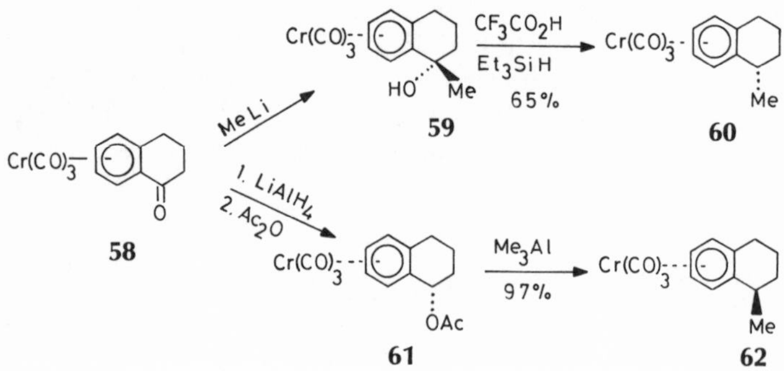

Scheme 17

Both *endo*- and *exo*-isomers 60 and 62 could be easily distinguished in the ^{1}H-NMR spectra. Free rotation of the $Cr(CO)_3$ residue should be more restricted in the *endo*-methyl complex 60 than the *exo*-methyl complex 62, because of an adverse interaction between the *endo*-substituent and $Cr(CO)_3$ groups. Since arene protons superimposed by the metal carbonyl bonds would be more deshielded than the alternate exposed protons,[48] the spread of

Table 4. Reaction results of Cr(CO)$_3$-stabilized carbonium ions with carbon nucleophiles.

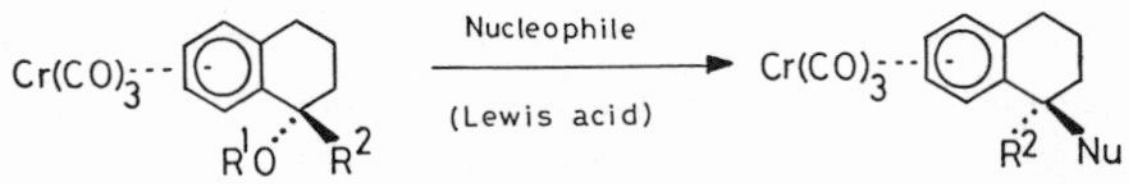

61 R^1 = Ac, R^2 = H   **62** R^2 = H, Nu = Me   **64** R^2 = H, Nu = Et
63 R^1 = R^2 = H   **65** R^2 = Nu = Me
59 R^1 = H, R^2 = Me   **66** R^2 = H, Nu = CH$_2$CH = CH$_2$
                  67 R^2 = H, Nu = CH = CH = CH$_2$

Entry	Complex	Nucleophile	Lewis acid	Product	Yield (%)
1	**61**	Me$_3$Al	none	**62**	99
2	**61**	Et$_3$Al	none	**64**	60
3	**63**	Me$_3$Al	TiCl$_4$	**62**	60
4	**59**	Me$_3$Al	TiCl$_4$	**65**	92
5	**61**	Et$_2$Zn	TiCl$_4$	**64**	62
6	**63**	Et$_2$Zn	TiCl$_4$	**64**	60
7	**61**	CH$_2$ = CHCH$_2$SiMe$_3$	TiCl$_4$	**66**	95
8	**61**	CH≡CCH$_2$SiMe$_3$	BF$_3$·OEt$_2$	**67**	76

δ values for the aromatic protons of the *endo*-isomer **60** (Δ 0.42 ppm) was greater than that of the *exo*-isomer **62** (Δ 0.15 ppm), reflecting the more restricted free rotation. The methyl signal (δ 1.40) of the *endo*-isomer **60** also appeared at a lower field than that (δ 1.30) of the *exo*-isomer **62** due to a similar anisotropic effect.

Although a free benzylic hydroxyl group of (η^6-arene)Cr(CO)$_3$ complexes could not be replaced by trialkylaluminum alone, the addition of TiCl$_4$ (which presumably forms alkyltitanium species[49]) gave smoothly *exo*-alkyl complexes (see Table 4). In the reaction with Et$_2$Zn, both the alcohol and the corresponding acetate complexes afforded *exo*-ethyl chromium complex only in the presence of TiCl$_4$ (entries 5 and 6). Therefore, two different alkyl groups can be introduced stepwise at the different benzylic positions depending on the reaction conditions and structure of the substrates. Thus, a benzylic acetoxyl group of a complex **68** was initially replaced by Et$_3$Al alone to give ethylated complex **69**, in which a tertiary benzylic alcohol was subsequently substituted with Me$_3$Al in the presence of TiCl$_4$ to afford a complex **70**.

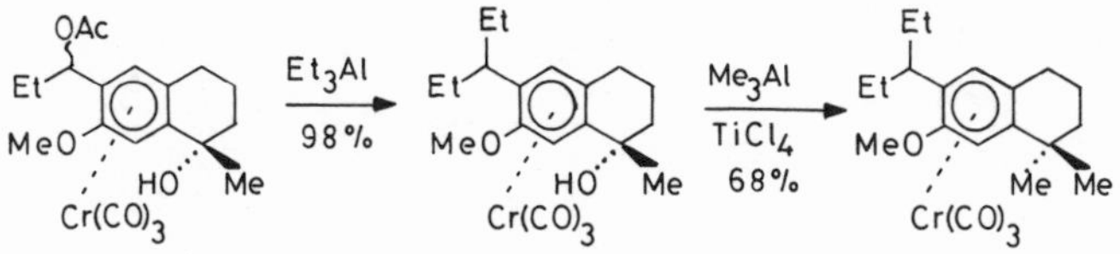

      68   **69**   **70**

Scheme 18

As other carbon nucleophiles, allyl- and propargyl trimethylsilanes[50] can be used for substitution of $Cr(CO)_3$-stabilized carbonium ions generated by the reaction with Lewis acids (Table 4). Among various Lewis acids, titanium tetrachloride and boron trifluoride etherate are most effective for these reactions with allyl- and propargyl trimethyl silanes. An important feature of these carbon–carbon bond forming reactions is the wide flexibility for the preparation of a variety of tricarbonyl(η^6-arene)chromium complexes. In particular, it is a very efficient method for the construction of quarternary centers without competitive elimination reaction. The stereochemistry of the quarternary carbon at the benzylic position depends on the order of introduction of two different nucleophiles. Complex **59** gave tricarbonyl(1-*exo*-allyl-1-*endo*-methyltetralin)chromium (**71**) by reaction with allyl trimethylsilane in the presence of $TiCl_4$. On the other hand, isomeric (1-*exo*-methyl-1-*endo*-allyltetralin)Cr(CO)$_3$ (**73**) was synthesized from **58** by the reaction with allylmagnesium chloride and subsequently with Me_3Al in the presence of $TiCl_4$.

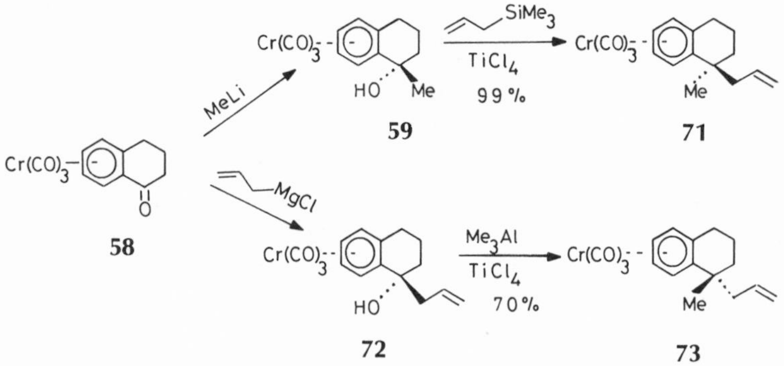

Scheme 19

Two groups reported[51,52] that the arene group of $Cr(CO)_3$-complexed *t*-benzylalcohols was exchanged with the electron-richer arenes in the presence of 2-methyl-1,3-cyclopentanedione and Triton B without carbon–carbon bond formation (Scheme 20). Although the mechanism of this ligand exchange reaction is uncertain, a bimolecular process or liberation of an active $Cr(CO)_3$ fragment via $Cr(CO)_3$-stabilized carbonium ion has been proposed.

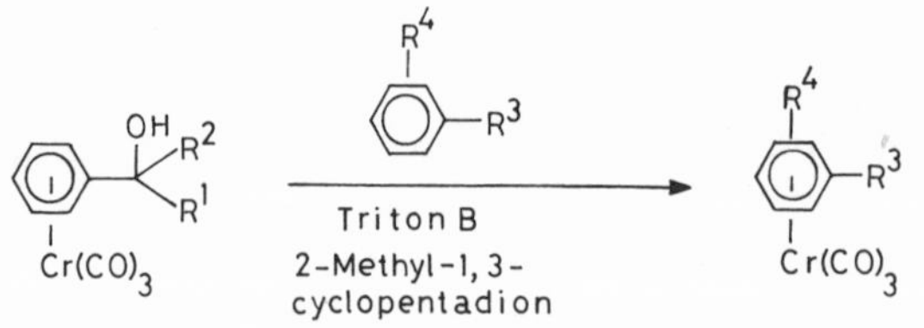

Scheme 20

We found[53] that such a carbonium ion generated, *in situ*, under acidic conditions instead of with Triton B as described above could be reacted with the electron rich aromatic or β-dicarbonyl compounds to afford carbon–carbon bond forming products. That is, simple addition of $HBF_4 \cdot OMe_2$ to a solution of *endo*-tetralol complex **63** and anisole in CH_2Cl_2 at 0°C gave an unseparable mixture of *ortho*- and *para*-methoxyphenyl compound **74** (Scheme 21). 1,2-Dimethoxybenzene and 1,2,3-trimethoxybenzene afforded single coupling products in good yields, respectively. Similarly, β-dicarbonyl compounds such as acetyl acetone and methyl acetoacetate were readily reacted with $Cr(CO)_3$-complexed carbonium ions to give the corresponding chromium complexes substituted at the benzylic position. While the primary and secondary benzylic hydroxyl groups can be easily replaced by electron rich arenes and β-dicarbonyl compounds, tertiary alcohol complexes gave only dehydrated complexes without the formation of coupling products under the same conditions.

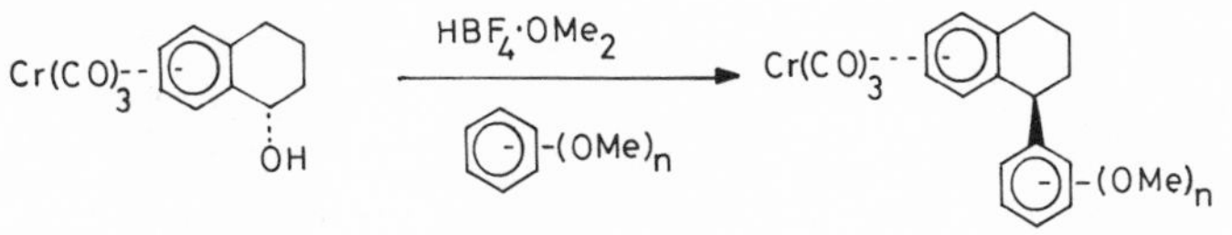

Scheme 21

*Preparation of tricarbonyl(1-*endo*-methyltetralin)chromium* (**60**). To a solution of the complex **59** (200 mg, 0.67 mmol) and triethylsilane (214 μL, 1.34 mmol) in dry CH_2Cl_2 (1.0 mL) was added trifluoroacetic acid (155 μL, 2.01 mmol) at room temperature under argon. The mixture was stirred at 30–40°C for 4 h, and quenched with water. The reaction mixture was extracted with CH_2Cl_2, and the combined organic layer was washed with saturated aqueous $NaHCO_3$ and brine, and dried over $MgSO_4$. Usual work up and purification by SiO_2 chromatography (ether/petroleum ether 1/3) afforded 122 mg of **60** (65%).

*Preparation of tricarbonyl(1-*exo*-methyltetralin)chromium* (**62**). To a solution of tricarbonyl(1-*endo*-acetoxytetralin)chromium (100 mg, 0.307 mmol) in dry CH_2Cl_2 (5 mL) was added Me_3Al (1.4 mL, 1.0 M in hexane, 1.4 mmol) at −78°C under argon. The mixture was stirred for 30 min at −78°C, and warmed to 0°C for 2 h. After addition of dilute HCl, the mixture was worked up as usual and purified to give *exo*-methyl complex **62** (82 mg, 97%).

B.   Stereoselective Synthesis of Chromium Complexes of *Ortho*- and *Meta*- Substituted Benzylalcohol Derivatives

As shown in Scheme 13, it is a further interesting problem with little precedent that each diastereomeric chromium complex of *o*- or *m*-substituted arenes having a chiral center on the side chain is stereoselectively synthesized. In particular, when chromium complexes possess an electron withdrawing group such as a nitrile on the side chain, an intramolecular nucleophilic addition of the stabilized carbanion to the arene ring can produce substituted fused- or spiro cyclic compounds stereoselectively. Since the benzylic acetoxyl and hydroxyl groups co-ordinated with $Cr(CO)_3$ could be easily substituted with a variety of carbon nucleophiles, it is necessary to synthesize stereoselectively each diastereomeric complex of benzylalcohol derivatives in the acyclic series for the achievement of this project.

The synthesis of each diastereomeric chromium complex of o-methoxy secondary benzylalcohol is achieved by the modification of the literature method[54] (Scheme 22). Reaction of (o-methoxybenzaldehyde)Cr(CO)$_3$ (75) with alkyllithium gave exclusively (S*,S*)-chromium[55] complex 76. On the other hand, (o-methoxyphenylalkyl ketone)chromium complexes 78 afforded predominantly diastereoisomeric (S*,R*)-complexes 77 by LiAlH$_4$ reduction. The extremely high diastereoselectivity of these reactions should result from the stereoselective exo addition of the nucleophile to the carbonyl group of the complexes 75 and 78, in which carbonyl oxygen is placed far from the o-methoxyl group because of electronic repulsion. However, this high stereoselectivity for nucleophilic addition to the carbonyl group is limited only in the chromium complexes having electron-donating methoxyl or methyl group at the ortho position to the carbonyl group.[56] Therefore, it becomes necessary to develop another useful method for the preparation of these type of diastereomeric chromium complexes.

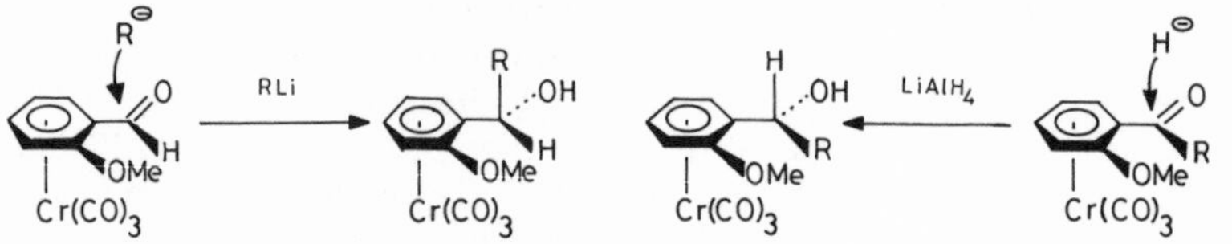

75	76(S*S*)	77(S*R*)	78
	a R = Me	b R = CHMe$_2$	

Scheme 22

Table 5.

Complex	Nucleophile	Ratio of 76:77	Yield (%)
75	MeLi	98:2	66
75	Me$_2$CHLi	99:1	41
78a	LiAlH$_4$	2:98	74
78b	LiAlH$_4$	1:99	84

We next attempted diastereoselective chromium complexation of the chiral arenes. Direct complexation[57] of 1-(*o*-methoxyphenyl)-ethylalcohol (**79**, R = Me) with $Cr(CO)_6$ in butyl ether, heptane and THF (10:1:1) at 130-140°C for 24 h afforded a mixture of (*S*,S**)-isomer **76** (R = Me) and *S*,R**-isomer **77** (R = Me) in a ratio of 70–80:30–20. The α-ethyl compound gave a higher ratio of the corresponding (*S*,S**)-complex. (The complexation by ligand transfer with (naphthalene)$Cr(CO)_3$ resulted in much better selectivity, as described later.) However, the complexation of the α-allyl compound with $Cr(CO)_6$ gave only a 1:1 mixture. This direct complexation to the arene nucleus may result from the same side[58] with the benzylic hydroxyl group via an interaction of chromium to oxygen atom in a conformation **80**, in which the steric hindrance between R and *o*-methoxyl groups is minimized. The alternative conformer **81** which leads to (*S*,R**)-complex has a severe steric interaction. In the case of allyl derivative **79** ($R = CH_2CH = CH_2$), the double bond can also be co-ordinated by chromium to give a 1:1 mixture.

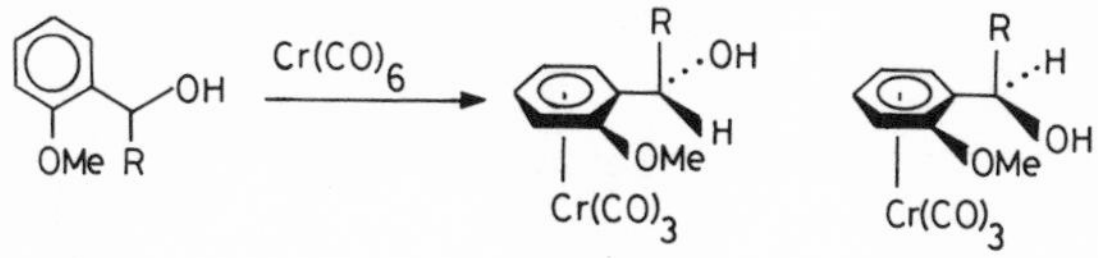

R	Ratio of **76:77**	Yield (%)
Me	78:22	65
Et	92:8	80
$CH_2CH = CH_2$	50:50	56

Scheme 23

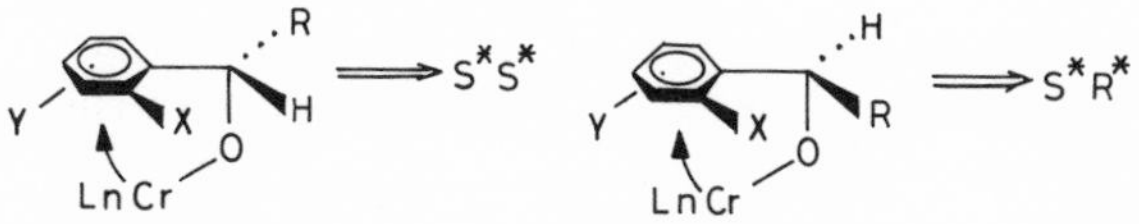

80   **81**

Scheme 24

An extension of this direct complexation can be applicable to stereoselective synthesis[57a] of the diastereomeric chromium complex of *meta* substituted secondary benzylalcohol derivatives, in which a sterically bulky and easily removable trimethylsilyl group is temporarily introduced at either *ortho* position to the hydroxyalkyl group allowing higher selectivity of the complexation reaction. For example, the reaction of 1-(2-trimethylsilyl-3-methoxyphenyl)-ethylalcohol (**82**, R = Me) with 1 equivalent of Cr(CO)$_6$ under the usual thermal conditions, followed by detrimethylsilylation, gave predominantly ($S*,S*$)-complex **83** as expected. On the other hand, 6-trimethylsilyl compound **85** afforded the diastereoisomeric ($S*,R*$)-complex **84** under the same reaction sequence. However, α-isopropyl-2-trimethylsilyl compound **82** (R = i-Pr) gave only a very low yield of the corresponding chromium complex, accompanied by other type of complexes (*vide infra*).

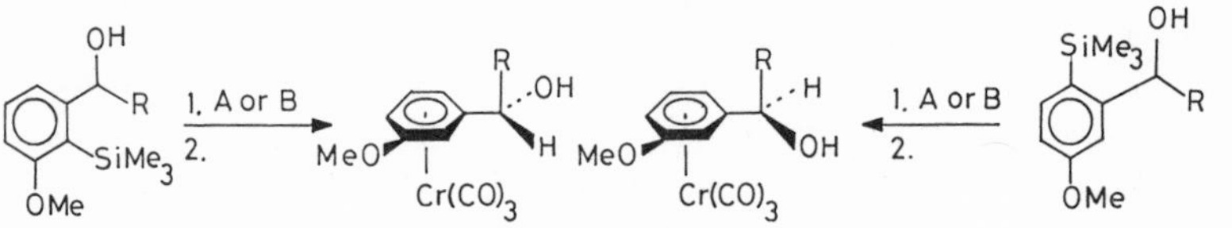

1.(A) 1eq Cr(CO)$_6$, (B) (naphthalene)Cr(CO)$_3$; 2. n-Bu$_4$N$^{\oplus}$F$^{\ominus}$

Scheme 25

Table 6.

Compound	Reagent	Ratio of **83:84**	Yield (%)
82a	A	98:2	69
82b	A	100:0	10
85a	A	5:95	66
86b	A	4:96	77
82a	B	100:0	97
82b	B	100:0	88
85a	B	2:98	82
85b	B	0:100	85

The thermal complexation of the *o*-trimethylsilylated secondary benzylic alcohol derivatives with Cr(CO)$_6$ sometimes gave different types of arene chromium complex in various ratios depending on the reaction conditions. For example, the reaction of **86** with an excess of Cr(CO)$_6$ over a longer reaction time gave unexpected cyclic siloxane chromium complexes **88** and **89** as major products (Table 7). The methyl signal at the benzylic position of the less polar *exo*-complex **89** appeared at a higher field (δ 1.45) than the corresponding signal (δ 1.60) of the more polar *endo*-isomer **88** as discussed in Section III.A. The stereochemistry of complexes **88** and **89** was confirmed by the following transformation. The complex **88** was converted to ($S*,S*$)-complex **83** (R = Me) by treatment with n-Bu$_4$N$^+$ F$^-$. The isomeric complex **89** gave the corresponding ($S*,R*$)-complex **84** (R = Me). Similarly, **82** (R = i-Pr) and **85** (R = Me) afforded the corresponding cyclic chromium complexes as a mixture of *endo*- and *exo*-alkyl substituted products with an excess of $Cr(CO)_6$ under severe reaction conditions. Also, treatment of the complex **87** with $Cr(CO)_6$ for 2 days at 140°C gave the complexes **88** and **89** as a mixture. The formation of cyclic siloxane chromium complexes **88** and **89** under these thermal conditions may result from a reaction path in which the initially formed chromium complex **87** was converted to *endo*-methyl complex **88** by an intramolecular attack of the benzylic alkoxide anion to the silicon atom, and then the *endo*-methyl complex **88** was equilibrated with more stable *exo*-methyl complex **89** under thermal conditions.[59]

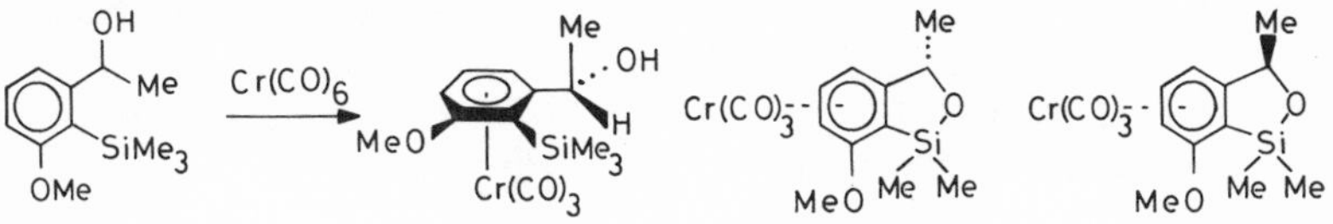

86 **87** **88** **89**

Scheme 26

Table 7.

Entry	Equivalents of Cr(CO)₆	Reaction time (h)	**87:88:89**	Yield (%)
1	1.3	24	78:20:2	85
2	1.7	40	25:67:8	90
3	2.0	50	10:65:25	85
4	3.0	60	5:55:40	80

Since this thermal complexation with $Cr(CO)_6$ reaches an equilibrium between the diastereomers over prolonged reaction times, a kinetic complexation under milder conditions is required to obtain higher diastereoselectivity and a better yield. Ligand transfer from (naphthalene)Cr(CO)₃ to another arene reported by Kündig *et al.*[60] proceeded in ether at low temperature and seemed to be promising for our purpose. The naphthalene chromium bond is labile, and the naphthalene ligand undergoes facile slippage freeing a co-ordination site ($\eta^6 \rightarrow \eta^4$) for an incoming ligand. This exchange reaction is accelerated by Lewis bases and donor solvents (e.g. THF). In fact, the reaction of **82** with (naphthalene)Cr(CO)₃ in ether containing 1 equivalent of THF at 70°C for 4 h in a sealed tube gave exclusively (S^*,S^*)-chromium complex **83** in better yield and with higher selectivity than the thermal complexation with $Cr(CO)_6$ after de-trimethylsilylation (Table 6). The more sterically bulky α-isopropyl derivative **82** (R = i-Pr) gave only **83** (R = i-Pr) without any diastereomeric contamination in higher yield than that with $Cr(CO)_6$. Similarly, this arene exchange reaction of regioisomeric trimethylsilyl compounds **85** gave the other diastereomeric complexes, (*S*,R**)-isomers **84** after desilylation.

Preparation of tricarbonyl(naphthalene)chromium. A mixture of naphthalene (9.2 g,72 mmol) and Cr(CO)$_6$ (8.4 g, 38 mmol) in di-butyl ether (240 mL) and THF (20 mL) was heated with stirring in 500 mL round-bottle flask fitted with a Liebig condenser (2.5 cm diameter, 60 cm length) at 150°C for 20 h under nitrogen atmosphere. After cooling to room temperature, solvents were removed *in vacuo*. The residue was dissolved with ether (300 mL), and the precipitate was removed by filtration. The red ether solution was evaporated *in vacuo*, and the crystalline residue was subjected to sublimation of naphthalene (70°C, 5 × 10^{-2} mmHg) to leave red (naphthalene)Cr(CO)$_3$ (5.5 g, 55%). In most cases, (naphthalene)Cr(CO)$_3$ obtained by the above method was used in the next experiments without further purification.

Preparation of(S*,S*)-tricarbonyl[1-(o-methoxyphenyl)-ethylalcohol]- chromium (83, R = Me) by ligand exchange with (naphthalene)Cr(CO)$_3$. 2-Trimethylsilyl compound **82** (R = Me) (224 mg, 1 mmol) and (naphthalene)Cr(CO)$_3$ (320 mg, 1.2 mmol) were placed in a heavy-wall glass tube equipped with a valve and gas inlet. After addition of ether (6.0 ml) and THF (100 μl, 1.2 mmol), the mixture was degassed by a freeze/pump/thaw cycle and then heated at 70°C for 4 h in a closed system. Filtration, evaporation *in vacuo* and SiO$_2$ chromatography gave the corresponding silylated (*S*,S**)-complex (348 mg; m.p. 127°C). To a solution of the above complex in THF (6 ml) was added 3.0 ml of *n*-tetrabutylammonium fluoride (0.4 M in THF, 1.2 mmol). The mixture was stirred at 0°C for 1 h, and worked up as usual to afford the (*S*,S**)-complex (278 mg).

C.   Stereospecific Carbon-Carbon Bond Formation via Cr(CO)₃-Stabilized Carbonium Ions in Acyclic Series

In the chromium complexes of cyclic compounds, carbon–carbon bond formation via Cr(CO)$_3$-stabilized carbonium ions proceeded stereoselectively to give *exo*-substituted complexes as discussed in Section III.A. In the acyclic systems, the stereochemical course of this S$_N$1 type substitution[57a,61] is a much more interesting problem. The reaction of (*S*,R**)-tricarbonyl(α- methyl-*o*-methoxy benzylacetate)chromium (**90**) with Et$_3$Al gave only one diastereomeric complex, [2-(*o*-methoxyphenyl)-butane]Cr(CO)$_3$ (**92**), in 89% yield. On the other hand, the diastereoisomeric acetate complex **91** with Et$_3$Al afforded the other diastereomeric coupling product **93** without formation of **92**. Similarly, the reaction of (*S*,S**)-(α-ethyl-*o*-methoxy benzylacetate)Cr-(CO)$_3$ (**94**) with Me$_3$Al gave a single substitution product **92**, and the stereoisomeric complex **95** afforded **93** as a sole product. These results indicate that this S$_N$1 type carbon–carbon bond forming reaction proceeded in stereospecific fashion.

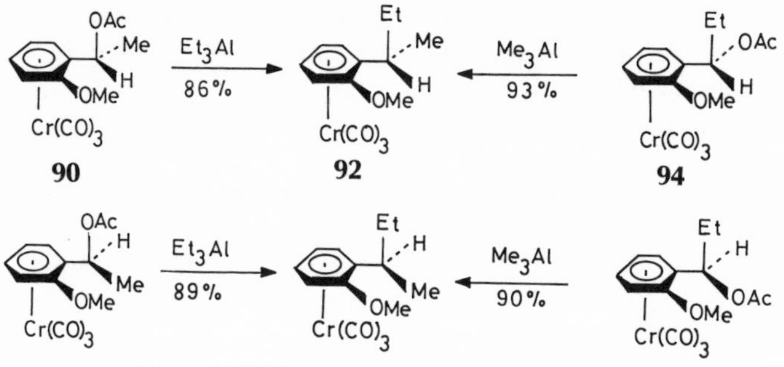

90 92 94 91 93 95

Scheme 27

In a similar manner, *m*-methoxy complexes **96** and **97** afforded stereospecifically **98** and **100** by the reaction with Et_3Al, or **99** and **101** by treatment with allyl trimethylsilane in the presence of $BF_3 \cdot OEt_2$, respectively. These results show that the coupling reaction via $Cr(CO)_3$-stabilized carbonium ions takes place stereospecifically even on the acyclic side chain. There is, however, a question whether this stereospecific reaction occurs at the benzylic position with retention or inversion.

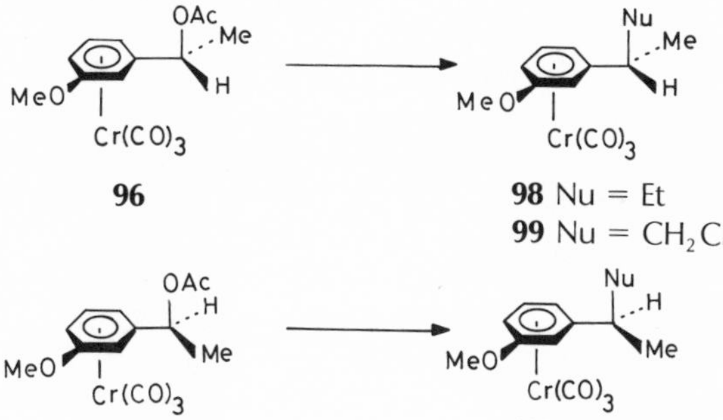

96

98 Nu = Et
99 Nu = $CH_2CH = CH_2$

97

100 Nu = Et
101 Nu = $CH_2CH = CH_2$

Scheme 28

Jaouen *et al.* reported[43c,d] that the replacement of α-carbonium ions with some hetero-atom nucleophiles proceeded with retention of configuration. This fact was ascertained by using a carbon nucleophile with an optically active chromium complex as follows (Scheme 29). Complexation of the optically active (*S*)-1-(*o*-methoxyphenyl)-ethylalcohol[62] (**102**) ($[\alpha]_D = -70°$, 96% ee) gave predominantly (*SS*)-chromium complex **103** ($[\alpha]_D = -188°$). After separation with SiO_2 chromatography, the major (*SS*)-complex **103** was converted via several steps into (*R*)-3-(*o*-methoxyphenyl)*n*-butyric acid

(106) ([α]$_D$ = − -16°, 95% ee).[63] The minor (RS)-complex 104 ([α]$_D$ = 369°) also afforded the same R-carboxylic acid 106. Therefore, the S$_N$1 coupling reaction at the benzylic position via Cr(CO)$_3$-stabilized carbonium ion takes place with retention of configuration.

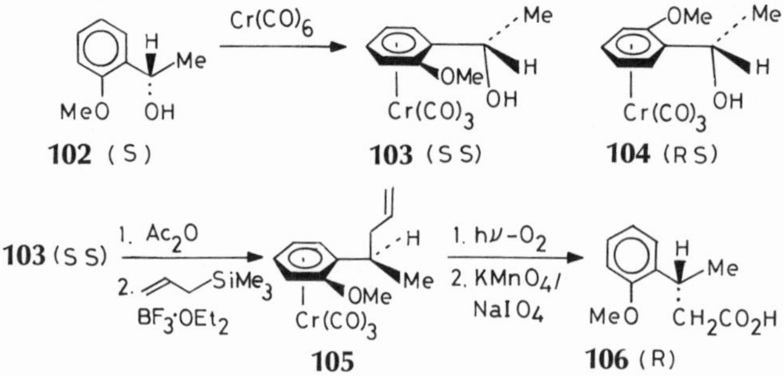

Scheme 29

The mechanism of the stereochemical retention in this S$_N$1 type reaction is illustrated in Scheme 30. The acetoxyl (or hydroxyl) group of diastereomeric chromium complexes 107 and 108 are stereoselectively eliminated from the exo-side owing to the steric effect and anchimeric assistance of Cr(CO)$_3$ group[64] to generate carbonium ion species 109 and 110, respectively. Nucleophiles attack the carbonium ions from the exo-side to give stereospecifically the coupling products 111 and 112, without equilibration of both carbonium ion species via a free rotation about the C(1)–C(Ar) bond. The difficulty of this interconversion between the carbonium ion species has been analyzed by NMR techniques.[65]

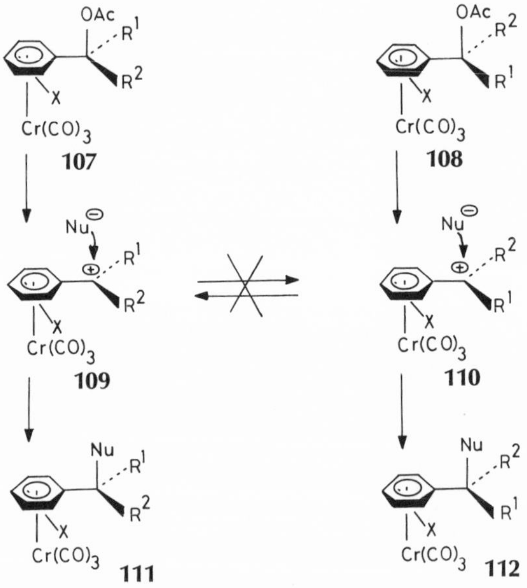

Scheme 30

The (S*,S*)-complex **113** bearing an α-isopropyl group was reacted with Me₃Al to give the stereospecific coupling product in lower yield (53%), compared with the corresponding (S*,R*)-complex **114** (74% yield). The lower yield in the (S*,S*)-complex may be attributed to the character of the intermediate carbonium ion, **109** (X = o-OMe, R¹ = H, R² = i-Pr), which has a severe steric interaction between *ortho* methoxyl and isopropyl groups. Similar tendency due to the steric effect was observed in other related complexes. For example, complex **115** did not react with Me₃Al alone. On the other hand, diastereoisomeric complex **116** afforded a smoothly methylated product stereospecifically under the same reaction conditions. However, complex **115** gave the methyl substitution complex with stereochemical retention in the presence of TiCl₄.

| 113 | 114 | 115 | 116 |

D. Diastereoselective Cr(CO)₃ Complexation Mediated by Remote Hydroxyl Group: Remote Diastereoselection

Diastereoselective Cr(CO)₃ complexation of *ortho*-substituted secondary benzylalcohol derivatives is described in Section III.B. The chromium complexes of *o*-alkoxyphenyl alkyl ketones bearing a chiral center at the C-*n* position in the alkyl side chain could be effectively employed to control the stereochemical relationship at C-1/C-*n* position, since the benzylic carbonyl group of these type of complexes can be converted into each diastereomeric alkyl substitution product via carbonium ions as discussed in Section III.C. In this chapter, we discuss the diastereoselective Cr(CO)₃ complexation of *ortho*-alkoxyphenyl alkane derivatives having two functional groups in the acyclic part, an acetal at C-1 (benzylic) position and a hydroxyl group at C-2, or C-3 position.[66]

Chromium complexation of 3-hydroxy-1-(*o*-methoxyphenyl)-butan-1-one was not succesful with either Cr(CO)₆ or (naphthalene)Cr(CO)₃. In the case of *ortho*-alkoxyphenyl ketone derivatives, the corresponding arene chromium complexes usually could be synthesized by an indirect route via the conversion of the benzylic ketone to acetal derivatives.[67] The complexation of ethylene acetal compound **117** with Cr(CO)₆ under the usual thermal conditions gave an easily separable diastereomeric mixture of (S*,R*)-complex **118** and (R*,R*)-complex **119** in a ratio of 55:45. This thermal complexation reaches equilibrium as mentioned in Section III.B. The ligand transfer with (naphthalene)Cr(CO)₃ resulted in much better selectivity as shown in Scheme 31.

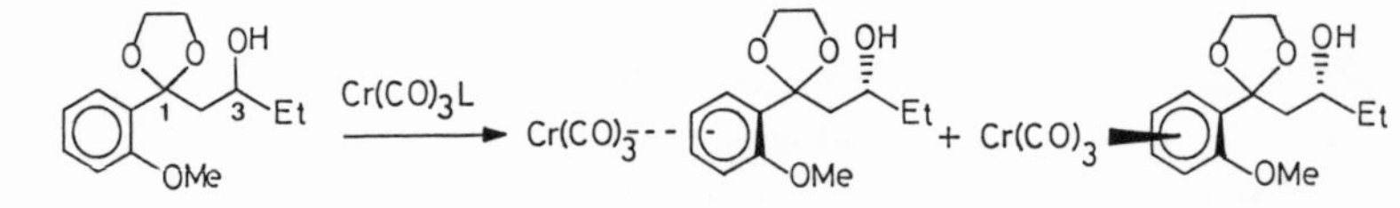

117   **118** (S^*R^*) **119** (R^*R^*)

Cr(CO)$_3$L	Reaction condition	**118:119**	Yield (%)
Cr(CO)$_6$	butyl ether; heptane; THF (10:1:1); 140°C; 24 h	55:45	75
(naphthalene)Cr(CO)$_3$	ethyl ether; 1 equivalent THF 70°C; 4 h	89:11	85

Scheme 31

In contrast to the ethylene acetal, the corresponding propylene acetal and 2,2-dimethylpropylene acetal derivatives, **120** and **121**, afforded no diastereo-selectivity even by the ligand transfer reaction. 1-(*o*-Methoxyphenyl)-3-hydroxybutane (**122**) without an acetal function at the benzylic position gave the corresponding diastereomeric complexes in a ratio of 61:39 by the ligand exchange reaction. However, gem-dimethyl compound **123** afforded pre-dominantly the corresponding (S^*,R^*)-complex in 92:8 ratio. The steric effect at the benzylic position is important for high diastereoselectivity.

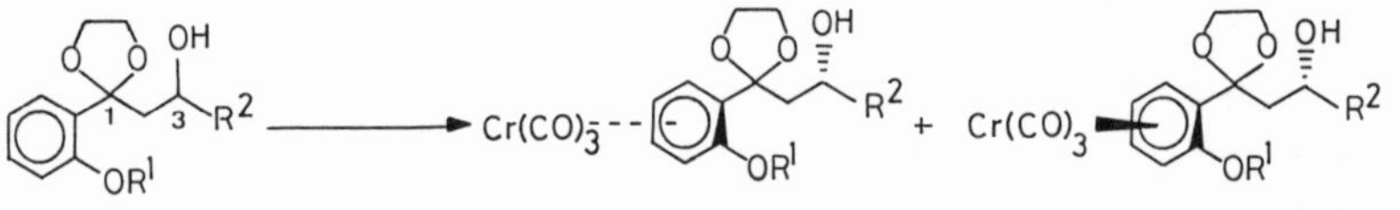

124   **125**   **126**

Scheme 32

Table 8.   Complexation with (Naphthalene)Cr(CO)$_3$

Entry	R^1	R^2	**125:126**	Yield (%)
1	Me	Et	89:11	85
2	Me	i-Pr	96:4	89
3	Me	(E) CH = CH (Me)	92:8	73
4	Me	(Z) CJ = CH (Me)	92:8	82
5	Me	(E) C (Me) = CH (Me)	98:2	85
6	Et	Et	94:6	95
7	i-Pr	Et	94:6	95
8	i-Pr	(E) CH = CH (Me)	94:6	80

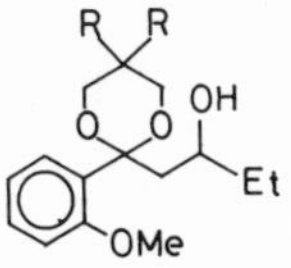

120 R = H (65:35) **122** (61:39) 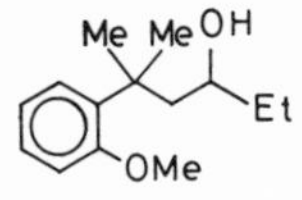 **123** (92:8)
121 R = Me (57:43)

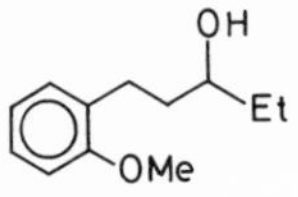

The reaction results of the ligand transfer with (naphthalene)Cr(CO)$_3$ are summarized in Table 8, and the selectivity of this ligand exchange increases with increasing steric bulkiness of both alkyl groups (R^1 and R^2). The Thorpe–Ingold effect[68] may be operative for the high diastereoselective complexation. The hydroxyl group on the side chain also plays an important function for high selectivity, because the corresponding acetoxyl compound gave only a product ratio of 65:35.

The appearance of high diastereoselective complexation on the ligand exchange reaction for compounds **124** to give chromium complexes **125** can be explained by Cr(CO)$_3$ complexation via an intermediate **127**, in which the ethylene acetal (or *gem*-dimethyl) group is moved away from the *o*-alkoxyl group by a rotation around the C(1)–C(Ar) bond due to the steric effect, and the R^2 group is positioned to avoid severe steric interactions with the hydrogen atoms of ethylene acetal (or *gem*-dimethyl) group. An alternative intermediate **128** to another diastereomeric complex **126** has significant steric interactions between R^2 and hydrogen atoms of acetal.

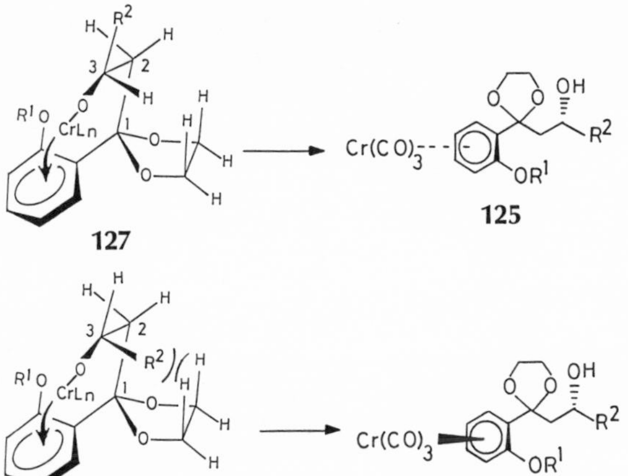

Scheme 33

The homologous compounds **129** with hydroxyl at the C-4 position gave no selectivity by the ligand transfer reaction[69] (Table 9). However, the compounds **130**, hydroxylated at the C-2 position, were reacted with (naphthalene)Cr(CO)$_3$ to afford predominantly (S^*,S^*)-chromium complexes **131**

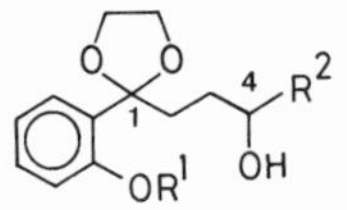

129

Table 9.   Complexation of **129** with (Naphthalene)Cr(CO)$_3$

R^1	R^2	Diastereomeric ratio	Yield (%)
Me	Et	50:50	75
Me	i-Pr	50:50	65
i-Pr	Et	50:50	60
i-Pr	i-Pr	56:44	70

(Table 10). The intermediate for complexation of these compound **130** may be similar to the compound **80** (Scheme 24), in which steric interactions between o-alkoxy group and alkyl groups of the side chain are diminished by the spacial arrangement of these groups.

These chromium complexes prepared by the above ligand transfer reactions can be employed to control stereochemistry in the acyclic part by the stereoselective transformation of the benzylic carbonyl group to each diastereomeric alkyl substitution product as mentioned in Sections III.B and III.C (Scheme 22). Protection of the hydroxyl group of the complex **125** (R^1 = Me, R^2 = Et) and subsequent hydrolysis of the ethylene acetal gave a complex **133** in 55% overall yield. Reduction of the carbonyl of the complex **133** followed by acetylation afforded (S*,R*,R*)-chromium complex **134**. The treatment of **134** with Me$_3$Al gave a single diastereomeric syn-complex **135**. On the other hand, reaction of **133** with MeLi gave a single methylated complex **136** which was converted into another isomer, anti-complex **137**.

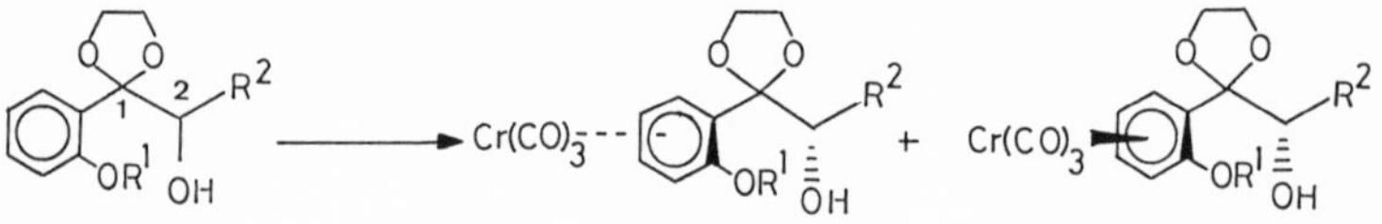

130   **131** (S*S*)   **132** (R*S*)

Scheme 34

Table 10.   Complexation with (naphthalene)Cr(CO)$_3$

Entry	R^1	R^2	Ratio of **131** and **132**	Yield (%)
1	Me	Me	85:15	80
2	Me	n-Bu	85:15	83
3	i-Pr	Me	92:8	88

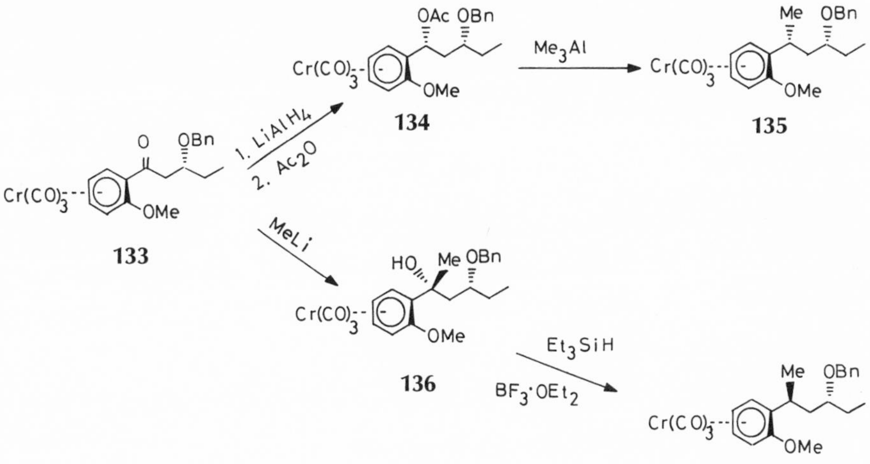

Scheme 35

The chromium complexes bearing an allylic alcohol function in the side chain can exhibit a stereocontrol between the C-1 and C-5 positions by a chirality transfer. Such a sequence would give a new method for stereoselective synthesis of 1,5-pattern of alkyl substituent; a relationship reminiscent of naturally occurring acyclic and macrocyclic systems such as a side chain of Vitamin E and K.[70] The acetate of (S*,S*)-chromium complex with E-double bond, 138, was treated with sodium dimethylmalonate in the presence of 10 mol% of Pd(0)[71] to give regio- and stereoselective coupling product, which was converted to the (S*,R*)-complex 139 in 76% overall yield after conversion of the acetal to carbonyl.[72] The coupling reaction via an intermediate π(allyl)Pd complex derived from the chromium complex with a malonate anion proceeded only at the C-5 position owing to a stereoelectronic effect of acetal and Cr(CO)₃-complexed aromatic ring. Although hydrolysis of the ethylene acetal of the resulting coupling product gave no satisfactory result under various acidic conditions, an oxidative deacetalization with trityl tetrafluoroborate[73] afforded the desired complex 139 without isomerization of the double bond to α,β-unsaturated ketone. The ketone complex 139 was converted into (S*,S*,S*)-complex 140 bearing 1,5-*syn* methyl in 56% overall yield by the reaction with an excess of MeLi and then hydrogenolysis. On the other hand, the complex 139 gave diastereomeric (S*,R*,S*)-complex 141 having 1,5-*anti* methyl in 72% yield by reduction, acetylation and methylation. Similarly, Z-double bond chromium complex 142 afforded stereoselectively two possible diastereomeric complexes 143 and 144, respectively, under the same reaction sequences (Scheme 37).

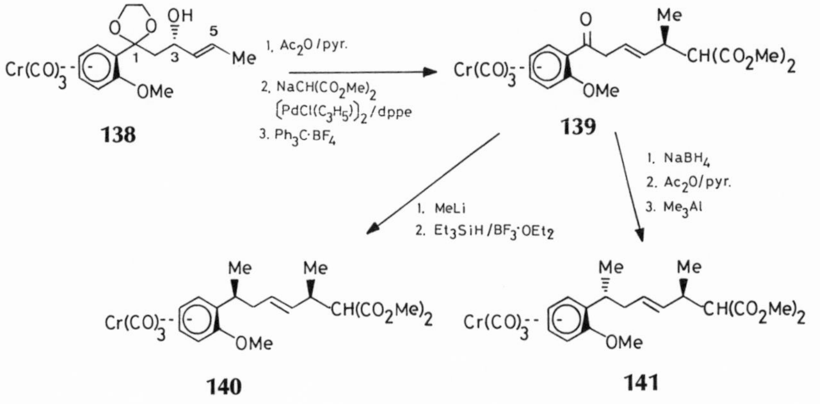

Scheme 36

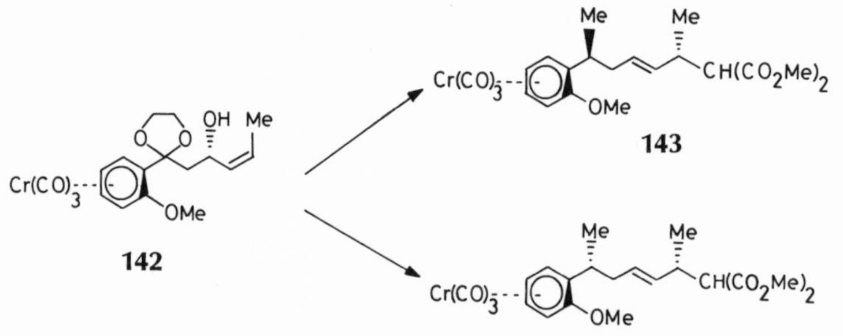

Scheme 37

The enolate Claisen rearrangement[72] of the complex **145**, prepared from **138**, gave stereoselectively 5,6-*syn* chromium complex **146** in 70% overall yield by usual reaction conditions.[74] The complex **146** could be converted to both 1,5-*syn*, 5,6-syn and 1,5-*anti*, 5,6-*syn* complexes by the above mentioned sequences. Similarly, the corresponding Z-complex **147** afforded 5,6-*anti* complex **148** by the same reaction conditions. These schemes incorporate significant flexibility with regard to the stereochemistry of the remote centers, since a simple variation of allylic olefin geometry would result in an effective stereocontrol of the remote chiral centers bearing various alkyl substituents.

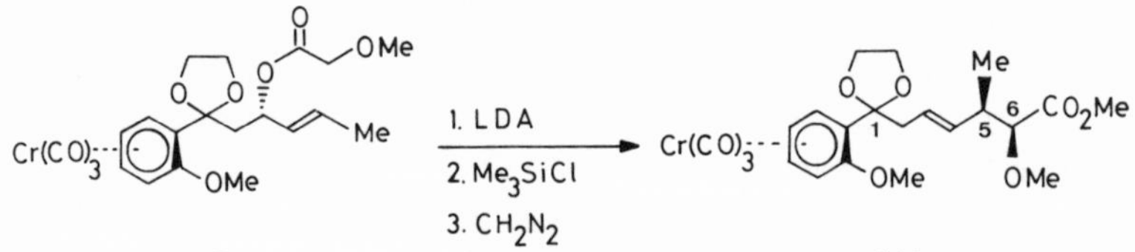

Scheme 38

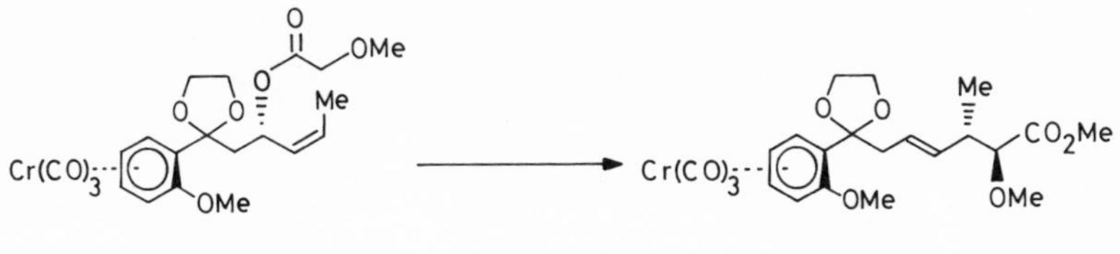

147   **148**

Scheme 38

E. Diastereoselective Addition of Crotyl-Metals to Cr(CO)$_3$-Complexed Aromatic Ketones

Over the last decade, considerable attention has been focused on the stereoselective carbon–carbon bond forming process in acyclic and other conformationally flexible molecules, particularly by the aldol addition[75] and the related reactions of crotylmetal reagents[76] with carbonyl compounds. Crotylmetal addition reactions are equal or even superior to the aldol reactions, since the adducts, being homoallylic alcohols, can be transformed into the aldol derivatives or other functional groups. Generally, a high degree of diastereoselectivity has been achieved in the reactions with aldehydes to form *anti-* and *syn*-adducts, respectively (Scheme 39).

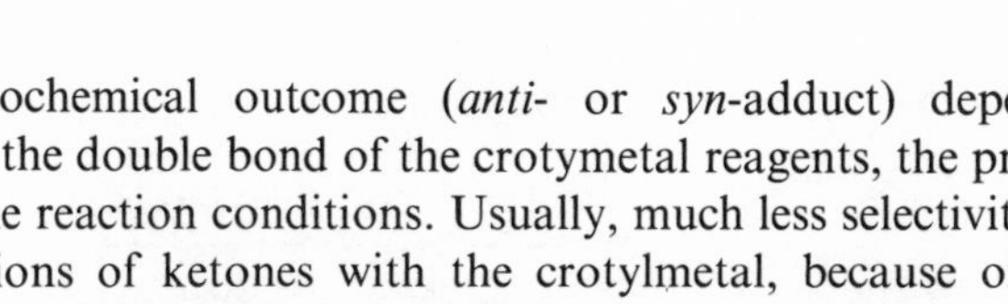

Scheme 39

The stereochemical outcome (*anti-* or *syn*-adduct) depends on the geometry of the double bond of the crotymetal reagents, the property of the metal and the reaction conditions. Usually, much less selectivity is observed in the reactions of ketones with the crotylmetal, because of the smaller difference in the steric size between both the groups attached to the carbonyl. Recently, Seebach,s[77] and Reetz,s[78] groups have reported that crotyltitanium reagents can be reacted with various ketones to yield *anti*-adducts with high selectivity (Scheme 40). Directing toward stereo- and regioselective synthesis of dihydroxyserrulatic acid (**205**), we have examined diastereoselective addition of crotylmetals to aromatic ketones. In order to obtain high selectivity, the aromatic part was modified temporarily to sterically bulkier group, e.g. the arene transition metal complexes.[79]

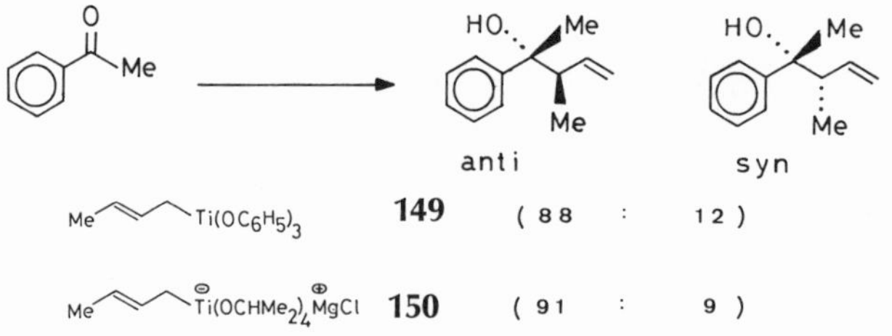

Scheme 40

The reaction of (5-methoxy-l-tetralone)Cr(CO)$_3$ (151) with crotylmag-nesium chloride gave no selectivity. Crotyltriphenoxytitanium (149) and an "ate" complex 150 also added to complex 151 with less selectivity. However, the reaction with crotyl Grignard or the lithium compound in the presence of 1 equivalent of trialkylaluminum afforded sufficiently high *anti*-selectivity without formation of the regioisomer by α-attack of the crotyl group (Scheme 41, Table 11). Addition of triethylaluminum showed particularly high *anti*-selectivity.[79] Since 1-tetralone itself without Cr(CO)$_3$ complexation showed a 67:33 ratio of *anti*- and *syn*-adducts under the same reaction conditions, the complexation apparently increases the selectivity on this reaction. The chromium complexes of acetophenone, propiophenone and isobutyro-phenone also gave *anti*-adducts predominantly with the aluminum "ate" complex (Table 12). The steric bulkiness of the alkyl group (R) does not exert a strong influence on the selectivity.

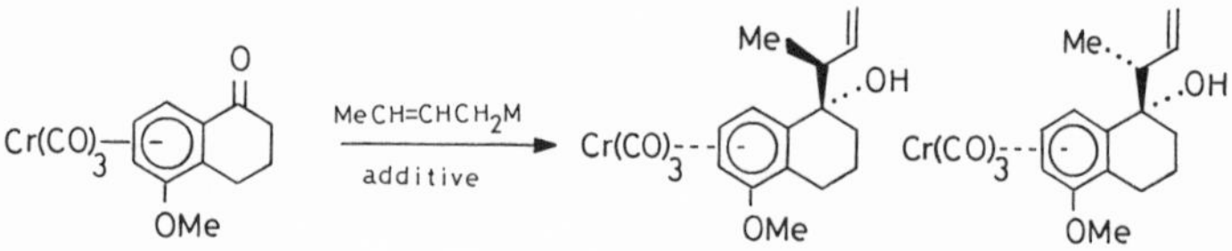

Scheme 41

Table 11.

Entry	M	Additive	152:153	Yield (%)
1	MgCl	none	50:50	75
2	MgCl	Me$_3$Al	80:20	75
3	MgCl	Et$_3$Al	93:7	73
4	MgCl	(i-Bu)$_3$Al	85:15	65
5	MgCl	5eq Et$_3$Al	66:33	80
6	MgCl	5eq (i-Bu)$_3$Al	30:70	40
7	MgCl	Et$_3$B	28:72	30
8	MgCl	(i-Am)$_3$B	90:10	30
9	MgCl	(i-Pro)$_4$Ti	64:34	30
10	Li	Et$_3$Al	94:6	60

Since the products lack vicinal hydrogens at the two chiral centers, the usual methods of *anti/syn* assignment by ^{1}H-NMR techniques cannot be applied to these compounds. Therefore, the stereochemistry in the (acetophenone)Cr(CO)$_3$ adduct was determined by comparison with an authentic sample, and by further transformation to the vicinal dimethyl compound as described later. For other adducts, the assignments are tentatively done based on the analogy to the proposed transition state topology.

This reaction presumably proceeds via a six-membered chair transition state, in which the smaller group on the ketone carbonyl function occupies a pseudo-axial position and the double bond of the aluminum "ate" complex exists as an E-form. Although the cyclic pentacoordinate transition state of aluminum "ate" complex may be curious, a number of other reactions have ben proposed via "ate" complexes of boron-80 and titanium.[78,81] Addition of an excess of trialkylaluminum as additive may support the above assumption to increase *syn* adduct. Thus, trialkylaluminum forms a co-ordination adduct, at first, with the carbonyl oxygen as Lewis acid, then addition to the carbonyl carbon takes place via the open-chain transition state.

Interestingly, the combination of Et$_3$B and crotyl Grignard reagent produced predominantly *syn*-adducts. The reactions of the ketones in the presence of R$_3$B are very selective. The use of triisoamylborane as an additive produced *anti*-adduct, and even with the use of triethylborane, crotyllithium instead of the corresponding Grignard reagent favored the *anti*-adduct (see Tables 11 and 12). An alternative transition state should be considered in the case of crotyl Grignard and Et$_3$B with Cr(CO)$_3$-complexed ketones. Also, Yamamoto *et al.* reported that crotyllithium was reacted with benzaldehyde in the presence of Et$_3$B to afford the *anti*-adduct with high selectivity[82] via a six-membered chair transition state, but crotylmagnesium chloride gave a 1:1 mixture of *anti*- and *syn*-adducts.

Table 12.

Entry	R	Additive	155:156	Yield (%)
1	Me	Et$_3$Al	89:11	95
2	Me	Et$_3$B	33:67	70
3	Me	5eq Et$_3$Al	69:31	70
4	Et	Et$_3$Al	85:15	96
5	CHMe$_3$	Et$_3$Al	84:16	65
6	H	Et$_3$Al	50:50	90

In contrast to the $Cr(CO)_3$-complexed aromatic ketones, the chromium complex of benzaldehyde exhibited no selectivity with the aluminum "ate" complex (Table 12, entry 6). Presumably, the addition of the "ate" complex to the aldehyde carbonyl is so rapid that carbon–carbon bond formation occurs prior to the formation of the above mentioned cyclic transition state. However, crotylchromium(II)[83] gave exclusively *anti-* adduct in 99% diastereoselectivity with (benzaldehyde)$Cr(CO)_3$.

In general, $Cr(CO)_3$-complexed aromatic ketones resulted in high *anti-*selectivity by the reaction with crotylaluminum "ate" complexes, as mentioned above. An electronic effect other than the steric effect of the $Cr(CO)_3$ group may also be operative for the high selectivity, in addition to the characteristics of crotylmetals and additives. Further investigations would be required for the discussion of the reaction mechanism.

The resulting benzylic hydroxyl group of the chromium complexes obtained by the above reactions could be further substituted with some carbon nucleophiles to lead to vicinal alkyl substituted compounds with stereochemical retention. For example, ionic hydrogenolysis of the complex **158** afforded the 1,2-dimethyl *syn* compound **159**. On the other hand, the treatment of the acetate complex **161** with Me_3Al gave the *anti* compound **162**. This method provides new stereoselective construction of both stereoisomers, in which two adjacent chiral centers are fully occupied by alkyl substituents.

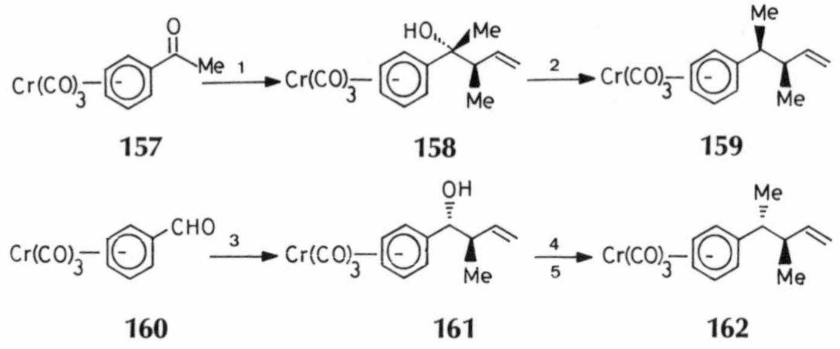

1) MeCH=CHCH$_2$MgCl/Et$_3$Al   2) Et$_3$SiH/CF$_3$CO$_2$H   3) MeCH=CHCH$_2$Br/CrCl$_2$
4) Ac$_2$O/pyr   5) Me$_3$Al

Scheme 42

Reaction of crotylmagnesium chloride with complex 151 in the presence of Et₃ Al.
In a solution of 1.62 mL of crotylmagnesium chloride (0.26 M in THF, 0.44 mmol) in dry THF (5 mL) was added 0.44 mL of triethylaluminum (1 M in hexane, 0.44 mmol) at $-78°C$ under an argon atmosphere. After stirring for 30 min, a solution of (5-methoxy-1-tetralone)$Cr(CO)_3$ **(151)** (130 mg, 0.40 mmol) in dry THF (4 ml) was injected to the above reaction mixture by a syringe at $-78°C$. The reaction mixture was warmed to $0°C$ for 3 h, and quenched with saturated aqueous ammonium chloride. The mixture was ex-

tracted with ether and worked up as usual. SiO$_2$ chromatography gave a yellow crystalline (110 mg) as a diastereomeric mixture of *anti-* and *syn-*adducts. The ratio of diastereomers was determined by ^{1}H-NMR.

IV. NATURAL PRODUCT SYNTHESIS UTILIZING (η^6-ARENE)Cr(CO)$_3$ COMPLEXES

A. Anthracyclinones

The clinical utility of the anthracycline antibiotics[84] has promoted considerable efforts to obtain them by either partial or total synthesis.[85] The members of the anthracycline family are in various stages of human cancer trials. The structure of the representative aglycones of "angular" and }"linear" type anthracyclines are indicated in Figure 4. Most of the variation in structures of the anthracycline family concerns the substitution pattern of the saturated A-ring.

These compounds appear to be ideal synthetic targets for the exploration of the desired lithiation of (η^6-arene)Cr(CO)$_3$ complexes. The development of general synthetic convergence to those molecules is gained by dissections a and b (Scheme 43) leading to fragments **174, 175** and **176, 177**, respectively. These reterosynthetic paths are mutually interactive in that aldehydes **174** and **177** are obtainable from **176** and **175**, respectively, by the reaction with DMF.

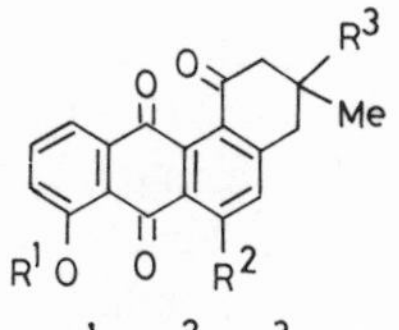

	R^1	R^2	R^3
163	H	OH	OH
164	H	OH	H
165	H	H	H
166	Me	H	H

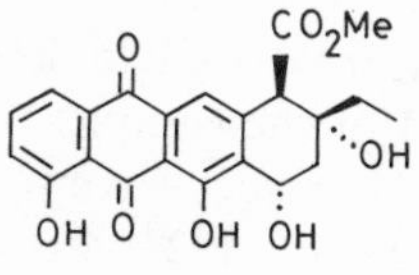

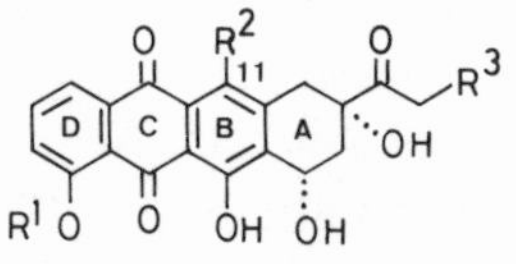

	R^1	R^2	R^3
167	Me	OH	OH
168	Me	OH	H
169	Me	H	OH
170	Me	H	H
171	H	H	H

172

Figure 4. Structure of representative aglycones of "angular" and "linear" type anthracyclines.

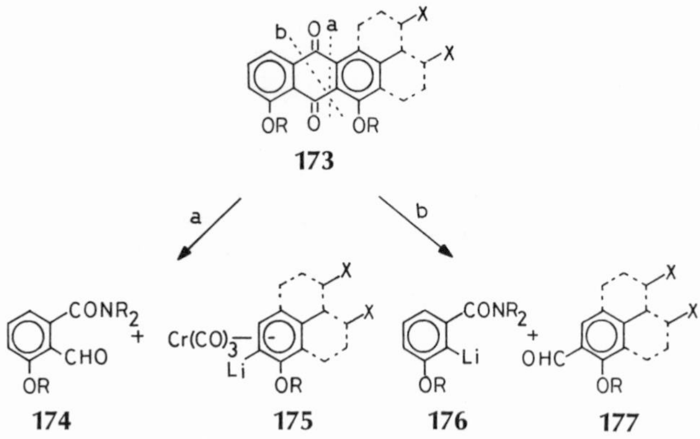

Scheme 43

1. Deoxyrabelomycin[86]

The benz[a]anthraquinones constitute a distinct class of antibiotics sub-stances isolated from cultures of several Streptomyces strains. Typical metabolites of this type of anthraquinone are rabelomycin[87] (**163**), X-I488IC[88] (**166**) and ochromycinone[89] (**165**). Although there are few synthetic studies[90] on this type, the group of structual "angular" analogues of the anthracycli-nones has been shown to exhibit weak antibacterial, enzyme inhibitory and antitumor activity.

(7-Methoxy-1-tetralone)Cr(CO)$_3$ (**178**) was methylated with MeI and NaH in DMF and benzene to give an easily separable 2-exo-methyl complex **179** (73%) and 2,2-dimethyl complex (14% yield). The mono methyl complex **179** was converted to (7-methoxy-2-exo-methyl-1-endo-tetralol)Cr(CO)$_3$ (**180**) by stereoselective reduction with LiAlH$_4$. Since the co-ordination by the Cr(CO)$_3$ group confers a three-dimensional structure which has stereochemi-cal consequences, electrophilic or nucleophilic attack at the reactive center always occurs stereospecifically in an exo-fashion. Directed lithiation of the complex **180**, followed by quenching with 2-formyl-3-methoxy-N,N-diethyl-benzamide[91] and subsequent decomplexation (exposure to sunlight), gave a diastereomeric mixture of phthalide **181** in 40% yield. Dehydration with KHSO$_4$ afforded an olefinic phthalide **183** in quantitative yield. This phthalide **183** was also obtained by the other sequence (path b in Scheme 43) in better yield. Treatment of di-lithio compound of the complex **180** with DMF, followed by demetallation and dehydration, afforded 6-formyl dihyd-ronaphthalene **182** in 92% yield. Condensation of **182** with di-lithio compound of 3-methoxybenzanilide[92] gave the phthalide **183** in 80% yield.

Reductive cleavage of the lactone with Zn dust, followed by ring closure with trifluoroacetic anhydride and trifluoroacetic acid under the usual methods, gave an anthrone **185** in 89% yield. The oxidation at the two benzylic positions of **185** was troublesome[93]. However, dihydroanthrone, obtained by catalytic hydrogenation of **185**, was smoothly converted into the anthraquinone **186** with CrO_3 in AcOH. Demethylation with $AlCl_3$ gave 3-deoxyrabelomycin **164**.

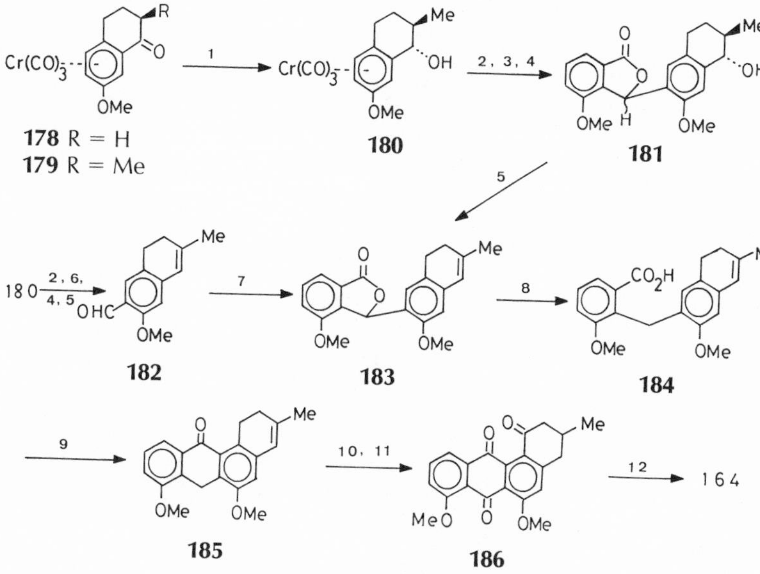

178 R = H
179 R = Me
180
181
182
183
184
185
186

1) LiAlH$_4$ 2) n-BuLi/TMEDA 3) 3-methoxy-2-formyl-N,N-diethylbenzamide
4) hv-O$_2$ 5) KHSO$_4$ 6) DMF 7) di-lithio of 3-methoxy-N-phenylbenzamide
8) Zn/CuSO$_4$/10% KOH/pyr 9) CF$_3$CO$_2$H/(CF$_3$CO)$_2$O 10) H$_2$/Pd-C 11) CrO$_3$/AcOH
12) AlCl$_3$

Scheme 44

2. Decarbomethoxyaklavinone

The synthetic strategy[86] to decarbomethoxyaklavinone is similar to that of rabelomycin. Mono-ethylation of (5-methoxy-1-tetralone) Cr(CO)$_3$ **(151)** at the C-2 position was not so facile as expected. Under the usual conditions, severe unseparable mixtures of mono- and diethyl derivatives were produced along with unreacted starting material. This problem could be surmounted by the reaction of the lithium enolate of **151** with 1 equivalent of triethanolamine borate[94] in DMSO, followed by trapping with EtI to give 70% yield of the mono-ethyl complex **187** and 20% of the starting complex. The directed lithiation of complex **188** under the usual conditions, followed by quenching with DMF gave a mixture of C-6 and C-8 formylated compounds in a ratio of 7:3. Undesirable lithiation at the C-8 position would be attributed to the co-ordination of the lithium to the benzylic alkoxide group. Regioselective introduction of the formyl group at the C-6 position can be easily achieved by protection of the benzyl alcohol as trimethylsilyl ether. Treatment of the lithio compound of complex **189** with DMF, followed by usual reaction sequence gave 6-formylated compound **190** without formation of the regioisomeric product in 73% overall yield after demetallation. Desilylation and dehydration afforded **191** which was converted into anthrone compound **193** in 64% overall yield by the same reaction sequence as mentioned above. The compound **193** was easily oxidized to the corresponding anthraquinone **194** which had already been converted to decarbomethoxyaklavinone (**195**) by Kende *et al.*[92] Although a similar synthetic sequence has been reported, the directed regioselective lithiation of (η^6-arene)Cr(CO)$_3$ provides a shorter route to the key intermediate.

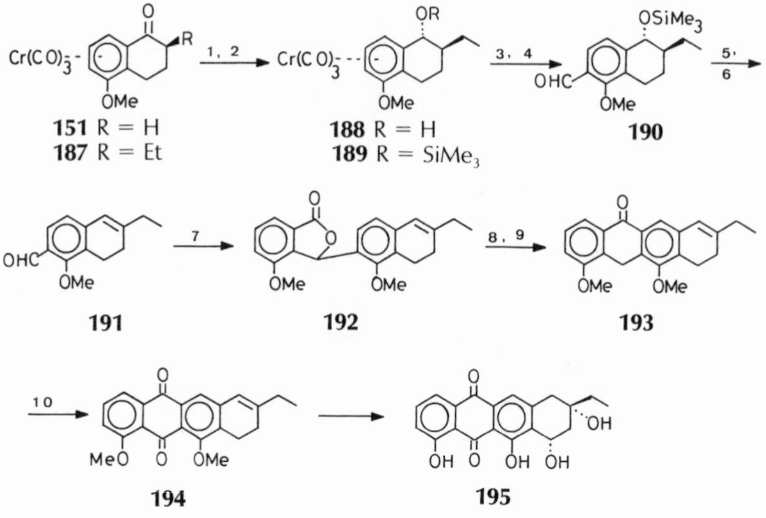

151 R = H
187 R = Et

188 R = H
189 R = SiMe$_3$

190

191   **192**   **193**

194   **195**

1) LiAlH$_4$ 2) Me$_3$SiCl/Et$_3$N 3) n-BuLi/TMEDA, then DMF 4) hν-O$_2$ 5) HCl
6) KHSO$_4$ 7) di-lithio of 3-methoxy-N-phenylbenzamide 8) Zn/CuSO$_4$/KOH/pyr
9) CF$_3$CO$_2$H/(CF$_3$CO)$_2$O 10) Triton B/O$_2$

Scheme 45

3.   11-Deoxydaunomycinon[95]

11-Deoxyanthracyclines, such as 11-deoxydaunomycin, 11-deoxyadriamycin and 11-deoxycarminomycin, are known to show significant anticancer activity and less cardiotoxity[96] than the 11-hydroxy analogues.

(η^6-Styrene)Cr(CO)$_3$ complexes are recognized as being equivalent to Michael acceptors, in which nucleophilic addition occurs at the β-position of the styrene ligand to stabilize the benzylic carbanions[97] (Scheme 46). Nucleophilic addition to this system can be used to introduce an acyl group at the C-9 position of anthracyclinones. The alternative method of carbanion addition to the carbonyl of the corresponding β-tetralone usually proceeds in low yield with a large excess of Grignard or organolithium reagents, because of easy enolization of the carbonyl groups.[98,99]

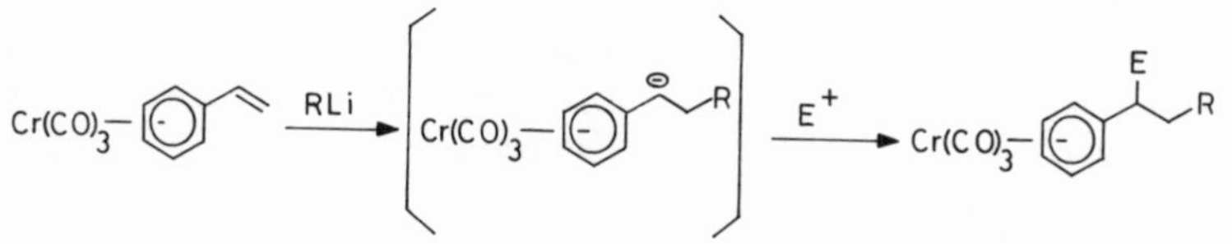

Scheme 46

(5-Methoxy-1-tetralone)Cr(CO)₃ (151) was converted to (3,4-dihydro- 5-methoxynaphthalene)Cr(CO)₃ (196) by reduction and subsequent dehydration. The reaction of 196 with the carbanion of a protected acetaldehyde cyanohydrin gave the 2-*exo*-substituted chromium complex 197 in 87% yield. Directed *ortho*-lithiation of 197, followed by quenching with 2-formyl-3-methoxy-*N,N*-diethylbenzamide and decomplexation, gave a condensation product, which was converted into a mixture of diastereomeric ketophthalide 198 by treatment with dilute acid and then base in 38% overall yield. The phthalide 198 was converted to anthraquinone via anthrone 199 by similar reaction sequences. The compound 200 has been already converted to 11-deoxydaunomycinon (170).[100]

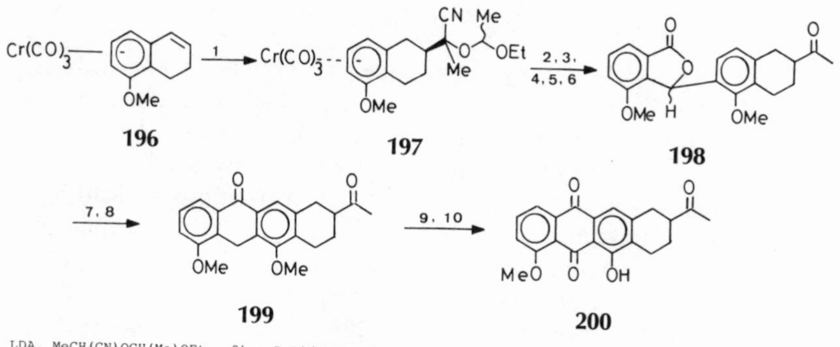

1) LDA, MeCH(CN)OCH(Me)OEt   2) n-BuLi/TMEDA   3) 3-methoxy-2-formyl-N,N-diethylbenzamide
4) hν-O₂   5) H₃O⁺   6) OH⁻   7) Zn/10% aq KOH/pyr   8) CF₃CO₂H/(CF₃CO)₂O   9) CrO₃/acetone/H₂SO₄
10) HBr/AcOH

Scheme 47

B. Hydroxycalamenenes

Phenolic cadinane-type sesquiterpenoids, *cis*-7-hydroxycalamenene[101] (201a), *cis*-7-hydroxycalamenal[102] (202a), *trans*-8-hydroxy- and 8-methoxy calamenenes[103] (**203b**, **204b**) and a prenylated analog, dihydroxyserrulatic acid[104] (**205**), have been isolated from plants, and some members of this class possess interesting physiological activities, such as phytoalexin and fish-poison. These compounds have 1,4,6-substituted tetrahydro-7 (or 8)-naphthol as a common structural unit. For the general synthetic method of these terpenoids, it is necessary to introduce two benzylic substituents (at C-1 and C-4) stereoselectively, and the aromatic substituents at the proper positions regioselectively, in the tetrahydronaphthol skeleton. It seems profitable for this purpose to employ (arene)Cr(CO)$_3$ complexes, as described above.

201 R = Me
202 R = CHO

203 R = H
204 R = Me

205

a, cis; b, trans

1.   7-Hydroxycalamenene[105]

(7-Methoxy-4-exo-isopropyl-1-tetralone)Cr(CO)$_3$ (**207**) is considered to be the most suitable compound for stereo- and regioselective total synthesis of both cis- and trans-7-hydroxycalamenenes. Complexation of methyl 4-(p-methoxyphenyl)-5-methylhexanoate with Cr(CO)$_6$ gave a yellow crystalline complex **206**, which was hydrolyzed with KOH to afford an acid in 72% overall yield. Although reaction of the acid complex with polyphosphoric acid[106] was not successful, cyclization of the corresponding acid chloride with AlCl$_3$ gave an 4-exo-isopropyl tetralone complex **207** in 60.5% yield, along with a trace of endo-isopropyl complex **208** (2.5% yield). This cyclization proceeds to avoid severe steric interaction between isopropyl and Cr(CO)$_3$ to give exclusively exo-isopropyl complex. Reaction of the exo-compound **207** with MeLi gave a single methylated complex which was converted to 6-formyl complex **209** by directed regioselective lithiation. Ionic hydrogenolysis of **209** resulted in a stereoselective hydride displacement via the benzylic carbonium ion, along with simultaneous exhaustive reduction of the formyl group, giving (1-endo-methyl-4-exo-isopropyl-6-methyl-7-methoxytetralin)-Cr(CO)$_3$ (**210**). Decomplexation, followed by demethylation, gave trans-7-hydroxycalamenene (**201b**).

On the other hand, *cis*-7-hydroxycalamenene (**201a**) was also stereoselectively synthesized from the major cyclization product, *exo*-isopropyl complex **207**. Reduction of the carbonyl, following acetylation and then treatment with Me₃Al afforded (1-*exo*-methyl-4-*endo*-isopropyl-7-methoxytetralin)-Cr(CO)₃ (**211**). Directed lithiation of **211** followed by quenching with DMF gave 6-formylated complex **212** without formation of the regioisomeric formylated compound. Decomplexation of **212**, hydrogenation of the formyl group with 5% Pd/C, and subsequent demethylation afforded *cis*-7-hydroxy-calamenene (**201a**).

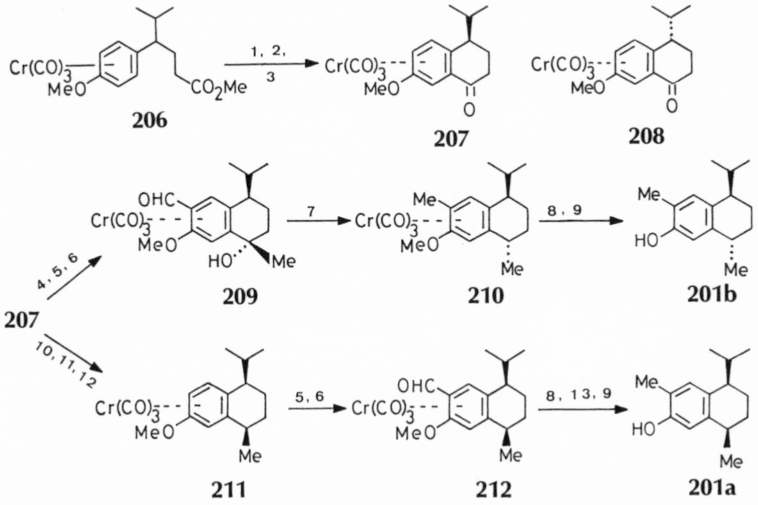

1) KOH/MeOH/H₂O 2) (COCl)₂ 3) AlCl₃ 4) MeLi 5) n-BuLi/TMEDA
6) DMF 7) CF₃CO₂H/Et₃SiH 8) hv-O₂ 9) BBr₃ 10) LiAlH₄
11) Ac₂O/pyr 12) Me₃Al 13) H₂/Pd-C

Scheme 48

2. 8-Hydroxycalamenene[107]

Both stereoisomers of 8-hydroxycalamenene were synthesized from a common intermediate, (8-methoxy-4-*exo*-isopropyl-1-tetralone)Cr(CO)₃ (**215**). 8-Methoxy-4-isopropyl-1-tetralone (**213**) was obtained from 4-(*m*-methoxyphenyl)-5-methyl-1-hexene by several steps. As discussed above, direct Cr(CO)₃ complexation of the compound **213** afforded no satisfactory results[67] owing to presumed co-ordination of the chromium by two oxygen atoms of ketone and ether functions. We therefore explored a conversion of the ketone **213** to an ethylene acetal compound with methyl orthoformate, ethylene glycol and catalytic amounts of *p*-TsOH at 40°C.[108] The ethylene acetal provided the corresponding arene complex **214** under the usual thermal conditions with Cr(CO)₆, which gave the desired 4-*exo*-isopropyl complex **215** in 82% yield by hydrolysis of the ethylene acetal, along with a trace of the *endo*-isopropyl complex. The complex **215** was stereoselectively converted into both stereoisomeric, 1-*endo*-methyl and 1-*exo*-methyl chromium complexes **216** and **219**, by the same reaction sequences as mentioned above. The ^{1}H-NMR signal of the benzylic methyl of the *exo*-complex **219** appeared at higher field (δ 1.19) than that of the *endo*-isomer **216** (δ 1.35). Also, the spread of δ values (Δ 0.47 ppm) of the aromatic protons in the *exo*-isomer was smaller than those of the *endo*-isomer (Δ 0.85 ppm) as discussed previously.

Since the next step requires functionalization at the C-6 position, the methoxyl group of the complex has to be converted into a triisopropylsilyloxy group for the *meta*-directed lithiation reaction. However, demethylation of **216** caused undesired decomplexation. Therefore, the nucleophilic addition method was adopted in the following manner. Reaction of 2-lithio-1,3-dithiane with the complex **216** and subsequent oxidative demetallation of an anionic chromium complex with I_2 gave a 6-dithianylated compound **217** (25%) and its hydrolyzed product **218** (36%), the desired substitution products at the *meta* position of methoxyl. *Trans*-8-methoxycalamenene was obtained from the compound **217** by desulfurization with Raney Ni, and from **218** by hydrogenolysis with Pd/C under a hydrogen atmosphere. Demethylation with BBr_3 afforded *trans*-8-hydroxycalamenene (**203b**).

On the other hand, *cis*-8-hydroxycalamenene (**203a**) was synthesized from **219** by the same reaction sequence.

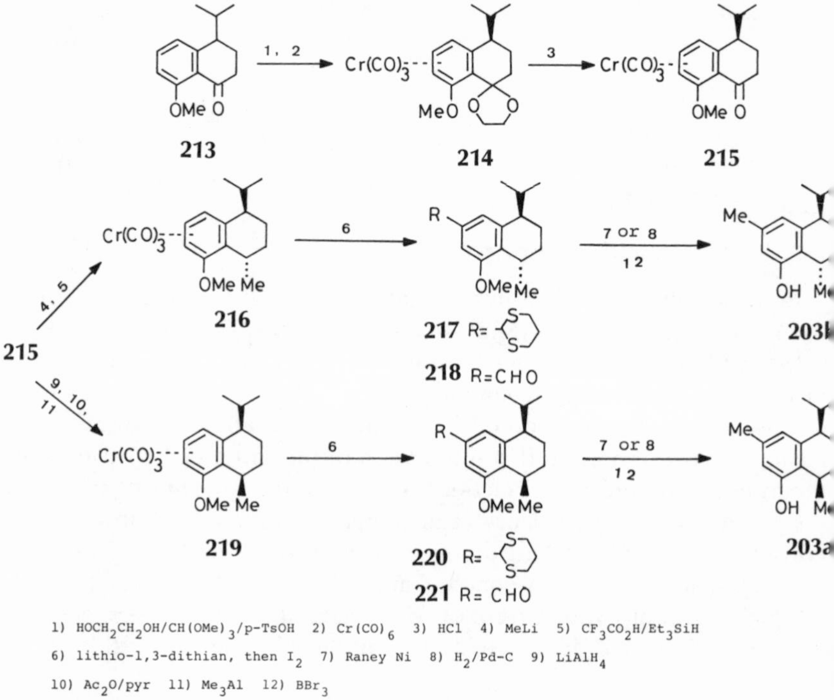
1, 2
3
Cr(CO)3
OMe O
MeO
213
214
215
4, 5
9, 10, 11
6
7 or 8
12
R
Me
OMe Me
OH Me
216
217 R=
218 R=CHO
203b
219
220 R=
221 R= CHO
203a
1) HOCH2CH2OH/CH(OMe)3/p-TsOH 2) Cr(CO)6 3) HCl 4) MeLi 5) CF3CO2H/Et3SiH
6) lithio-1,3-dithian, then I2 7) Raney Ni 8) H2/Pd-C 9) LiAlH4
10) Ac2O/pyr 11) Me3Al 12) BBr3

Scheme 49

C. Acorenone and Acorenone B[61]

Acorenone (**228**) and acorenone B (**226**) are members of the natural spiro[4,5]decane system. The synthetic strategies for stereospecific generation of the quarternary carbon have been an attractive challenge for organic synthesis. In most of the previous synthesis[109] of spirosesquiterpenoids, the strategy has involved the formation of *gem*-disubstituted monocyclic ring systems and then construction to the second ring. Semmelhack and Yamashita reported[110] new synthetic methods for the synthesis of acorenone and acorenone B utilizing (η⁶-arene)Cr(CO)₃ complexes, in which the spirocyclopentane unit was stereoselectively formed by an intramolecular *exo*-nucleophilic addition at the *meta*-position of Cr(CO)₃-complexed anisole derivative (Scheme 50). However, the direct chromium complexation of the starting material **222** with Cr(CO)₆ under thermal conditions gave a mixture of diastereomeric chromium complexes **223** and **224** in a ratio of 6:4, along with the corresponding complexes in which a Cr(CO)₅ unit attached to the nitrile group. Diastereoselective synthesis of each complex of this type is an interesting problem with little precedent. This problem has easily been solved by the methods of diastereoselective Cr(CO)₃ complexation and subsequent stereospecific carbon–carbon bond formations as described in Sections III.B and C.

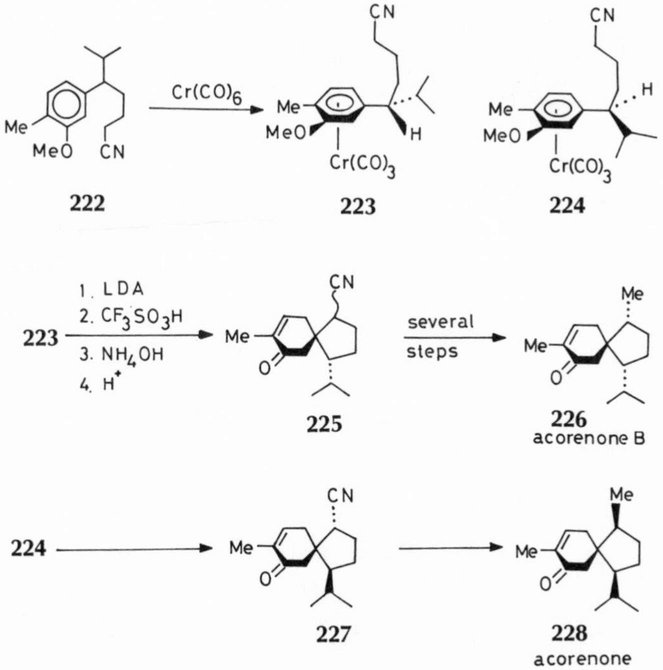

Scheme 50

6-Trimethylsilyl-3-methoxybenzaldehyde was reacted with isopropyl lithium to afford **229**. The ligand exchange reaction of (naphthalene)Cr(CO)$_3$ with the compound **229**, followed by detrimethylsilylation gave exclusively the (S^*,R^*)-complex **230** as described above. Regioselective introduction of methyl at the C-4 position was achieved by directed lithiation and subsequent quenching with 1 equivalent of MeI and HMPA in 95% yield without formation of regioisomeric methylated products. After acetylation, the complex **231** was reacted with ethyl 2-trimethylsilyl-3-butenoate[111] in the presence of $BF_3 \cdot OEt_2$, and subsequent hydrogenation of a double bond of the coupling product with PtO_2 afforded pure diastereomeric chromium complex **232** in 66% overall yield. The conversion of the ester of the complex **232** to nitrile **223** was achieved by treatment with dimethylaluminum amide[112] in refluxing xylene in a 45% yield, along with 20% yield of an amide complex **233**. The amide complex **233** was transformed into the nitrile complex **223** by reaction with Me_2AlNH_2 under the same conditions. Complex **223** has already been converted to acorenone B by the previous method.[110]

Similarly, acorenone was stereoselectively synthesized through complexes **235** and **224** from the regioisomeric 2-trimethylsilyl compound **234** by the same reaction sequence.

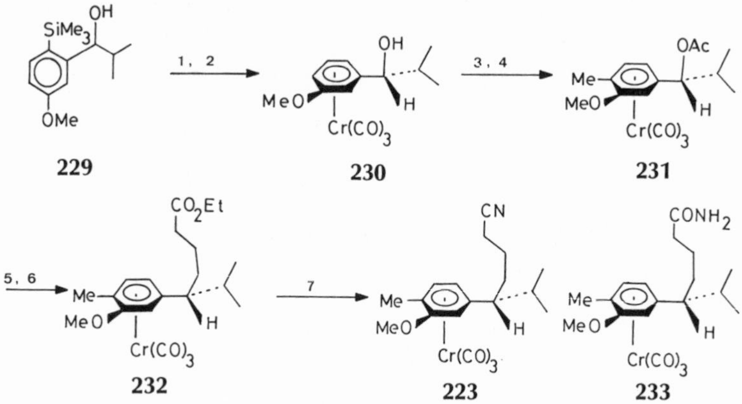

229 **230** **231**

232 **223** **233**

1) (naphthalene)Cr(CO)$_3$ 2) n-Bu$_4$N$^+$F$^-$ 3) n-BuLi/TMEDA, then MeI/HMPA 4) Ac$_2$O/pyr 5) CH$_2$=CHCH(SiMe$_3$)CO$_2$Et/BF$_3$·OEt$_2$ 6) H$_2$/PtO$_2$ 7) Me$_2$AlNH$_2$

Scheme 51

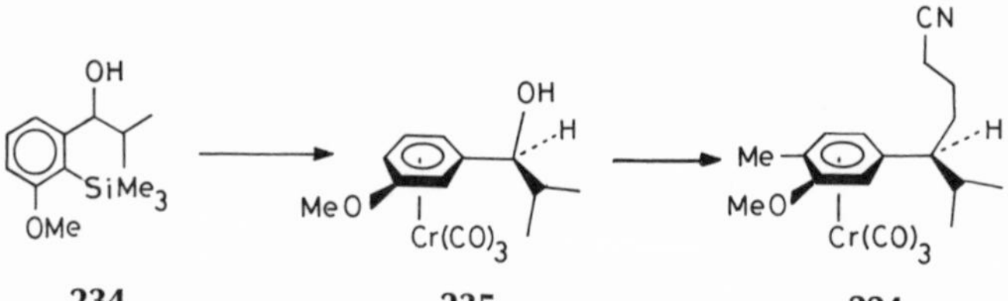

234 **235** **224**

ACKNOWLEDGEMENTS

I would appreciate the intellectual contributions of the following students: Naomi Nishikawa, Kazuhiko Take, Kazuo Isobe, Tatsuya Minami and Toshio Kobayashi. I would also like to thank Professor Yuji Hayashi for helpful discussions. This work was supported by a Grant-in-Aid from the Ministry of Japanese Education.

REFERENCES AND NOTES

1. For reviews: (a) Semmelhack, M. F., *Ann. N. Y. Acad. Sci.*, 1977, *295*, 36. (b) Jaouen, G., Arene Complexes in Organic Synthesis, in "*Transition Metal Organometallics in Organic Synthesis*", Alper, H., Ed.; Academic Press: New York, 1978; Vol 2. (c) Collman, J. P.; Hegedus, L. S., "*Principles and Applications of Organotransition Metal Chemistry*", University Science Books: Mill Valley, 1980; Chapter 14. (d) Davis, R.; Kane-Maquire, L. A. P., Chromium Compounds with η^2-η^8 Carbon Ligands, in "*Comprehensive Organometallic Chemistry*", Wilkinson, G.; Ston, F. G. A.; Abel, E. W., Eds.; Pergamon Press: Oxford: 1982; Vol. 3. Chapterh 26-2. (e) Watts, W. E., The Organic Chemistry of Metal-Coordinated Cyclopentadienyl and Arene Ligand in: "*Comprehensive Organometallic Chemistry*", Wilkinson, G.; Ston, F. G. A.; Abel, E. W., Eds.; Pergamon Press: Oxford, 1982; Vol 8. Chapter 59. (f) Davies, S. G., "*Organotransition Metal Chemistry*", Pergamon Press: Oxford, 1982. (g) Pearson, A. J., "*Metallo-Organic Chemistry*", John Wiley: New York, 1985; ch 9.

2. (a) Nicholls, N.; Whiting, M. C., *J. Chem. Soc.* 1959, 551. (b) Fisher, E. O.; Öfele, K., *Z. Naturforsch. Teil B* 1958, *13*, 458. (c) Fisher, E. O.; Öfele, K.; Esseler, H.; Frohlich, W.; Mortensen, J. P.; Semmlinger, W., *Chem. Ber.* 1958, *91*, 2763. (d) Natta, G.; Ercoli, R.; Calderazzo, F.; Santambrogio, E., *Chim. Ind. (Milan)* 1958, *40*, 1003.

3. (a) Siverthorn, W. E., *Adv. Organomet. Chem.* 1975, *13*, 47. (b) Rausch, M. D., *J. Org. Chem.* 1974, *39*, 1787. (c) Anderson, W. P.; Hsu, N.; Stanger Jr, C. W.; Munson, B., *J. Organomet. Chem.* 1974, *69*, 249. (d) Moser, G. A.; Raush, M. D., *Synth. React. Inorg. Metal. Org. Chem.* 1974, *4*, 38.

4. Top, S.; Jaouen, G., *J. Organomet. Chem.* 1979, *182*, 381.

5. Mahaffy, C. A. L.; Pauson, P. L., *Inorg. Synth.* 1979, *19*, 154.

6. Strohmeier, W., *Chem. Ber.* 1961, *94*, 2490.

7. This figure is adopted from a excellent literature; Semmelhack, M. F.; Clark, G. R.; Garcia, J. L.; Harrison, J. J.; Thebtaranonth, Y.; Wulff, W.; Yamashita, A., *Tetrahedron*, 1981, *37*, 3957.

8. Kündig, E. P., *Pure Appl. Chem.* 1985, *57*, 1855.

9. (a) Elschenbroich, C., *J. Organomet. Chem.* 1968, *14*, 157. (b) Rausch, M. D., *Pure Appl. Chem.* 1972, *30*, 523.

10. (a) Semmelhack, M. F.; Harrison, J.J.; Thebtaranonth, Y., *J. Org. Chem.* 1979, *44*, 3275. (b) Kündig, E. P.; Simmons, D. P., *J. Chem. Soc., Chem. Commun.* 1983, 1320. (c) Kündig, E. P.; Desobry, V.; Simmons, D. P., *J. Am. Chem. Soc.* 1983, *105*, 6962.

11. (a) Beck, H. J.; Fisher, E. O.; Kreiter, G. C., *J. Organomet. Chem.* 1971, *26*, C41. (b) Fisher, E. O.; Stückler, P.; Beck, H. J.; Kreisel, F. R., *Chem. Ber.* 1976, *109*, 3089.

12. (a) Semmelhack, M. F.; Bisaha, J.; Czarny, M., *J. Am. Chem. Soc.* 1979, *101*, 768. (b) Card, R. J.; Trahanovsky, W. S., *J. Org. Chem.* 1980, *45*, 2555. (c) Semmelhack, M. F.; Zask, A., *J. Am. Chem. Soc.* 1983, *105*, 2034. (d) A transmetallation reaction between bis(η^6-

phenyltricarbonylchromium)mercury and n-butyllithium afforded (η^6-phenyllithium)tricarbonyl chromium in moderate yield; Rausch, M. D.; Gloth, R. E., *J. Organomet. Chem.* 1978, *153*, 59.

13. (a) Sandilands, L. M.; Lock, C. J. L.; Faggiani, R.; Hao, N.; Sayer, B. G.; Quilliam, M. R.; Mccarry, B. E.; Mcglinchey, M. J., *J. Organomet. Chem.* 1982, *224*, 267. (b) Semmelhack, M. F.; Ullenius, C., *Ibid.* 1982, *235*, C10.

14. Ghavshou, M.; Widdowson, D. A., *J. Chem. Soc., Perkin Trans. 1* 1983, 3065.

15 Card, R. J.; Trahanovsky, W. S., *J. Org. Chem.* 1980, *45*, 2560.

16. Lepley, A. P.; Khan, W. A.; Guinamini, A. B.; Guinamini, A. G., *J. Org. Chem.* 1966, *31*, 2047.

17. (a) Masters, N. F.; Widdowson, D. A., *J. Chem. Soc., Chem. Commun.* 1983, *955*. (b) Beswick, P. J.; Leach, S. J.; Masters, N. F.; Widdowson, D. A., *Ibid.* 1984, *46*. (c) Clough, J. M.; Mann, I. S.; Widdowson, D. A., *Tetrahedron Lett.* 1987, *28*, 2645.

18. (a) Fukui, M.; Ikeda, T.; Oishi, T., *Tetrahedron Lett.* 1982, *23*, 1605. (b) Idem., *Chem. Pharm. Bull.* 1983, *31*, 466.

19. (a) Carter, O. L.; Mcphail, A. T.; Sim, G. A., *J. Chem. Soc. (A).* 1967, 1619. (b) Meurs, F. van; Toorn, J. M.; Bekkum, H., *J. Organomet. Chem.* 1976, *113*, 341.

20. Jackson, W. R.; Jennings, W. B.; Rennison, S. C.; Spratt, R., *J. Chem. Soc. (B).* 1969, 1214.

21. Albright, T. A.; Hoffmann, P.; Hoffmann, R., *J. Am. Chem. Soc.* 1977, *99*, 7546.

22. Semmelhack, M. F.; Garcia, J. L.; Cortes, D.; Farina, R.; Hong, R.; Carpenter, B. K., *Organometallics*, 1983, *2*, 467.

23. Nechvatal, G.; Widdowson, D. A.; Williams, D. J., *J. Chem. Soc. Chem. Commum.* 1981, 1260.

24. Nechvatal, G.; Widdowson, D. A., *J. Chem. Soc. Chem. Commun.* 1982, 467.

25. For reviews: (a) Gshwend, H. W.; Rodriguez, H. R., *Org. React.* 1979, *26*, 1. (b) Beak, P.; Snieckus, V., *Acc. Chem. Res.* 1982, *15*, 306. (c) *Tetrahedron Symposia-in Print*, No 9.; *Tetrahedron*, 1983, *39*, 1955–2091.

26. (a) Uemura, M.; Tokuyama, S.; Sakan, T., *Chem. Lett.* 1975, 1195. (b) Uemura, M.; Nishikawa, N.; Tokuyama, S.; Hayashi, Y., *Bull. Chem. Soc. Jpn.* 1980, *53*, 293. (c) House, H. O.; Strincland, R. C.; Zaiko, E. J., *J. Org. Chem.* 1976, *41*, 2401. (d) Trost, B. M.; Rivers, G. T.; Gold, J. M., *Ibid.* 1980, *45*, 1835. (e) Winkle, M. R.; Ronald, R. C., *Tetrahedron*, 1983, *39*, 2031.

27. Uemura, M.; Nishikawa, N.; Hayashi, Y., *Tetrahedron Lett.* 1980, *21*, 2069.

28. (a) Uemura, M.; Nishikawa, N.; Take, K.; Ohnishi, M.; Hirotsu, K.; Higuchi, T.; Hayashi, Y., *J. Org. Chem.* 1983, *48*, 2349. (b) Uemura, M.; Take, K.; Isobe, K.; Minami, T.; Hayashi, Y., *Tetrahedron Symposia-in-Print*, No 24.; *Tetrahedron*, 1985, *41*, 5771.

29. Blagg, J.; Davies, S. G.; Goodfellow, C. L.; Sutton, K. H., *J. Chem. Soc., Chem. Commun.* 1986, 1283.

30. *exo*-Hydroxyl complex was prepared from the *endo*-hydroxyl complex **26** via $Cr(CO)_3$-stabilized carbonium ion.

31. Under forcing conditions, nuclear lithiation of benzylalcohol can be effected well: (a) Panetta, C. A.; Dixt, A. S., *Synthesis*, 1981, 59. (b) Meyer, N.; Seebach, D., *Chem. Ber.* 1980, *113*, 1304.

32. (a) Wittig, G.; Davis, P.; Koeenig, G., *Ber.* 1951, *84*, 627. (b) Benzylalkyl ethers and sulfides co-ordinated by $Cr(CO)_3$ group allowed α-substitution via corresponding α-carbanion by suppression of the Wittig rearrangement; Davies, S. G.; Holman, N. J.; Laughton, C. A.; Mobbs, B. E., *J. Chem. Soc. Chem. Commun.* 1983, 1316.

33. Wakefield, B. J., *"The Chemistry of Organolithium Compounds"*, Pergamon Press: New York, 1974, p. 203.

34. Uemura, M.; Hayashi, Y., unpublished results.

35. Gill, J.C.; Marples, B. A.; Traynor, J. R., *Tetrahedron Lett.* 1987, *28*, 2643.

36. Gilday, J. P.; Widdowson, D. A., *Tetrahedron Lett.* 1986, *27*, 5525.

37. Gilday, J. P.; Widdowson, D. A., *J. Chem. Soc. Chem. Commun.* 1986, 1235.
38. (a) Blagg, J.; Davies, S. G.; Mobbs, B. E., *J. Chem. Soc. Chem. Commun.* 1985, 619. (b) Blagg, J.; Davies, S. G., *Ibid.* 1986, 492.
39. (a) Jaouen, G.; Vessières, A., *Pure Appl. Chem.* 1985, *57*, 1865. (b) Top, S.; Vessieres, A.; Abjean, J. P.; Jaouen, G., *J. Chem. Soc., Chem. Commun.* 1984, 428.
40. Abbayes, H. Boudeville, M. A., *J. Org. Chem.* 1977, *42*, 4104.
41. (a) Holmes, J. D.; Jones, D. A. K.; Pettit, R., *J. Organomet. Chem.* 1965, *4*, 324. (b) Gubin, S. P.; Khandkrarova, V. S.; Kreindlin, A. Z., *Ibid.* 1974, *64*, 229.
42. (a) Davies, R. E.; Simpson, H. D.; Griece, N.; Pettit, R., *J. Am. Chem. Soc.* 1971, *93*, 6688. (b) Bly, R. S.; Strickland, R. C.; Swindell, R. T.; Veazey, R. L., *Ibid.* 1970, *92*, 3722. (c) Bly, R. S.; Maier, T. L., *J. Org. Chem.* 1978, *43*, 614.
43. (a) Top, S.; Meyer, A.; Jaouen, G., *Tetrahedron Lett.* 1979, 3537. (b) Top, S.; Jaouen, G., *J. Chem. Soc., Chem. Commun.* 1979, 224. (c) Idem., *J. Org. Chem.* 1981, *46*, 78. (d) Top, S.; Jaouen, G.; McGlinchey, M., *J. Chem. Soc. Chem. Commun.* 1980, 1110.
44. Reetz, M. T.; Sauerwald, M., *Tetrahedron Lett.* 1983, *24*, 2837.
45. Uemura, M.; Isobe, K.; Hayashi, Y., *Tetrahedron Lett.* 1985, *26*, 767.
46. Kursanov, D. N.; Parnes, Z. N.; Lion, N. M., *Synthesis*, 1974, 633.
47. Nucleophilic or electrophilic attack at the reactive center adjacent to the aromatic moiety allows the approach of reagents from the opposite side to the bulky Cr(CO)₃ group: see Ref. 1.
48. Gracey, D. E. F.; Jackson, W. R.; Jennings, W. B.; Rennison, S. C.; Sprat, R., *J. Chem. Soc. (B)*. 1969, 1210.
49. Reetz, M. T., *"Organotitanium Reagents in Organic Synthesis"*, Springer-Verlag: Berlin, 1986.
50. Uemura, M.; Kobayashi, T.; Hayashi, Y., *Synthesis*, 1986, 386.
51. Goasmat, F.; Dabard, R. Patin, H., *Tetrahedron Lett.* 1975, 2359.
52. Meyer, A.; Jaouen, G., *J. Organomet. Chem.* 1975, *97*, C21.
53. Uemura, M.; Minami, T.; Hayashi, Y., *J. Organomet. Chem.* 1986, *299*, 119.
54. (a) Meyer, A.; Dabard, R., *J. Organomet. Chem.* 1972, *36*, C38. (b) Besancon, J.; Tirouflet, J.; Card, A.; Dausausoy, Y., *Ibid.* 1973, *59*, 267. (c) Solladie-Cavallo, A.; Suffert, J., *Tetrahedron Lett.* 1984, *25*, 1897. (d) Idem., *Synthesis*, 1985, 659 and references cited therein.
55. The symbol (*S*,S**) represents an enantiomeric mixture of (*S,S*)- and (*R,R*)-chromium complex. The first symbol indicates a configuration of the aromatic part co-ordinated by Cr(CO)₃, and the second one shows that of the chiral center at the benzylic position.
56. Reaction of Grignard reagents or organolithium compounds with (*m*-methoxybenzaldehyde)Cr(CO)₃ gave a 1:1 mixture of diastereomeric chromium complexes.
57. (a) Uemura, M.; Kobayashi, T.; Minami, T.; Hayashi, Y., *Tetrahedron Lett.* 1986, *27*, 2479. (b) Brocard, J.; Lebibi, J.; Pelinski, L.; Mahmoud, M., *Ibid.* 1986, *27*, 6335.
58. Complexation of 1-hydroxyindane with (trispyridine)Cr(CO)₃ in the presence of boron trifluoride etherate gave *endo*-complex in very low yield: Gracey, D. E. F.; Jackson, W. R.; Jennings, W. B.; Mitchell, T. R. B., *J. Chem. Soc. (B)* 1969, 1204.
59. Gracey, D. E. F.; Jackson, W. R.; McMullen, C. H.; Thompson, N.J., *J. Chem. Soc. (B)*. 1969, 1197.
60. (a) Kündig, E. P.; Perret, C.; Spichiger, S.; Bernardinelli, G., *J. Organomet. Chem.* 1985, *286*, 183. (b) Kündig, E. P.; Desobry, V.; Grivet, C.; Rudolph, B.; Spichiger, S., *Organometallics*, 1987, *6*, 1173. (c) Traylor, T. G.; Goldberg, M. J., *Ibid.* 1987, *6*, 2531. (d) Traylor, T. G.; Stewart, K. J.; Goldberg, M. J., *J. Am. Chem. Soc.* 1984, *106*, 4445. (e) Cais, M.; Kohn, D. H.; Lapid, A.; Tatarsky, D.; Dabard, R.; Jaouen, G.; Simmonlaux, G., *J. Organomet. Chem.* 1980, *154*, 91.
61. Uemura, M.; Kobayashi, T.; Isobe, K.; Minami, T.; Hayashi, Y., *J. Org. Chem.* 1986, *51*, 2859.

62. Optically active (S)-alcohol **102** was prepared by fractional recrystallization of brucine salt of an adduct from the (DL)-compound and phthalic anhydride. Optically purity was determined by ^{1}H-NMR of Mosher ester; (a) Yamaguchi, S., nuclear magnetic resonance analysis using chiral derivatives in "*Asymetric Synthesis*". Morrison, J. D., Ed., Academic Press: New York, 1983, Vol. 1, Chapter 7. (b) Meyers, A. I.; Kendall, R. M., *Tetrahedron Lett.* 1974, 1337.

63. Optically purity was determined by 1H-NMR of amide derivative from (R)-carboxylic acid **106** and L-phenylalaninol; Meyers, A. I.; Smith, R. K.; Whintten, C. E., *J. Org. Chem.* 1979, *44*, 2250.

64. Trahanovsky, W. S.; Card, R. J., *J. Am. Chem. Soc.* 1972, *94*, 2897.

65. Acampora, M.; Ceccon, A.; Dalfarra, M.; Giacometti, G.; Rigatti, G., *J. Chem. Soc. Perkin Trans. 2.* 1977, 483.

66. Uemura, M.; Minami, T.; Hayashi, Y., *J. Am. Chem. Soc.* 1987, *109*, 5277.

67. Although 5- or 7-methoxy-1-tetralone gave the corresponding arene chromium complex in good yield by the complexation with $Cr(CO)_6$, 8-methoxy-1-tetralone and 5,8-dimethoxy1-tetralone afforded no chromium complexation products under the same conditions. This failure of direct complexation reaction of these compounds may be attributed to intramolecular co-ordination of the chromium with two oxygen atoms of carbonyl and methoxyl groups.

68. (a) DeTar, D. F.; Luthra, N. P., *J. Am. Chem. Soc.* 1980, *102*, 4505. (b) Exon, C.; Magnus, P., *Ibid.* 1983, *105*, 2477.

69. Uemura, M.; Nishimura, H.; Hayashi, Y., unpublished results.

70. (a) Chan, K-K.; Cohen, N.; De Nobel, J. P.; Specian, Jr. A. C.; Saucy, G., *J. Org. Chem.* 1976, *41*, 3497. (b) Cohen, N.; Eichel, W. F,; Lopresti, R. J.; Neukom, C.; Saucy, G., *Ibid.* 1976, *41*, 3505. (c) Cohen, N.; Eichel, W. F.; Lopresti, R. J.; Neukom, C.; Saucy, G., *Ibid.* 1976, *41*, 3512. (d) Trost, B. M.; Klun, T. P., *J. Am. Chem. Soc.* 1979, *101*, 6756.

71. For reviews: (a) Trost, B. M., *Acc. Chem. Res.* 1980, *13*, 385. (b) Tsuji, J., "*Organic Synthesis with Palladium Compounds*", Springer-Verlag: New York, 1980. (c) Trost, B. M.; Verhoeven, T. R., In "*Comprehensive Organometallic Chemistry*", Wilkinson, G.; Stone, F. G. A.; Abel, E. W., Eds.; Pergamon Press: New York, 1982, Vol. 8. p. 799. (d) Tsuji, J., *Pure Appl. Chem.* 1982, *54*, 197.

72. Uemura, M.; Minami, T. Hayashi, Y., *J. Org. Chem.* 1989, *54*, 469.

73. Barton, D. H. R.; Magnus, P. D.; Smith, G.; Zurr, D., *J. Chem. Soc., Chem. Commun.* 1971, 861.

74. Gould, T. J.; Balestra, M.; Wittman, M. D.; Gary, J. A.; Rossano, L. T.; Kallmerten, J., *J. Org. Chem.* 1987, *52*, 3889 and references cited therein.

75. For reviews: (a) Mukaiyama, T., *Org. React.* 1982, *28*, 203. (b) Braun, M., *Angew. Chem. Int. Ed. Engl.* 1987, *26*, 24. (c) Masamune, S.; Choy, W.; Petersen, J. S.; Sita, L. R., *Ibid.* 1985, *24*, 1. (d) Nogradi, M., "*Stereoselective Synthesis*", VCH: Weinheim, 1986, p 193.

76. (a) Hoffman, R. W., *Angew. Chem. Int. Ed. Engl.* 1982, *21*, 555. (b) see ref. 75(d) p. 176.

77. Seebach, D.; Widler, L., *Helv. Chim. Acta.* 1982, *65*, 1972.

78. Reetz, M. T.; Steinbach, R.; Westermann, J.; Peter, R.; Wenderoth, B., *Chem. Ber.* 1985, *118*, 1441.

79. Uemura, M.; Minami, T.; Isobe, K.; Kobayashi, T.; Hayashi, Y., *Tetrahedron Lett.* 1986, *27*, 967.

80. Yamamoto, Y.; Yatagai, H.; Maruyama, K., *J. Am. Chem. Soc.* 1981, *103*, 1969.

81. Reetz, M. T.; Wenderoth, B., *Tetrahedron Lett.* 1982, *23*, 5259.

82. Yamamoto, Y.; Yatagai, H.; Maruyama, K., *J. Chem. Soc., Chem. Commun.* 1980, 1072.

83. Hiyama, T.; Okuda, Y.; Kimura, K.; Nozaki, H., *Bull. Chem. Soc. Jpn.* 1982, *55*, 561.

84. Arcamone, F., in "*Anticancer Agents Based On Natural Products Models*", Cassady, J. M.; Douros, J. D., Eds.; Academic Press: New York, 1980, 1.

85. For reviews on synthesis of anthracyclinones; (a) Krohn, K., *Angew. Chem. Int. Ed. Engl.*

1986, *25*, 790. (b) *Tetrahedron Symposia-in Print*, No 17, Kelly, T. R., Ed.; *Tetrahedron*, 1984, *49*, 4539-4793.

86. Uemura, M.; Take, K.; Hayashi, Y., *J. Chem. Soc. Chem. Commun.* 1983, 858.
87. Liu, W. C.; Parker, W. L.; Slusarchyk, D. S.; Greenwood, G. L.; Graham, S. F.; Meyers, E., *J. Antibiot.* 1970, *23*, 437.
88. Sezaki, M.; Kondo, S.; Maeda, K.; Umezawa, H.; Ohno, M. *Tetrahedron*, 1970, *26*, 5171.
89. Inamura, N.; Kakinuma, K.; Ikekawa, N.; Tanaka, H.; Omura, S., *J. Antibiot.* 1982, *35*, 602.
90. Angular anthracyclinones, ochromicinone and X-14881C, have been synthesized: (a) Katsuura, K.; Snieckus, *Can. J. Chem.* 1987, *65*, 124. (b) Guingant, A.; Barreto, M. M., *Tetrahedron Lett.* 1987, *28*, 3107.
91. This compound was obtained from 3-methoxy-*N,N*-diethylbenzamide by the literature method: de Silva, O.; Watanabe, M.; Snieckus, V., *J. Org. Chem.* 1979, *44*, 4802.
92. (a) Kende, A. S.; Rizzi, P., *Tetrahedron Lett.* 1981, *22*, 1779. (b) Kende, A. S.; Boettger, S. D., *J. Org. Chem.* 1981, *46*, 2799.
93. The A-ring aromatized compound was obtained as a major product.
94. Rathke, M. W.; Lindert, A., *Synth. Commun.* 1978, 9.
95. Uemura, M.; Minami, T.; Hayashi, Y., *J. Chem. Soc. Chem. Commun.* 1984, 1193.
96. (a) Arcamone, F.; Cassinelli, G.; Dimatteo, F.; Forenza, S.; Ripamonti, M. C.; Rivoḷa, G.; Vigevani, A.; Clardy, J.; McCabe, T., *J. Am. Chem. Soc.* 1980, *102*, 1462. (b) Cassinel, G.; Rivola, G.; Ruggieri, D.; Arcamone, F.; Grein, A.; Merli, S.; Spalla, C.; Cazza, A. M.; Marco, A. D.; Pratesi, G., *J. Antibiot.* 1982, *35*, 176.
97. Semmelhack, M. F.; Seufert, W.; Keller, L., *J. Am. Chem. Soc.* 1980, *102*, 6584.
98. (a) Gesson, J. P.; Mondon, M., *J. Chem. Soc., Chem. Commun.* 1982, 421. (b) Yadav, J.; Corey, P.; Hsu, C.-T.; Perlman, K.; Sih, C. J., *Tetrahedron Lett.* 1981, *22*, 811. (c) Kelly, T. R.; Vaya, J.; Ananthasubramanian, L., *J. Am. Chem. Soc.* 1980, *101*, 5983.
99. This problem was solved by using organocerium compound: (a) Imamoto, T.; Sugiura, Y.; Takiyama, H., *Tetrahedron Lett.* 1984, *25*, 4233. (b) Imamoto, T.; Kusumoto, T.; Tawarayama, Y.; Sugiura, Y.; Mita, T.; Hatanaka, Y.; Yokoyama, M., *J. Org. Chem.* 1984, *49*, 3904. (c) Imamoto, T.; Takiyama, N.; Nakamura, K., *Tetrahedron Lett.* 1985, *26*, 4763.
100. Kimball, S. D.; Walt, D. R.; Johnson, F., *J. Am. Chem. Soc.* 1981, *103*, 1561.
101. (a) Rowe, J. W.; Toda, J. K., *Chem. Ind. (London)* 1969, 922. (b) Burden, R. S.; Kemp, S. M., *Phytochemistry*, 1983, *22*, 1039.
102. Lindgre, O. B.; Svahn, C. M., *Phytochemistry*, 1968, *7*, 1407.
103. (a) Kashman, Y., *Tetrahedron*, 1979, *35*, 263. (b) Nishizawa, M.; Inoue, A.; Sastrapradja, S.; Hayashi, Y., *Phytochemistry*, 1983, *22*, 2083.
104. (a) Croft, K. D.; Ghisalberti, E. L.; Jefferies, P. R.; Raston, C. L.; White, A. H.; Hall, S. R., *Tetrahedron*, 1977, *33*, 1475. (b) Bunko, J. D.; Ghisalberti, E. L.; Jefferies, P. R., *Aust. J. Chem.* 1981, *34*, 2237.
105. Uemura, M.; Isobe, K.; Take, K.; Hayashi, Y., *J. Org. Chem.* 1983, *48*, 3855.
106. (a) Dabard, R.; Jaouen, G., *Bull. Soc. Chim. Fr.* 1974, 1639. (b) Jaouen, G.; Dabard, R., *Ibid.* 1974, 1646.
107. Uemura, M.; Isobe, K.; Hayasji, Y., *Chem. Lett.* 1985, 91.
108. Under usual azeotropic conditions with p-TsOH in ethylene glycol and benzene under refluxing, **213** gave no satisfactory result; Broadhurst, M. J.; Hassal, C. H.; Thomas, G. H., *J. Chem. Soc., Perkin Trans.* 1982, 2239.
109. (a) Marshall, J. A.; Brady, S. F.; Anderson, N. H., *Fortschr. Chem. Org. Naturst.* 1974, *31*, 283. (b) Krapcho, A. A., *Synthesis*, 1978, 77.
110. Semmelhack, M. F.; Yamashita, A., *J. Am. Chem. Soc.* 1980, *102*, 5924.
111. Albaugh-Robertson, P.; Katzenellenbogen, J. A., *J. Org. Chem.* 1983, *48*, 5288.
112. Wood, J. L.; Khatri, N. A.; Weinreb, S. M., *Tetrahedron Lett.* 1979, 4907.

π-BOND HYBRIDIZATION IN TRANSITION METAL COMPLEXES
A STEREOELECTRONIC MODEL FOR CONFORMATIONAL ANALYSIS

William E. Crowe and Stuart L. Schreiber

OUTLINE

Advances in Metal-Organic Chemistry, Volume 2, pages 247–267
Copyright © 1991 JAI Press Ltd
All rights of reproduction in any form reserved
ISBN: 0-89232-948-3

I. INTRODUCTION

Numerous theoretical[1] and structural[2] studies point to the fact that π-bonding plays an important role in determining the geometry of transition metal complexes. Conformational preferences of π-bonding ligands are often rationalized by simple consideration of the relative energies of the metal d-orbitals which can overlap with the ligand in question. For example, in d8 trigonal bipyramidal complexes π-acid ligands favor equatorial positions and adopt geometries which allow π-overlap with the $d_{x^2-y^2}$ or d_{xy} orbitals (metal-centered HOMOs) lying in the equatorial plane. Likewise d^6 octahedral and pseudo-octahedral complexes possessing a single carbonyl (or nitrosyl) ligand have a well-defined metal-centered HOMO lying in the plane perpendicular to the carbonyl (or nitrosyl) ligand (Figure 1). Gladysz has shown that stereoelectronic interactions involving the metal-centered HOMO in CpRe(NO)(PPh$_3$)R complexes dramatically influence the structure and reactivity in a wide variety of these chiral molecules.[3] Thus, electronic interactions involving ligand π-orbitals and metal-centered HOMOs or LUMOs, which we will term primary stereoelectronic interactions, are often reliable predictors of ligand conformation.[4]

Ordinarily, assessment of primary stereoelectronic interactions plays a key role in the conformational analysis of transition metal complexes. Nevertheless, ligand alignment is seldom uniquely determined by primary stereoelectronic interactions alone as there are often two ligand alignments that maximize π-bonding or primary stereoelectronic interaction. Considering the chiral metal acyl complexes studied by Gladysz,[5] Davies[6] and Liebeskind[7] we note that the observed alignment of the acyl carbonyl antiperiplanar to the π-bonding CO (or NO) ligand maximizes π-backbonding to the acyl ligand from the metal-centered HOMO. However, alignment of this acyl carbonyl syn to the CO (or NO) ligand also maximizes π-backbonding. This raises the question: What forces govern the preferred antiperiplanar orientation of the acyl carbonyl?

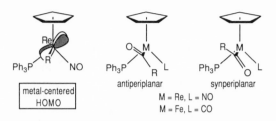

Figure 1. Alternative acyl ligand conformations that maximize primary stereoelectronic interactions.

The influence of steric interactions on the *syn–anti* isomerism of acyl ligands in $CpFe(CO)(PPh_3)(RC=O)$ and $CpRe(NO)(PPh_3)(RC=O)$ complexes has been discussed at length by Davies.[8] While steric interactions are certainly important in these systems, there are some trends which indicate that *syn–anti* isomerism in these complexes is not adequately rationalized by consideration of steric interactions alone. Notably, formyl complexes (R = H) show the same antiperiplanar conformational preference as acyl complexes possessing bulky alkyl (e.g. R = sec-butyl) substituents (vide infra).

As a number of metal acyl complexes with cis π-bonding ligands show a preference for alignment of the acyl oxygen antiperiplanar to the π-bonding ligand, and since evaluation of steric interactions does not adequately explain this alignment in all of the cases, we sought another explanation for the phenomenon. In this paper it is proposed that syn-anti isomerism in transition metal acyl complexes possessing cis π-bonding ligands is influenced by *secondary* stereoelectronic interactions. Secondary stereoelectronic interactions result from the mixing of metal-centered d and p orbitals which occurs during π bonding. This dp mixing leads to hybridization of π-bonding molecular orbitals toward the π-bonding ligand. Interaction of the C–O σ* orbital of the acyl ligand with the π-bonding orbitals of the *cis* CO (or NO) ligand leads to a stereoelectronic preference antiperiplanar orientation of the acyl ligand.

Secondary stereoelectronic interactions also influence the alignment of *cis* π-bonding ligands in systems other than the acyl complexes mentioned above. Based on the presence of orbital interactions similar to those found for acyl complexes, we propose that ligand orientations that place a ligand acceptor orbital (e.g. σ*) *syn* or a ligand donor orbital (e.g. heteroatom lone pair) anti to a *cis* π-bonding ligand are stereoelectronically favored (Figure 2). The theoretical basis for the conformational model depicted in Figure 2 will be presented below (Section III.A). First, however, we consider the structural data relevant to *syn–anti* isomerism in metal acyl complexes and the conformational model proposed by Seeman and Davies.[10]

Figure 2.   Ligand conformations favored by secondary stereoelectronic interactions.

II.   SYN–ANTI ISOMERISM IN METAL ACYL COMPLEXES

A.   Steric Interactions in Metal Acyl Complexes: The Seeman–Davies Model

Certain reactions of organic ligands attached to stereogenic transition metal centers in complexes of the type $CpML_3$ have been shown to proceed with a remarkable degree of asymmetric induction. Prompted by this observation, Seeman and Davies recently published conformational models for alkyl and acyl complexes containing the organometallic fragment CpFe-(CO)(PPh$_3$).[9,10] Their models were based on the results of extended Hückel and *ab initio* calculations on model systems where PPh$_3$ was replaced with simpler ligands. Their principal conclusions were that iron serves primarily as a template for the ligands and that alkyl or acyl ligand conformation is governed by steric interaction with the PPh$_3$ ligand.

Having noted that the conformation of triphenylphosphine is remarkably constant in a variety of complexes of the type CpFe(CO)(PPh$_3$)R, Seeman and Davies based their model for the conformational analysis of acyl complexes on the geometry depicted in Figure 3.[10] They concluded that steric interaction between the acyl ligand and one of the phenyl groups of PPh$_3$ is responsible for the acyl group preferring to be coplanar with the CO ligand. The preference for alignment of the acyl oxygen *anti* to the M–CO bond was attributed to steric interaction between the acyl ligand and the *ortho* C-H bond of a second phenyl group. Based on this analysis it was suggested that the anti conformational preference will be more pronounced for acyl ligands

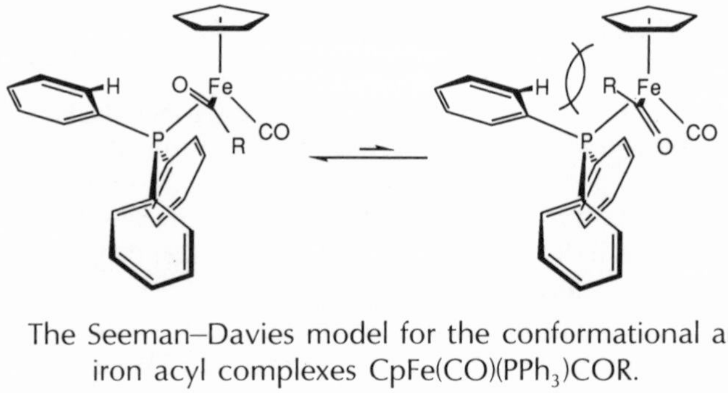

Figure 3.   The Seeman–Davies model for the conformational analysis of iron acyl complexes CpFe(CO)(PPh$_3$)COR.

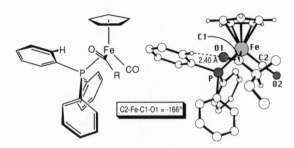

Figure 4. Structure and C_{CO}–Fe–C_{acyl}–O_{acyl} torsional angle of CpFe(CO)
(PPh$_3$)COCH(Me)Et (**1**).

with larger substituents. Consistent with the Seeman–Davies model, all
crystallographically characterized iron acyl complexes of the type CpFe-
(CO)(PPh$_3$)(COR) have acyl oxygens aligned roughly antiperiplanar to the
CO ligand. Moreover, the C_{CO}–Fe–C_{acyl}–O_{acyl} torsional angle for these complexes varies with the size of the acyl substituent indicating that steric
interactions play a role in determining acyl ligand conformation: C_{CO}–Fe–
C_{acyl}–O_{acyl} = − 166 for R = sec-butyl (**1**, Figure 4)[11] whereas C_{CO}–Fe–C_{acyl}–
O_{acyl} = − 159 for R = Me.

Rhenium acyl complexes of the type CpRe(NO)(PPh$_3$)(COR) which are
isostructural with the iron acyl complexes discussed above also exhibit antiperiplanar acyl ligand conformations. For example, CpRe(NO)(PPh$_3$)-
COCH(Me)CH$_2$Ph (2) depicted in Figure 5 has a N–Re–C_{acyl}-O_{acyl} torsional
angle of 179°.[12] In contrast to observations made for iron acyl complexes, this
rhenium acyl has nearly perfectly antiperiplanar acyl ligand alignment. This
is probably indicative of stronger primary stereoelectronic interactions (π–
backbonding) in rhenium acyls than in iron acyls. Stronger π-backbonding for
rhenium acyls is also suggested by acyl ligand carbonyl stretching frequen-
cies: v_{CO} = ca. 1560 cm^{-1} for acyl ligands attached to rhenium whereas v_{CO} =
ca. 1600 cm^{-1} for acyl ligands attached to iron.

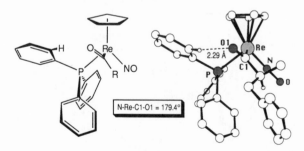

Figure 5. Structure and N-Re–C_{acyl}–O_{acyl} torsional angle of CpRe(NO)
(PPh$_3$)COCH(Me)CH$_2$Ph (**2**).

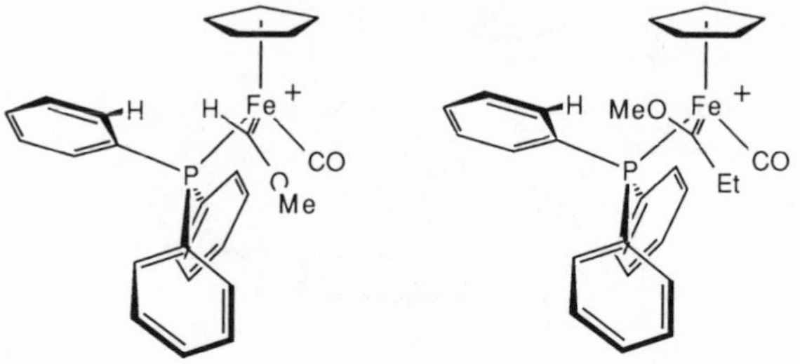

Figure 6. Structures of Fischer carbene complexes [CpFe(CO)(PPh$_3$) CHOMe]$^+$ (**3**) and [CpFe(CO)(PPh$_3$)CEtOMe]$^+$ (**4**).

Evidence for the steric interactions proposed in the Seeman–Davies model is further provided by the observed conformations of carbene ligands attached to the [CpFe(CO)(PPh$_3$)]$^+$ metal fragment. For cationic carbene complexes [CpFe(CO)(PPh$_3$)(CHOMe)]$^+$ and [CpFe(CO)(PPh$_3$)(CEtOMe)]$^+$ depicted in Figure 6[8a] the observed orientation of the carbene ligand is that which places the smaller substituent *anti* to the CO ligand. Similar observations have been reported for benzylidene ligands attached to the isosteric rhenium fragment [CpRe(NO)(PPh$_3$)]$^+$.[13] For these rhenium complexes the anti isomer is clearly the thermodynamic product as the less stable *syn* isomer is the exclusive kinetic product (Figure 7).

B. Acyl Ligand Conformations not Explained by the Seeman–Davies Model

If Seeman–Davies-type steric interactions are the only forces responsible for the observed conformational preferences for metal acyl complexes then

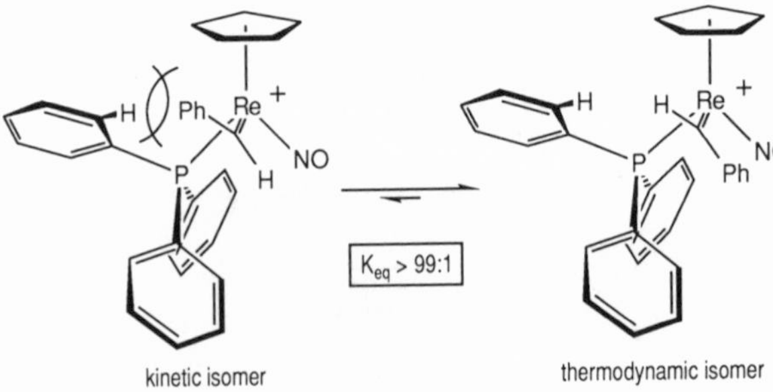

kinetic isomer   thermodynamic isomer

Figure 7. Kinetic (**5k**) and thermodynamic (**5t**) isomers of [CpRe(NO) (PPh$_3$)CHPh]$^+$.

formyl complexes should — due to the smaller size of H vs. O on the formyl ligand — show a preference for alignment of the acyl oxygen *syn* to the M–CO bond. *This is not the case.* The structure of CpRe(NO)(PPh$_3$)(CHO), has been determined by Gladysz[14] and this formyl complex has the formyl oxygen oriented antiperiplanar (*N*-Re–C$_{acyl}$–O$_{acyl}$ torsional angle = 176°)[15] to the M–NO bond (Figure 8). This conformation is not predicted by the Seeman–Davies model. Gladysz has noted that the observed conformation allows for a stabilizing electronic interaction between the highest occupied d orbital on the rhenium and the π*$_{CO}$ orbital on the acyl ligand. The rhenium formyl complex also shows a very short Re–C$_{formyl}$ bond (2.06 Å) and a low formyl carbonyl stretching frequency (v_{co} = 1558 cm^{-1}); both of these observations are in accord with a ubstantial amount of Re to formyl electron donation. However, synperiplanar orientation of the formyl ligand would allow for the same electronic stabilization. There are no obvious intramolecular steric interactions between the other ligands and the formyl ligand; in fact the triphenylphosphine conformation is as observed by Davies and Seeman in iron acyl complexes. Intermolecular steric interactions (crystal packing forces) are unlikely to play a major role in determining the formyl ligand conformation due to the smallness of this ligand relative to other ligands on the complex. Clearly, some other force must be acting to align the formyl ligand in the observed antiperiplanar orientation.

Based on the examination of some other metal acyl complexes with cis CO ligands that do not possess Seeman–Davies-type steric interactions (Figures 9–11), it is tempting to speculate that there is an intrinsic electronic preference for alignment of the acyl oxygen *anti* to the CO ligand. In Herndon's iron acetyl complex **7**[16] (Figure 9) the bulky triphenylphosphine ligand central to the Seeman–Davies model is replaced by the smaller dimethylphenylphosphine ligand. Although there is no steric interaction between the acyl

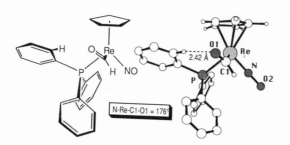

Figure 8.   Structure and *N*-Re–C$_{acyl}$–O$_{acyl}$ torsional angle of CpRe(NO) (PPh$_3$)CHO (**6**).

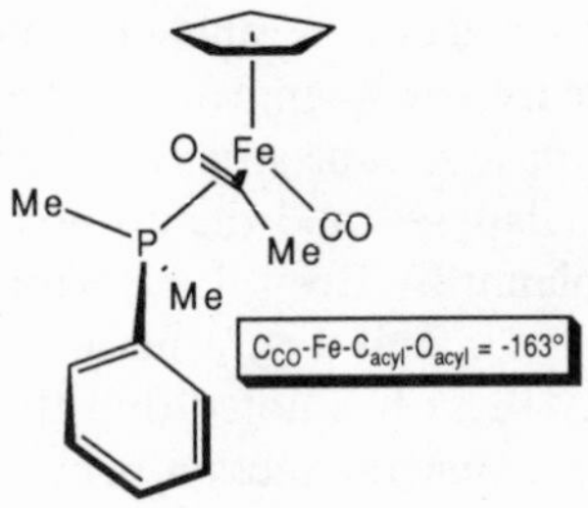

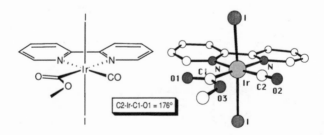

Figure 9.   Solid state geometry of CpFe(CO)(PMe$_2$Ph)COMe (**7**).

ligand and the ortho C–H bond of a phenyl group, the acetyl ligand adopts an antiperiplanar alignment. The carbomethoxy ligand in Albano's octahedral iridium complex **8**[17] (Figure 10) certainly is in a very different steric environment than found in pseudooctahedral complexes of the type CpFe-(CO)(PPh$_3$)(COR), yet the carbonyl oxygen of the carbomethoxy ligand is nevertheless aligned antiperiplanar to the CO ligand. Berke's octahedral rhenium complex **9**[18] depicted in Figure 11 is another example of a rhenium formyl complex where the formyl oxygen is aligned *anti* to the *cis* π-bonding ligand (cf. Gladysz's complex, Figure 8) despite steric interactions which should favor *syn* alignment.

Electronic interactions that lead to preferential stabilization of the antiperiplanar rotational isomer of the metal acyl complexes discussed above are depicted in Figure **12**. Primary stereoelectronic interaction is optimized when the acyl ligand is coplanar with the M–CO bond as this allows for maximum overlap of the π*$_{CO}$ orbital of the acyl ligand with the metal-centered HOMO (cf. Figure 1). Secondary stereoelectronic interaction is optimized when the acyl ligand is antiperiplanar to M–CO bond as this allows for maximum overlap of the σ*$_{CO}$ orbital of the acyl ligand with the M–CO π–bonding orbital which is metal-centered and hybridized toward the CO ligand. Similar

Figure 10.   Structure and C$_{CO}$–Ir–C$_{acyl}$–O$_{acyl}$ torsional angle of Ir(bipy)I$_2$(CO)CO$_2$Me (**8**).

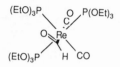

Figure 11. Solid state geometry of Re(CO)$_2$[P(OEt)$_3$]$_3$CHO (**9**).

electronic effects have been invoked to explain the stability of the Z conformation of carboxylic esters.[19] The theoretical basis for π-bond hybridization and secondary stereoelectronic interactions is presented in the next section. We close this section with a discussion of some metal acyl complexes with *cis* CO ligands where the acyl ligands are not oriented antiperiplanar to CO ligands.

Ruthenium formyl complex **10** (Figure 13), prepared by Nelson,[20] has a C_{CO}–Ru–C_{formyl}–O_{formyl} torsional angle of $-28°$. This complex possesses several unusual structural features including an exceptionally short C–O bond length (1.106 Å) and a large Ru–C–O bond angle (140°) for the formyl ligand — the C–O bond of the formyl ligand is shorter than that of the carbonyl ligand present the same complex (C–O = 1.134 Å for the carbonyl ligand). The reasons for the unusual structural features present in **10** are unclear. However, a comparison between Nelson's rhutenium formyl complex (**10**: C–O_{formyl} = 1.11 Å, v_{formyl} CO = 1601 cm^{-1}) and Gladysz's rhenium formyl complex (**6**: C–O_{formyl} = 1.22 Å, v_{formyl} CO = 1554 cm^{-1}) indicates that there is substantially less backbonding to and hence less electronic stabilization of the formyl ligand in the rhuthenium complex. In Nelson's complex, the stereoelectronic preference for antiperiplanar alignment of the formyl ligand is presumably overridden by the steric demand of the bulky pentamethylcyclopentadienyl ligand.

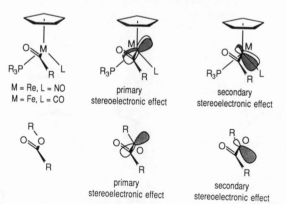

Figure 12. Stereoelectronic interactions in metal acyl complexes and carboxylic esters.

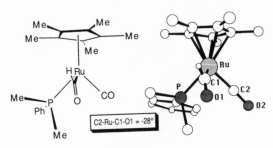

Figure 13. Structure and C_{CO}–Ru–C_{formyl}–O_{formyl} torsional angle of Cp*Ru
(CO)(PMe$_2$Ph)CHO (**10**).

Polar groups such as the acyl C–O bond in metal acyl complexes are often involved in electrostatic interactions. Figure **14** shows two acyl complexes whose solid state structures are determined by electrostatic interactions. The orientation of the acetyl ligand in iron complex **11** is controlled by a strong hydrogen bond to the NH group attached to the phosphine ligand.[21] Orientation of acetyl ligands in the binuclear rhodium complex **12** is determined by dipole–dipole interaction.[22] The observed conformation leads to the most favorable alignment of Rh$^{\delta+}$–I$^{\delta-}$ and C$^{\delta+}$–O$^{\delta-}$$_{acyl}$ dipoles.

Although not all acyl complexes with *cis* π-bonding ligands have acyl oxygens aligned antiperiplanar to the π-bonding ligand, the structural data presented above suggests that antiperiplanar alignment is an intrinsic preference. The origin of this preference is not apparent from the available experimental data. However, consideration of rhenium formyl complexes **6** and **9** suggests that antiperiplanar alignment is not explained by analysis of steric interactions alone. We proposed above (Figure 12) that the intrinsic preference for antiperiplanar acyl ligand alignment is due to a secondary stereoelectronic interaction. In the next section the theoretical basis of this stereoelectronic model is presented.

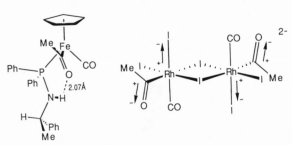

Figure 14. Structures CpFe(CO)(PPh$_2$NHCHMePh)COMe (**11**) and {[RhI$_3$(CO)COMe]$_2$}$^{2-}$ (**12**) where electrostatic interactions determine acyl ligand conformation.

III. π-BOND HYBRIDIZATION

A. Second-Order Perturbation Theory and π-Bond Hybridization

Perturbation theory provides a means by which the energies and shapes of molecular orbitals can be correlated to the orbitals of interacting fragments. Estimation of interaction energies and mixing coefficients can be done quantitatively in molecular orbital calculations, or qualitatively by relating the interaction integrals to orbital overlap. The qualitative assessment of orbital energies and shapes is beautifully illustrated in a recent text by Albright et al.[23]

Evaluation of the orbital mixing which occurs when a π-bonding ligand interacts with a transition metal center requires the assessment of second-order mixing which takes place when two or more orbitals on the same fragment (i.e. metal d and p orbitals) overlap with the same orbital of an interacting ligand (i.e. ligand p orbital). The result of this second-order mixing is that bonding and antibonding orbitals are hybridized toward the ligand involved in the bonding interaction, and nonbonding orbitals are hybridized away from the ligand.[24] This is illustrated in Figure 15 for the interaction of a π-bonding ligand with a transition metal center. Since this result is central to our description of π-bond hybridization it will be elaborated further below.

Represented below are the equations of perturbation theory. Equation 1 shows that interaction with a higher energy orbital ($e_j > e_i$) is stabilizing ($E_{int} < 0$), and interaction with a lower energy orbital ($e_j < e_i$) is destabilizing ($E_{int} > 0$). The shapes of molecular orbitals are determined by the mixing coefficients as indicated in Eqs. 2–5.

$$E_{int} = -(H_{ij} - e_i S_{ij})^2 / (e_j - e_i) \quad \text{interaction energy} \quad (1)$$

$$\Psi_i^M = c_{ii}\phi_i^F + c_{ji}\phi_i^L + c_{ki}\phi_k^F \quad \text{molecular orbital} \quad (2)$$

$$c_{ji} = -(H_{ij} - e_i S_{ij})/(e_j - e_i) \quad \text{first-order mixing coefficient} \quad (3)$$

$$c_{ki} = (H_{ij} - e_i S_{ij})(H_{jk} - e_i S_{jk})/(e_j - e_i)(e_k - e_i)$$

$$\text{second-order mixing coefficient} \quad (4)$$

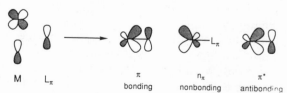

M   L$_\pi$ π   n$_\pi$ π*
                        bonding   nonbonding   antibonding

Figure 15. Orbital mixing leading to π-bond hybridization.

$$c_{ii} = 1 - S_{ij}c_{ji} - 1/2(c_{ji})^2 \quad \text{normalization coefficient} \qquad (5)$$

where S_{ij}, S_{jk} are overlap integrals; H_{ij}, H_{jk} are interaction integrals; e_i, e_j, e_k are orbital energies (Coulomb integrals); ϕ_i^F, ϕ_j^L are fragment and ligand interacting orbitals; and ϕ_k^F is the fragment hybridizing orbital.

The second order mixing coefficient can be written as a product of the first order coefficient and a second term. Hence Eq. 2 can be rewritten as Eq. 7. Note that c'_{ki} has the form of a mixing coefficient (Eq. 8).

$$c_{ki} = c_{ji} \cdot c'_{ki} \qquad (6)$$

$$\Psi_i^M = c_{ii}\phi_i^F + c_{ji}[\phi_j^L + c'_{ki}\phi_k^F] \qquad (7)$$

$$c'_{ki} = -(H_{jk} - e_i S_{jk})/(e_k - e_i) \qquad (8)$$

The shapes of the molecular orbitals can be determined qualitatively by the signs of the mixing coefficients. These can be simply related to spatial overlap between the interacting orbitals. Equation 9 is a generalized version of the Wolfsberg–Helmholz formula which relates interaction integrals (H_{ij}) to orbital overlap. Using this approximation we note that the mixing coefficient c_{ji} and the pseudo-mixing-coefficient c'_{ki} are proportional to orbital overlap and inversely proportional to energy separation (Eqs. 10 and 11). An important assumption made here is that the interaction integral is the dominant term in the numerator of the equations defining the mixing coefficients (Eq. 12).

$$H_{ij} = -f(e_i, e_j) \cdot S_{ij} \quad \text{Wolfsberg-Helmholz formula}^{25} \qquad (9)$$

$$c_{ji} \propto S_{ij}/(e_j - e_i) \qquad (10)$$

$$c'_{ki} \propto S_{jk}/(e_k - e_i) \qquad (11)$$

$$|H_{ij}| > |e_i S_{ij}| \quad \text{"normal" orbital mixing}^{24} \qquad (12)$$

If the phases of the interacting orbitals are arranged so that all overlap integrals are positive, the sign of the coefficients c_{ji} and c'_{ki} will be solely determined by the relative energies of the interacting orbitals. Interaction with a higher energy orbital yields a positive coefficient; interaction with a lower energy orbital yields a negative coefficient. Four cases may be considered: (1) e_j and $e_k > e_i$; (2) e_j and $e_k < e_i$; (3) $e_j > e_i$ and $e_k < e_i$; (4) $e_k > e_i$ and $e_j < e_i$. Cases 1 and 2 correspond to bonding and antibonding orbitals, respectively. Cases 3 and 4 correspond to nonbonding orbitals. Case 4 is illustrated in Figure 16.

Hybridization is determined by the second order mixing coefficient, c_{ki}. The molecular orbital is hybridized toward the ligand (or ligand set) when c_{ki} is positive and hybridized away from the ligand when c_{ki} is negative. The second order mixing coefficient is the product of two terms (Eq. 6). C_{ki} is positive when both of these terms have the same sign and negative when they

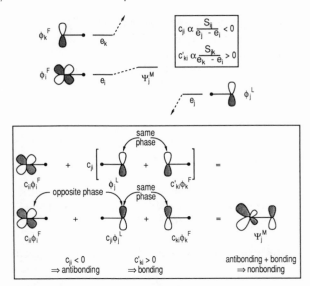

Figure 16. Second-order mixing illustrated for the hybridization of a non-bonding orbital.

have the opposite sign. In cases 1 and 2 above the second order mixing coefficient is positive and in cases 3 and 4 it is negative. Hence, *bonding and antibonding orbitals are hybridized toward the ligands and nonbonding orbitals are hybridized away from the ligands.*

B. Angular Distortions that Suggest π-Bond Hybridization

Ligands that are *cis* to a π-bonding ligand almost always bend away from the π-bond. Ibers has attributed the bending back of cis ligands to steric

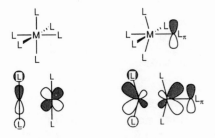

Figure 17. Ligands that are *cis* to a π-bonding ligand bend away from the π bond.

interactions caused by the shortness of the metal-ligand multiple bond,[26] whereas Hoffmann has rationalized this angular distortion on electronic grounds.[27] Regardless of the cause, bending away the cis ligands is concomitant with hybridization of the π-bonding orbital toward and the non-bonding n_π orbital away from the π-bonding ligand (Figure 17).

C. Stereoelectronic Interactions that Influence Syn–Anti Isomerism

Inspection of Figure 2 indicates that there are three types of stereoelectronic interactions whch could influence the syn–anti isomerism of a ligand bound *cis* to a π-bonding ligand. These are:

(a) $\pi \leftrightarrow$ acceptor (e.g. σ^*)—a stabilizing two-center–two–electron interaction.

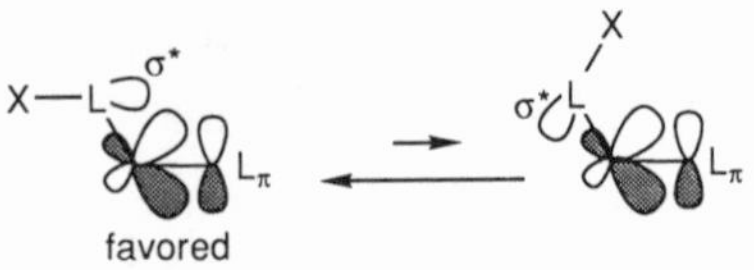

(b) $\pi \leftrightarrow$ donor (e.g. lone pair)—a destabilizing two-center–four–electron interaction.

(c) $n_\pi \leftrightarrow$ donor (e.g. lone pair)—a stabilizing two-center–two–electron interaction.

The relative importance of the three types of stereoelectronic interactions depicted above will, of course, depend on the nature of the metal–L_π bonding interaction.[28] However, in all cases ligand conformations which place a ligand acceptor orbital *syn* or a donor orbital *anti* to a metal ligand π bond are stereoelectronically favored (Figure 2).

IV. SYN–ANTI ISOMERISM OF LIGANDS OTHER THAN η^1-ACYL

A. Bent Nitrosyl Ligands

$$
\underset{\text{favored}}{\underset{\text{M}}{\text{OC}}\diagdown\overset{\overset{\overset{O}{\parallel}}{N}}{}}\quad\xrightarrow{}\quad\longleftarrow\quad\underset{\text{M}}{\text{OC}}\diagdown\overset{\overset{(\cdot\cdot)}{N}}{}\diagup{O}
$$

(see Figure 2)

X-ray crystal structures of bent-nitrosyl complexes [Ir(P Ph$_3$)$_2$(CO)(NO)Cl]$^+$ (**13**)[29] and [Ru(PPh$_3$)$_2$(NO)$_2$Cl]$^+$ (**14**)[30] are shown in Figures 18 and 19. In both cases the bent nitrosyl ligand adopts the stereoelectronically preferred conformation (see Figure 2) that places its lone pair antiperiplanar to the *cis* π–bonding ligand.

Bent nitrosyl complexes are related to acyl complexes by the isoelectronic substitution N for CR. As this substitution replaces C–R bond with a nitrogen lone pair, the stereoelectronically preferred alignment of the bent nitrosyl N=O bond is different from that of the acyl C=O bond (Eq. 13).

$$
\underset{\substack{\text{M}-\text{CO}\\\text{acyl complex}}}{\overset{O}{\diagup}\diagdown^R}\qquad\underset{\substack{\text{M}-\text{CO}\\\text{bent nitrosyl complex}}}{\overset{O}{\diagdown}\!\!=\!\!N^{\cdot\cdot}}\;\underset{\longleftarrow}{\longrightarrow}\;\underset{\text{M}-\text{CO}}{\overset{\cdot\cdot}{N}\diagup^O}
$$

It is interesting to compare bent nitrosyl and acyl complexes to thier organic analogs: nitrite and carboxylic esters. As nitrite esters are related to carboxylic esters by the same isoelectronic substitution (N for CR), the stereoelectronically preferred alignment of the nitrite N=O bond should be different from that of the carboxylic C=O bond. In fact, this appears to be

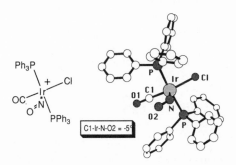

Figure 18. Structure and C_{CO}–Ir–N–O torsional angle for [Ir(PPh$_3$)$_2$ (CO)(NO)Cl]$^+$ (**13**).

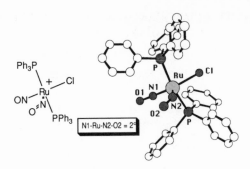

Figure 19. Structure and N_{linear}–Ir–N_{bent}–O torsional angle of [Ru(PPh$_3$)$_2$
(NO)$_2$Cl]$^+$)**14**).

the case: In both methyl nitrite[31] and methyl formate[32] the *Z* conformer is
preferred, but to a very different extent (Eqs. 14 and 15). This difference is

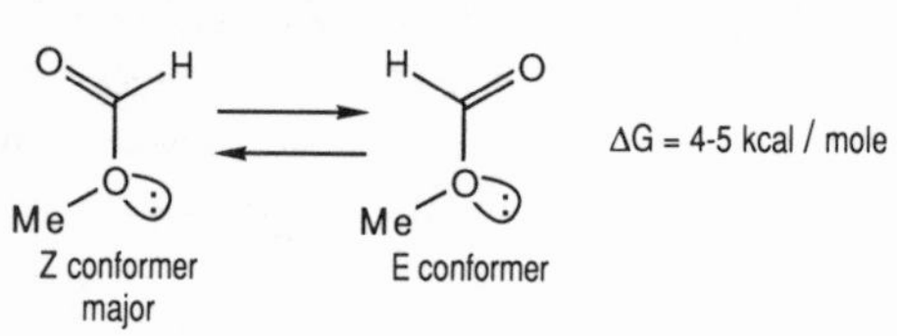

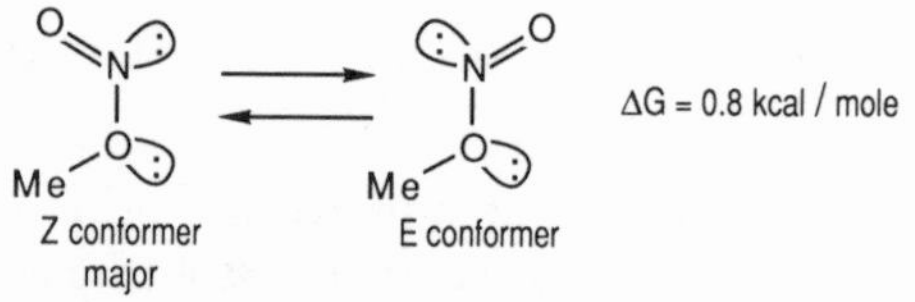

$$\Delta G = 4\text{-}5 \text{ kcal / mole}$$

Z conformer
major

E conformer

$$\Delta G = 0.8 \text{ kcal / mole}$$

Z conformer
major

E conformer

understood if the forces governing these conformational preferences are
viewed as a superposition of electrostatic (dipolar) and stereoelectronic
interactions (Figure 20). Electrostatic forces favor the *Z* conformation for
both nitrite and carboxylic esters. Stereoelectronic forces, however, favor the
Z conformation for carboxylic esters and the *E* conformation for nitrite
esters. In the case of methyl formate stereoelectronic and electrostatic forces
act in unison to produce a energetically substantial conformational prefer-

electronic
and
dipolar
forces

electronic
forces

dipolar
forces

carboxylic   ester
cooperative   forces
⇒ large ΔG

nitrite   ester
opposing   forces
⇒ small ΔG

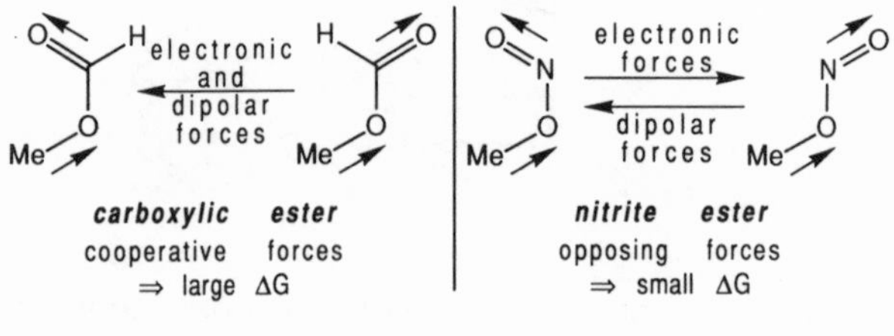

Figure 20. Electronic and dipolar forces in organic esters.

ence, whereas in the case of methyl nitrite these forces largely cancel out to produce a small conformational preference.

B. Acetylene Ligands

Figures 21–23 shows some ynolate and ynol ether complexes of tungsten (15–17) which are derived from carbyne–CO coupling reactions.[33] In each case the alkyne is aligned parallel to the M–CO bond axis and its oxygen substituent is oriented *syn* to the carbonyl ligand. The electronic interactions that cause the acetylene ligands to be parallel to the carbonyl ligand have been discussed at length by Templeton.[34] The acetylene is thus aligned so that it may receive electron density from a filled tungsten-centered orbital and donate electron density to an empty one. The *syn* orientation of the oxygen substituent is not, however, explained by these interactions. The π^*_{CC} orbital of an ynolate or ynol ether is hybridized toward the carbon bearing the oxygen substituent and the filled metal-centered orbital with which it interacts is polarized toward the carbonyl ligand (Figure 24). Thus, the observed conformation of the acetylenes is that which maximizes the overlap between these two orbitals.

C. Bent Imido Ligands

(see Fig. 2)

X-ray crystal structures of bis-imido complexes $Mo(S_2CNEt_2)_2(NPh)_2$ (**18**)[35] and $OsO_2(N-t-Bu)_2$ (**19**)[36] are shown in Figures 25 and 26. In both

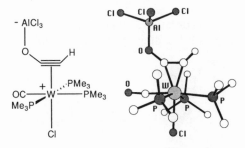

Figure 21. Structure of W(HC≡COAlCl₃)(CO)(PMe₃)₃Cl (**15**).

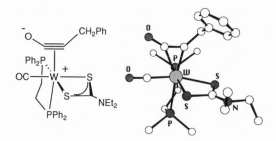

Figure 22.   Structure of W(PhCH$_2$C≡CO)(CO)(S$_2$CNEt$_2$)$_2$ (**16**).

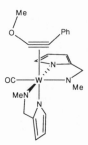

Figure 23.   Solid state structure of W(PhC≡COMe)(CO)[2-(MeNCH$_2$) pyrrole]$_2$ (**17**).

cases the bent imido ligand adopts the stereoelectronically preferred conformation (see Figure 2) that places its lone pair antiperiplanar to the metal nitrogen triple bond of the linear imido ligand.

Interestingly, both **18** and **19** adopt conformations which place the imido subtituent in a sterically more hindered environment. In these cases secondary stereoelectronic effects override steric preferences.

V.   CONCLUSION

Syn–anti ligand conformational preferences in a number of transition metal complexes possessing *cis* π-bonding ligands are not adequately explained by

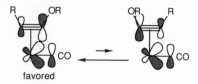

Figure 24.   Preferred acetylene alignment for maximum π-backbonding in complexes **15–17**.

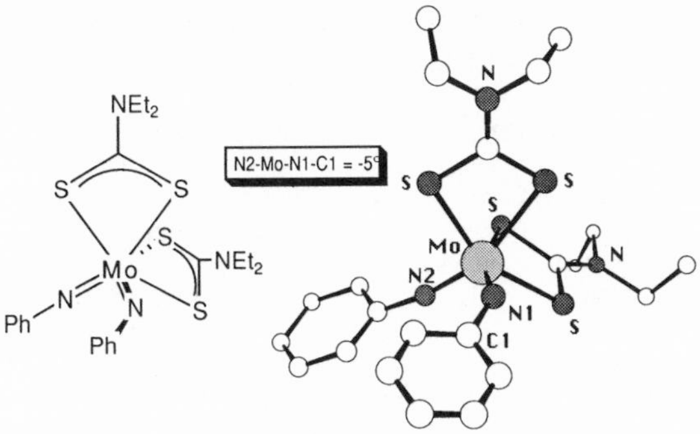

Figure 25. Structure and N_{linear}–Mo–N_{bent}–C_{ipso} torsional angle of Mo(S$_2$ CNEt$_2$)$_2$(NPh)$_2$ (**18**).

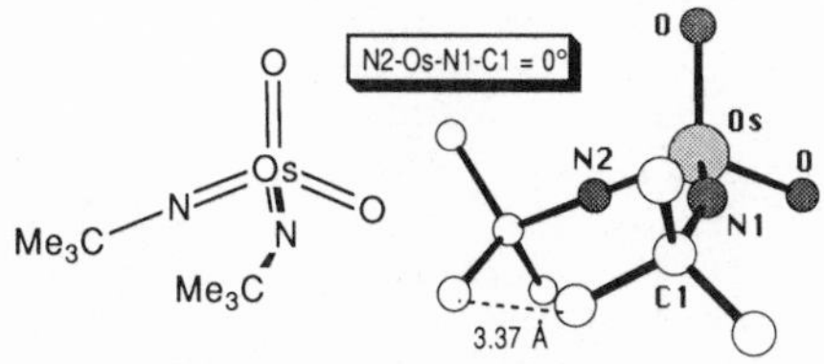

Figure 26. Structure and N_{linear}–Mo–N_{bent}–C torsional angle of OsO$_2$(N-t-Bu)$_2$ (**19**).

consideration of steric and primary stereoelectronic interactions alone. A stereoelectronic model has been developed here which does successfully rationalize these preferences.

Notably, the cases where stereoelectronic *syn–anti* preferences override clear steric preferences (complexes **6, 9, 18** and **19**) occur for metals (Re, Mo and Os) that have been noted to show a predilection for forming metal-ligand multiple bonds.[37] It may be that stereoelectronic interactions (primary and secondary) are more important for these metals than for metals, such as Fe, which do not show a tendency to form multiple bonds. It will be interesting to see if the stereoelectronic model described herein finds application in the conformational analysis of newly discovered organometal systems.

REFERENCES AND NOTES

1. (a) Albright, T. A. *Acc. Chem. Res.* 1982, *15*, 149. (b) Schilling, B. E. R.; Hoffmann, R.; Faller, J. W. *J. Am. Chem. Soc.* 1979, *101*, 592. (c) , N. M.; Fenske, R. F. *Organometallics* 1982, *1*, 974. (d) , N. M.; Fenske, R. F. *J. Am. Chem. Soc.* 1982, *104*, 3879. (e) Hall, M. B.; Fenske, R. F. *Inorg. Chem.* 1972, *11*, 768. (f) Rappé, A. K.; Goddard, III, W. A. *J. Am. Chem. Soc.* 1982, *104*, 3287. (g) Rappé, A. K.; Goddard, III, W. A. *J. Am. Chem. Soc.* 1982, *104*, 448. (h) Carter, E. A.; Goddard, III, W. A. *J. Am. Chem. Soc.* 1986, *108*, 4746.
2. (a) Nugent, W. A.; Mayer, J. M. "Metal–Ligand Multiple Bonds"; Wiley-Interscience:

New York, 1988. (b) Schrock, R. R. *Acc. Chem. Res.* 1979, *12*, 98. (c) Templeton, J. L.; Winston, P. B.; Ward, B. C. *J. Am. Chem. Soc.* 1981, *103*, 7713.

3. Georgiou, S.; Gladysz, J. A. Tetrahedron 1986, *42*, 1109 .

4. Since the evolution of the frontier orbitals that participate in primary stereoelectronic interactions is fairly well understood and has been discussed at length elsewhere (see, for example, Ref. 1a) it will not be considered further here.

5. Bodner, G. S.; Patton, A. T.; Smith, D. E.; Georgiou, S.; Tam, W.; Wong, W. -K.; Gladysz, J. A. *Organometallics* 1987, *6*, 1954 .

6. Davies, S. G.; Dordor-Hedgecock, I. M.; Sutton, K. H.; Walker, J. C. *Tetrahedron* 1986, *42*, 5123 .

7. Liebeskind, L. S.; Welker, M. E.; Fengl, R. W. *J. Am. Chem. Soc.* 1986, *108*, 6328.

8. (a) Blackburn, B. K.; Davies, S. G.; Whittaker, M. in *"Stereochemistry of Organometallic and Inorganic Compounds"*; Bernal, I. (Ed.), Elsevier Science: Amsterdam, 1988, Vol. 3. (b) Blackburn, B. K.; Davies, S. G.; Sutton, K. H.; Whittaker, M. *Chem. Soc. Rev.* 1988, *7*, 147.

9. (a) Seeman, J. I.; Davies, S. G. *J. Chem. Soc., Chem. Commun.* 1984, 1019. (b) Seeman, J. I.; Davies, S. G. *J. Am. Chem. Soc.* 1985, *107*, 6522. (c) Davies, S. G.; Dordor-Hedgecock, I. M.; Sutton, K. H.; Whittaker, M. *J. Am. Chem. Soc.* 1987, *109*, 5711.

10. Seeman, J. I.; Davies, S. G.; Williams, I.H . *Tetrahedron Lett.* 1986, *27*, 619.

11. Baird, G. J.; Bandy, J. A.; Davies, S. G.; Prout, K. *J. Chem. Soc., Chem. Commun.* 1983, 1202.

12. Smith, D. E.; Gladysz, J. A. *Organometallics* 1985, *4*, 1480.

13. (a) Kiel, W. A.; Lin, G- .Y.; Constable, A. G.; McCormick, F. B.; Strouse, C. E.; Eisenstein, O.; Gladysz, J. A. *J. Am. Chem. Soc.* 1982, *104*, 4865. (b) Kiel, W. A.; Buhro, W. E.; Gladysz, J. A. *Organometallics* 1984, *3*, 879.

14. Wong, W-. K.; Tam, W.; Strouse, C. E.; Gladysz, J. A. *J. Chem. Soc., Chem. Commun.* 1979, 530.

15. Structures from the Cambridge Crystallographic Data Base were examined using software developed by Simon K.Kearsley at Yale University.

16. Herndon, J. W.; Wu, C.; Ammon, H. L. *J. Org. Chem.* 1988, *53*, 2873.

17. Abano, V. G.; Bellon, P. L.; Sansoni, M. Inorg. Chem. 1969, *8*, 298.

18. Sontag, C.; Orama, O.; Berke, H. *Chem. Ber.* 1987, *120*, 559.

19. Deslongchamps, P. "Stereoelectronic Effects in Organic Chemistry"; Pergamon: Oxford, 1983; Chapter 3.

20. Nelson, G. O.; Sumner, C. E. *Organometallics* 1986, *5*, 1983.

21. Korp, J. D.; Bernal, I. *J. Organomet. Chem.* 1981, *220*, 355.

22. Adamson, G. W.; Daly, J. J.; Forster, D. *J. Organomet. Chem.* 1974, *71*, C17.

23. Albright, T. A.; Burdett, J. K.; Whangbo, M-. H. *Orbital Interactions in Chemistry*"; Wiley-Interscience: New York, 1985.

24. The assumption here is that $|H_{ij}| > |e_i S_{ij}|$. In cases where this relationship does not hold counterintuitive orbital mixing (COM) occurs. Most of the time the effects of COM are small and do not affect the frontier orbitals of the molecule. For further discussion, see: Whangbo, M.-H.; Hoffmann, R. *J. Chem. Phys.* 1978, *68*, 5498.

25. In the original Wolfsberg–Helmholz formula $-f(e_i, e_j) = K (e_i + e_j)$, where $K =$ a proportionality constant.

26. Bright, D.; Ibers, J. A. *Inorg. Chem.* 1969, *8*, 709.

27. DuBois, D. L.; Hoffmann, R. *Nouv. J. Chim.* 1977, *1*, 479.

28. For example, if L_π is a π-acidic ligand on a low-valent metal center the M–L_π π bonding orbital will be metal-centered, whereas if L_π is a π-basic ligand on a high-valent metal center the M–L_π bonding orbital will be ligand-centered. Hence, stereoelectronic interactions involving M–L_π π bonding orbitals should be less important for π-basic ligands, due to poorer orbital overlap with other ligands.

29. (a) Hodgson, D. J.; Ibers, J. A. Inorg. Chem. 1968, 7, 2345. (b) Hodgson, D. J.; Ibers, J. A. Inorg. Chem. 1969, 8, 1282.

30. Pierpont, C. G.; Van Derveer, D. G.; Durland, W.; Eisenberg, R. J. Am. Chem. Soc. 1970, 92, 4760.

31. Turner, P. H.; Corkill, M. J.; Cox, A. P. J. Phys. Chem. 1979, 83, 1473.

32. (a) Ruschin, S.; Bauer, S. H. J. Phys. Chem. 1980, 84, 3061. (b) Blom, C. E.; Günthard, Hs. H. Chem. Phys. Lett. 1981, 84, 267.

33. (a) Churchill, M. R.; Wasserman, H. J.; Holmes, S. J.; Schrock, R. R. Organometallics 1982, 1, 766. (b) Birdwhistell, K. R.; Tonker, T. L.; Templeton, J. L. J. Am. Chem. Soc. 1985, 107, 4474. (c) Mayr, A.; McDermott, G. A.; Dorries, A. M.; Van Engen, D. Organometallics 1987, 6, 1053.

34. (a) Templeton, J. L.; Winston, P. B.; Ward, B. C. J. Am. Chem. Soc. 1981, 103, 7713. (b) Tatsumi, K.; Hoffmann, R.; Templeton, J. L. Inorg. Chem. 1982, 21, 466. (c) Winston, P. B.; Burgmayer, S. J. N.; Templeton, J. L. Organometallics 1983, 2, 167.

35. Haymore, B. L.; Maata, E. A.; Wentworth, R. A. D. J. Am. Chem. Soc. 1979, 101, 2063.

36. Nugent, W. A.; Harlow, R. L.; McKinney, R. J. J. Am. Chem. Soc. 1979, 101, 7265.

37. The formation of metal-ligand multiple bonds seems to be most favorable along a diagonal from vanadium to osmium: see Nugent, W. A.; Mayer, J. M. Metal–Ligand Multiple Bonds; Wiley-Interscience: New York, 1988; pp. 29–32.

38. Some of the illustrations used here have been used in a later review article appearing in: Shembayati, S.; Crowe, W. E.; Schreiber, S. L. Angew. Chem. Int. Ed. Eng. 1990, 29, 256.

PALLADIUM-MEDIATED METHYLENECYCLOPROPANE RING OPENING
APPLICATIONS TO ORGANIC SYNTHESIS

William A. Donaldson

OUTLINE

Advances in Metal-Organic Chemistry, Volume 2, pages 269–293
Copyright © 1991 JAI Press Ltd
All rights of reproduction in any form reserved
ISBN: 0-89232-948-3

I. INTRODUCTION

The hydrometallation,[1] carbometallation,[2] halometallation,[3] and oxometallation[4] of olefins and polyenes are of great importance to synthetic organic chemistry. Likewise, addition of an M–R or an M–X bond across allenes 1 and/or methylenecyclopropanes ("homo-allenes", 2) has also proven to be of interest. The reactivity of these latter substrates 2 is a central focus of our research efforts.

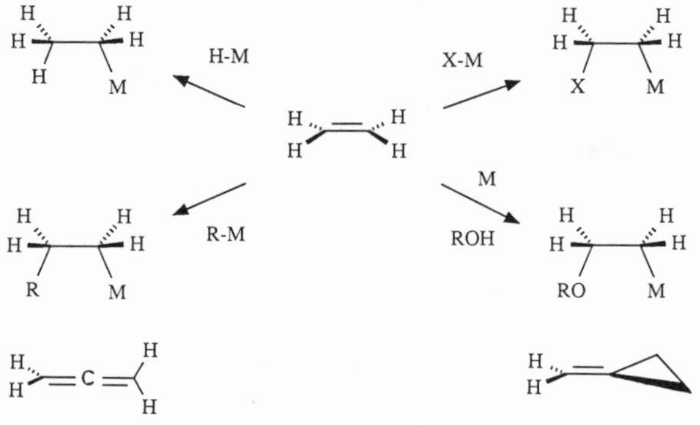

1   2

II. DEVELOPMENT OF A RING HOMOLOGATION
METHODOLOGY

The chloropalladation of 1 affords a π-allyl complex 3 containing two allene molecules coupled at the central carbons.[5] The chloropalladation of

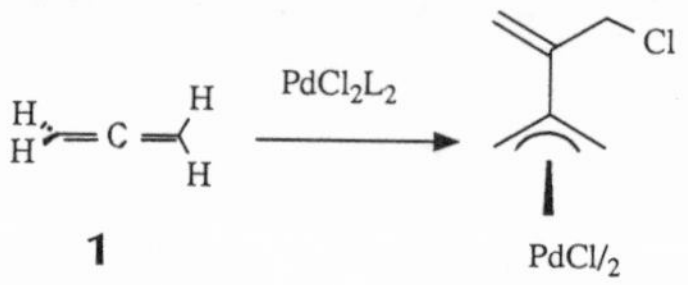

methylenecyclopropane 2a yields a π-allyl complex 4 containing a single rearranged unit of the starting material, via cleavage of the C-1–C-2 bond.[6] Similarly, methylenecyclopropanes bearing electron-withdrawing substituents (2b, c) undergo chloropalladation via cleavage of C-1–C-2 bond.[7] These reactions are believed to occur via initial insertion of 2 into the Pd–Cl bond, followed by a cyclopropylcarbinyl-to-homoallyl rearrangement to afford **5**. Subsequent β-hydride elimination and insertion of the diene into the resultant Pd–H bond affords the product **4**.

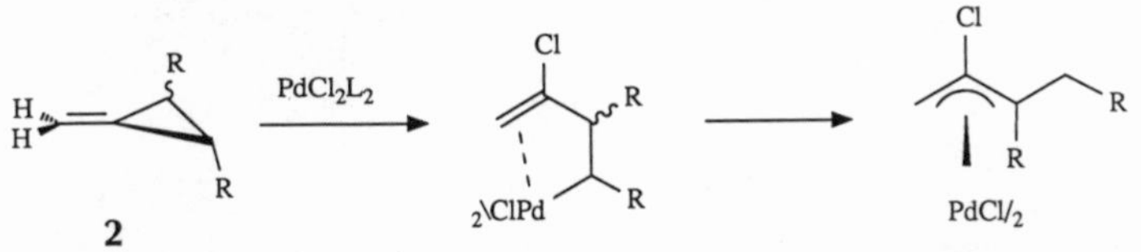

a, R = H; b, R = CO2Me (trans); c, R = CO2Me (cis)　**5**　**4**

In comparison, alkyl- and aryl-substituted methylenecyclopropanes undergo chloropalladation via cleavage of the C-2–C-3 bond to generate π-allyl **6a** and **b**.[6,8] A theoretical and experimental investigation of the reaction has lead to the following proposed mechanism:initial coordination

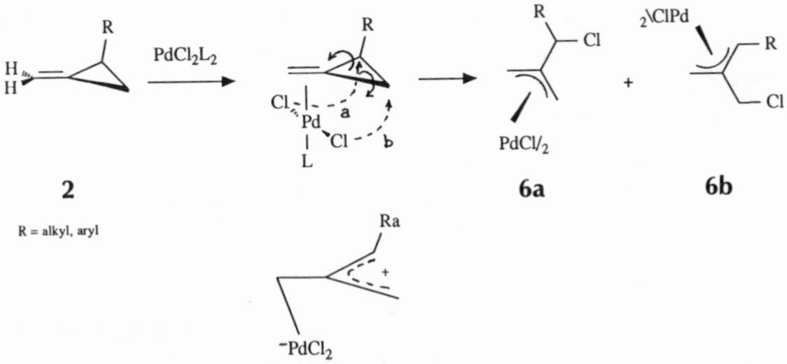

R = alkyl, aryl

Scheme 1

of PdCl$_2$ to the less hindered face of **2**, followed by disrotatory ring opening and suprafacial transfer of chloride ligand to either C-2 or C-3.[9] Extended Hückel calculations indicate that the Pd-metal slips away from the central carbon ($\eta 2 \rightarrow \eta 1$) prior to, or during, chloride transfer.[9a] While the timing of these two movements is not entirely known, the authors indicate that "little, if any, additional activation energy" is required for the Cl transfer. These authors also found no evidence for the intermediacy of the zwitterionic trimethylenemethane species **7**.[9b] Our own investigation of the chloropalladation of 7-methylene-1-phenylbicyclo[4.1.0]heptane (**8**), in methanol, yielded a product **9** which can only arise via attack of methanol exclusively at the phenyl-substituted cyclopropane carbon.[10] This result strongly implies that the intermediate in the ring opening may involve some carbocation character. Further examination of the nature of this intermediate/transition state is in progress, and it appears that the timing and regioselectivity of Cl transfer

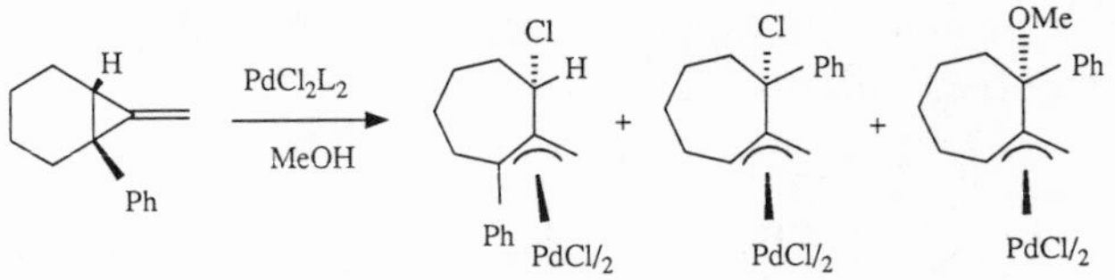

8   **9**

may be effected by both electronic and steric factors as well as by additional ring strain.[11]

Due to the great potential for exploitation of the π-allyl moiety,[12] we became particularly interested in the chloropalladation of Ω-methylenebicyclo[n.1.0]alkanes **10**. The requisite substrates **10** may be prepared from the corresponding cyclic olefins **11** in two[13] or three steps.[14] Since the chloropalladation of **10** must occur via cleavage of the C-2–C-3 bond, these steps constitute an overall ring homologation methodology. We have explored the generality of this ring homologation methodology by preparing a series of (3-chloro-2-methylenecycloalkyl)palladium chloride dimers (**12**, $n = 6,7,8,9,11,13,16$) from the corresponding cyclic olefins; all in excellent isolated chemical yield.[15,16]

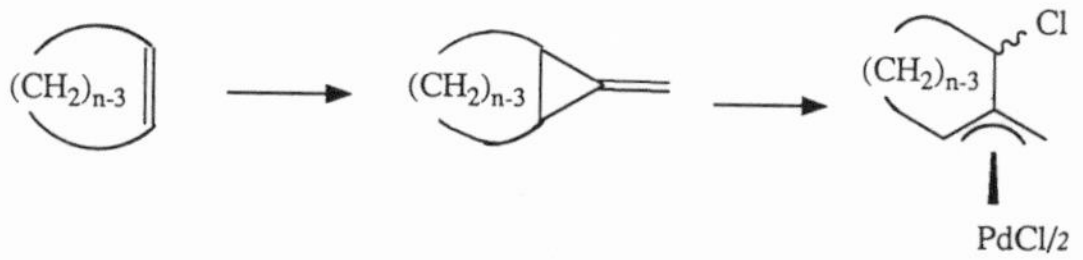

11   **10**   **12**

Scheme 2

Preparation of (3-chloro-2-methylenecyclooctyl)palladium chloride dimer (**12**, $n = 8$). To a stirred solution/suspension of bis(acetonitrile)palladium chloride (0.28 g, 0.74 mmol) in CH_2Cl_2 (20 mL) was added 8-methylenebicyclo[5.1.0]octane (0.09 g, 0.74 mmol). The orange–red solution rapidly became yellow and was stirred for an additional 30 min. The solvent was removed under reduced pressure and dried *in vacuo* to afford a yellow powder, 99%.

Generally, terminal substituents on acyclic π-allyls tend to occupy the *syn* position;[17] however, this trend can be reversed due to steric interactions with adjacent bulky substituents.[18] Examination of the larger ring π-allyls (**12**,

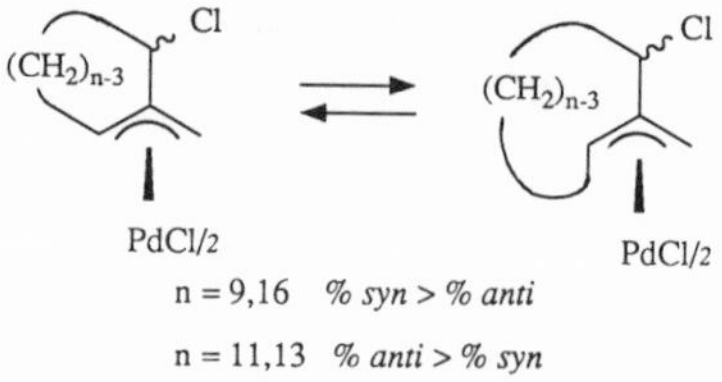

$n = 9,16$ % *syn* > % *anti*

$n = 11,13$ % *anti* > % *syn*

$n \geq 9$) indicate that they exist in an equilibrium of *syn* and *anti* isomers. The ratio of geometric isomers is dependent on the size of the carbocyclic ring. Apparently, factors such as ring strain, torsional strain and transannular interactions must also be important to the relative stabilities of the *syn* and *anti* isomers of cyclic π-allyl complexes.

The chemical consequences of ring conformation on the reactivity of macrocyclic compounds has been experimentally demonstrated and computationally modeled.[19] As will be later shown, the conformation and configuration of complexes **12**, plays an important role in their reactivity. The recent development of energetic parameters for coordinated transition metals[20] has opened the possibility for molecular mechanics analysis of the (3-chloro-2-methylenecycloalkyl)palladium complexes **12**. For example, the (3-chloro-2-methylenecycloheptyl)palladium complex is found to adopt a single diastereomer/conformer in solution as well as in the crystalline state.[21] Of the four *possible* chair structures (**A, B, C, D**) the conformer with lowest calculated energy (**A**) is identical to that found in the crystalline state, and the calculated bond lengths and angles for **A** are similar to those reported for the crystal structure.

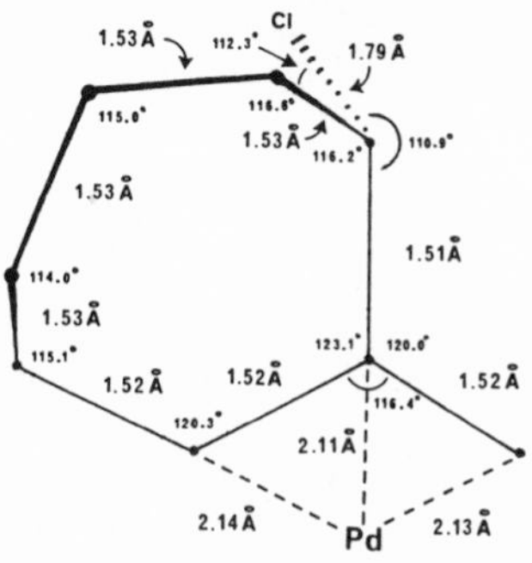

A

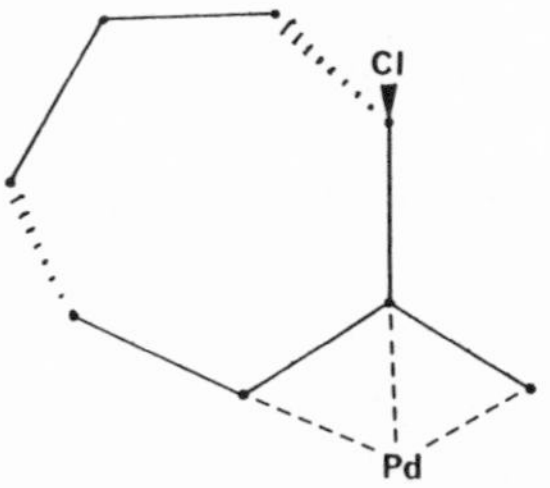

B 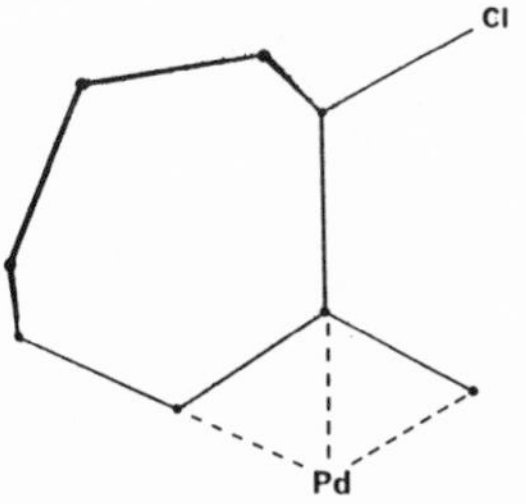   **C** 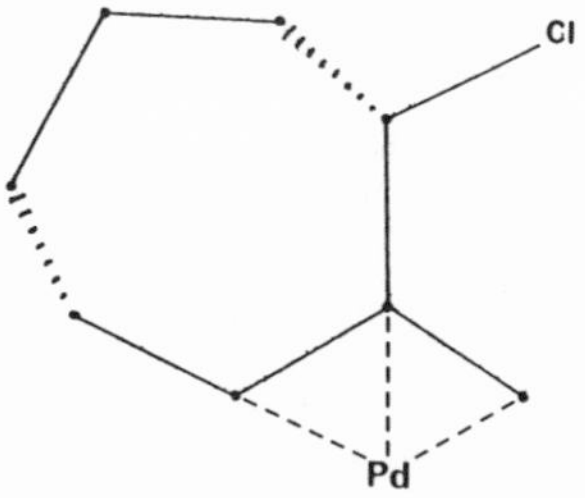   **D**

III. REACTIVITY OF CYCLIC π-ALLYL COMPLEXES (12)

The base mediated cleavage of Pd–allyl complexes is perhaps the oldest reported (and simplest) reaction which liberates the organic ligand.[22] Treatment of the (3-chloro-2-methylenecycloalkyl)palladium chloride dimers 12 with alkaline methanol results in the formation of a mixture of the α-methoxyolefin 13 and the corresponding cyclic olefin 14, in good yield.[23] This latter product represents not only cleavage of the π-allyl complex but a formal reduction of the C3–C1 bond, as well. We have proposed the following mechanism: initial cleavage of the π-allyl 12 generates the allylic chloride 15 and finely divided palladium metal. Oxidative addition of Pd(0) into the

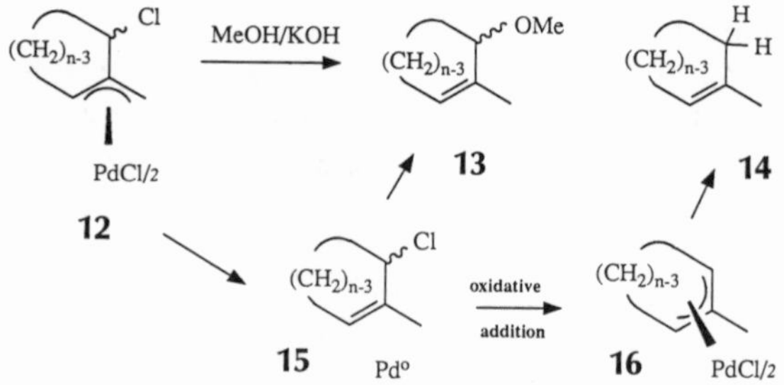

Scheme 3

allylic halide bond[24] affords a new π-allyl 16 which upon subsequent cleavage under the excess alkaline methanol conditions gives the cyclic olefin product (14). Alternatively, solvolysis of the allylic halide 15 gives the α-methoxyolefin product 13. The ratio of α-methoxyolefins to cyclic olefins (13:14) decreases with increasing size of the cyclic ring. Since the products do not interconvert under the reaction conditions, the ratio of 13:14 must represent the relative rates of solvolysis versus oxidative addition of the allylic chloride 15. It is well known that the rate of solvolysis of chlorocycloalkanes decreases with increasing ring size, due to internal strain (I-strain) generated on changing the hybridization of the center undergoing solvolysis.[25] In comparison, oxidative addition of vapor deposited *metal atoms* has been proposed to occur via an initial electron transfer.[26] Since initial electron transfer is anticipated to occur with little change in the geometry of allylic halide 15, we have proposed that the rate of oxidative addition of 15 is essentially independent of ring size. Thus, it is not surprising that the percentage of α-methoxyolefin 13 in the product mixture decreases with increasing ring size.

We have utilized this reactivity in the synthesis of (±)-13-methyltridecanolide (17), the major macrolide constituent (0.03%) of *Galbanum* resin, starting with commercially available cyclododecene (Scheme 4).[27,28]

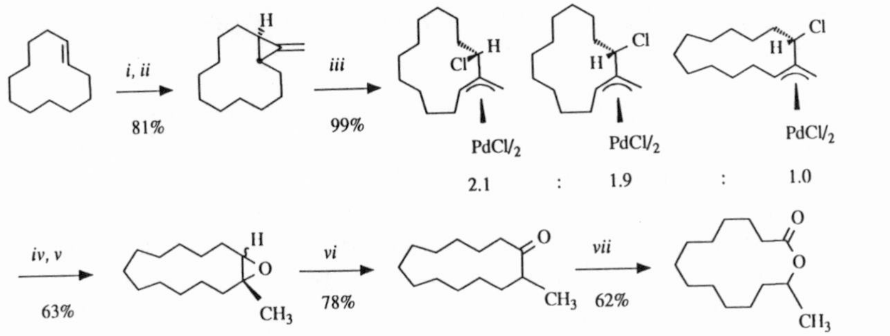

81% 99%

PdCl/2 PdCl/2 PdCl/2

2.1 : 1.9 : 1.0

iv, v vi vii

63% CH₃ 78% CH₃ 62% CH₃

i, Cl₂HCCH₃, *n*-BuLi; *ii*, *t*BuOK; *iii*, PdCl₂L₂; *iv*, MeOH/KOH; *v*, *m*CPBA; *vi*, BF₃·Et₂O; *vii*, CH₃CO₃H

Scheme 4

While the cleavage of (3-chloro-2-methylenecycloalkyl)palladium complexes **12** may not be a good general route for the homologation of cyclic olefins **11**, it does highlight the point that compounds **12** may act as "doubly activated" π-allyls under certain reaction condition. Perhaps of greater synthetic interest is the reaction of π-allyl palladium complexes with stabilized carbon nucleophiles in the presence of excess phosphine ligands.[29]

The reaction of (3-chloro-2-methylenecycloheptyl)palladium chloride dimer (**18**) with two equivalents of malonate anion afford the tetraester product **19** in excellent isolated yields.[30] In addition, reaction of **18** with a single equivalent of malonate anion affords the functionalized cycloheptadiene **20** as product. Thus, under the appropriate conditions, complex **18** may react as either a "trimethylenemethane dication" synthon (**21**)[31] or as an "isoprenyl monocation" synthon (**22**).

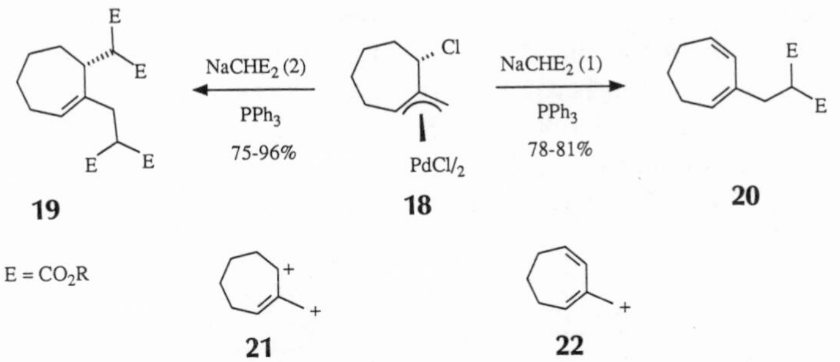

NaCHE₂ (2)   NaCHE₂ (1)

PPh₃   PPh₃

75-96%   PdCl/2   78-81%

19   **18**   **20**

E = CO₂R

21   **22**

Preparation of 2-(2′,2′-(dicarboethoxy)ethane]-3-(dicarboethoxy)methane-1-cycloheptene (**19**). To a solution of (3-chloro-2-methylenecycloheptyl)palladium chloride dimer (250 mg, 0.875 mmol) and triphenylphosphine (918 mg, 3.50 mmol) in dry THF (25 mL) under N₂, was added a solution of sodio diethylmalonate (1.75 mmol, freshly prepared from xs. NaH and diethylmalonate) in dry THF (10 mL). The pale yellow solution turned orange–yellow upon addition. After 24 h the solvent was removed, the residue taken up in CH₂Cl₂, washed 1 × with H₂O and the solvent removed to afford the crude product. Purification by "flash" chromatography[32] (elution with hexane, followed by elution with hexanes:ethyl acetate 28:1) and distillation (kugelrohr) afforded the product as a clear oil. Yield: 90%.

These results clearly indicate that initial nucleophilic attack occurs at the unsubstituted allylic terminus to afford the allylic chloride **23**. The allylic halide **23** oxidatively adds to Pd(0) with inversion at C-3 to give a new π-allyl **24**.[33] In the absence of an additional equivalent of nucleophile **24** undergoes β-hydride elimination to afford the cycloheptadiene product **20**.[34] Conversely, nucleophilic attack on **24** gives the product **19** in which two new carbon–carbon bonds have been formed.

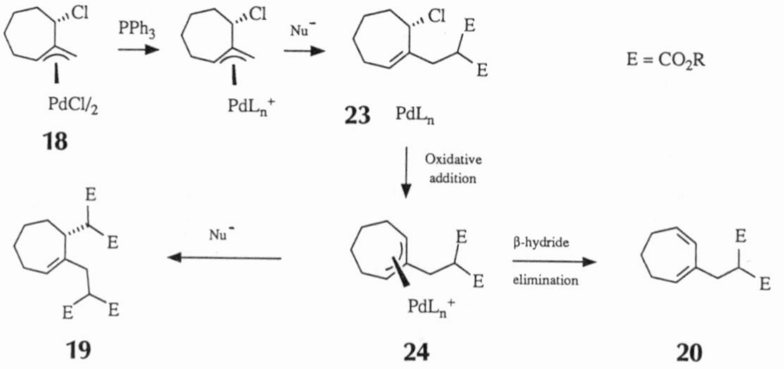

Scheme 5

Notably, the new π-allyl **24** is a symmetrical intermediate. The use of chiral chelating phosphine ligands in place of triphenylphosphine would destroy this symmetry and could result in asymmetric induction in the attack of the second nucleophile.[35] Surprisingly, the reaction of racemic π-allyl **18** with two equivalents of malonate anion in the presence of the chiral phosphine *s,s*-DIOP gave only the product from β-hydride elimination.[20] In sharp contrast, reaction of the six membered cyclic π-allyl **25** gave a mixture of **26**, **27a** and **27b**. While chiral lanthanide NMR shift reagents failed to separate the enantiomeric resonance signals of **26** the enantiomeric signals for the *dl* compound **27a** were separable and appeared in a 3:2 ratio.

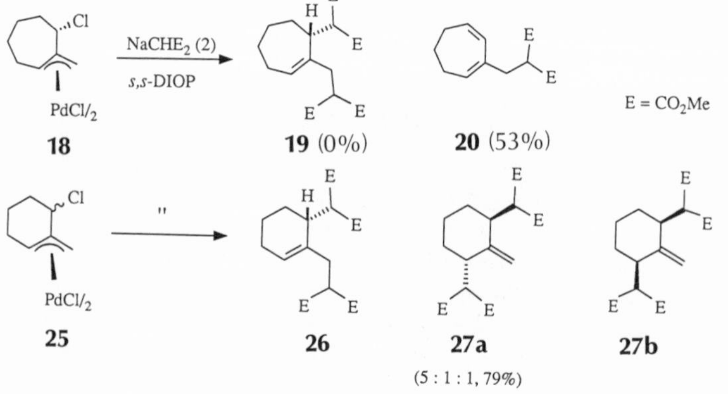

$(5:1:1, 79\%)$

The remarkable difference in reactivity between the (cycloheptyl)- and the (cyclohexyl)palladium allyls might be rationalized by examination of the structures of intermediates **24** and **28**. Only in the boat conformer of each is the proper orientation for *β*-hydride elimination achieved. By analogy to the corresponding hydrocarbons, the barrier for chair–boat interconversion should be lower for the seven-membered ring (**24**) than for the six-membered ring **28**. More importantly, the axial *β*-hydrogens of **24′**, are closer to the Pd metal than the axial *β*-hydrogens of **28′**. Thus, the relative rates of*β*-hydride elimination versus nucleophilic attack should be greater for **24** than for **28**.

24 **24′**

28 $E = CO_2Me$ **28′**

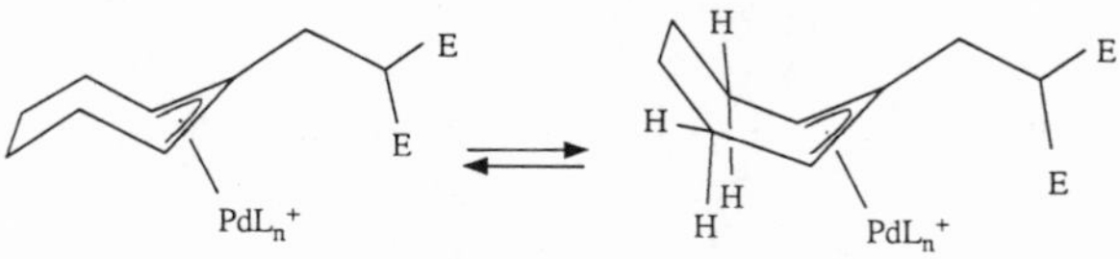

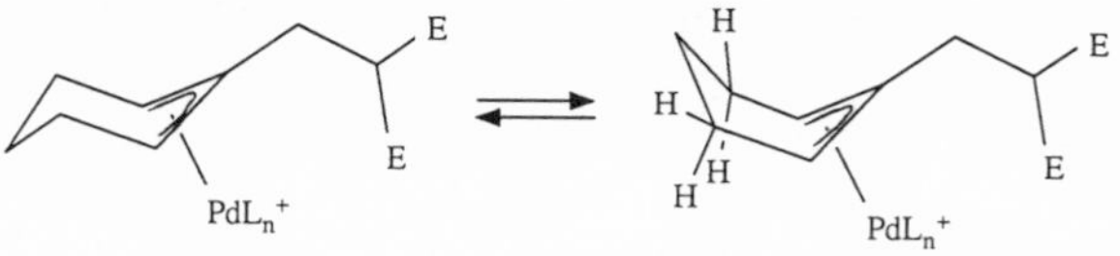

In addition, reaction of the cyclooctyl Pd-allyl **29** with two equivalents of malonate, in the presence of triphenylphosphine, gives *only* a 1,3-cyclooctadiene product, **30**.[30] Examination of the intermediate **31** indicates that the *β*-hydrogens are even closer to the Pd-metal than in the seven-membered π-allyl intermediate **24**.

29 **30** (71%) **31**

$E = CO_2Me$

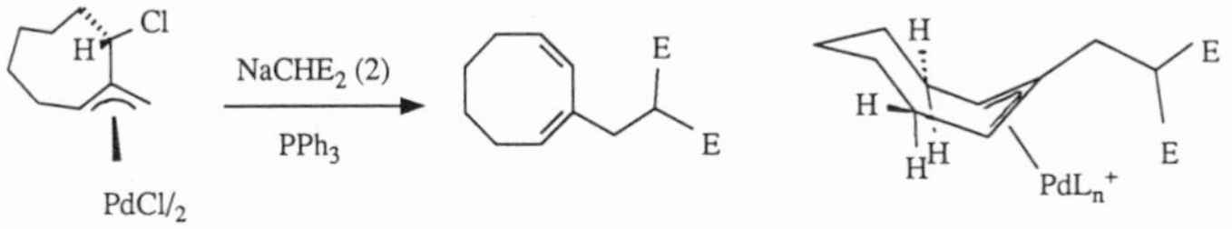

It has previously been noted from the fluxional behavior of cationic cyclic π-allyl-μ-hydride iron complexes **32**, that the *β*-hydrogens become considerably closer to the metal center with an increase in the carbocyclic ring size. Thus the barrier for the exchange of agostic hydrogens in **32** decreases from $E_a = 10.4\,\text{kcal mol}^{-1}$ for the cyclohexenyl complex ($n = 6$) to $E_a < 5\,\text{kcal mol}^{-1}$ for the cyclooctenyl complex ($n = 8$).[35]

L = P(OMe)$_3$
Ref. 35

32a   **32b**

In the above discussion, we have indicated that the ability of complexes **12** ($n = 6, 7, 8$) to react as a "trimethylenemethane dication" synthon or as an "isoprenyl monocation" synthon is dependent on the size of the carbocyclic ring, the nature of the phosphine ligand and on the amount of malonate anion present. More recently we have found that the reactivity of **18** is also dependent on the nature of the nucleophile.[36] The reaction of **18** with two equivalents of sodio methyl(phenylsulfonyl)acetate gives good yields of the product derived from synthon **21** (i.e. **33**, albeit as a mixture of diasteromers). In comparison, the reaction of **18** with two equivalents of lithio bis-(phenylsulfonyl)methane gives only the product derived from synthon **22** (i.e. **34**). We speculate that attack of the second equivalent of this bulky nucleophile on the sterically crowded π-allyl intermediate **35** is sufficiently slow compared to β-hydride elimination.

E = CO$_2$Me, S = SO$_2$Ph

Palladium allyl complexes are also known to react with "harder" carbon nucleophiles. The reaction of **18** with phenylmagnesium chloride, or phenyl-lithium (in the presence of triphenylphosphine) gave biphenyl as the only isolated product. This nucleophile coupling may result from an initial electron transfer from the phenyl Grignard reagents to the C3-chloro sub-stituent. In comparison, the reaction of **18** with the milder sodium tetra-phenylborate (2 equivalents) gave the product of addition of two phenyl nucleophiles (**36**) and the (2-benzylcycloheptyl)palladium chloride dimer **37**. *Isolation of this new π-allyl provides substantial evidence for our proposal that oxidative addition of Pd(0) into the allylic halide bond forms this intermediate in all of the reactions of π-allyl **18** with 2 equivalents of nucleophile.* In addition, this result demonstrates that the addition of two *different* nucleo-philes, in specific order, may be accomplished.[37]

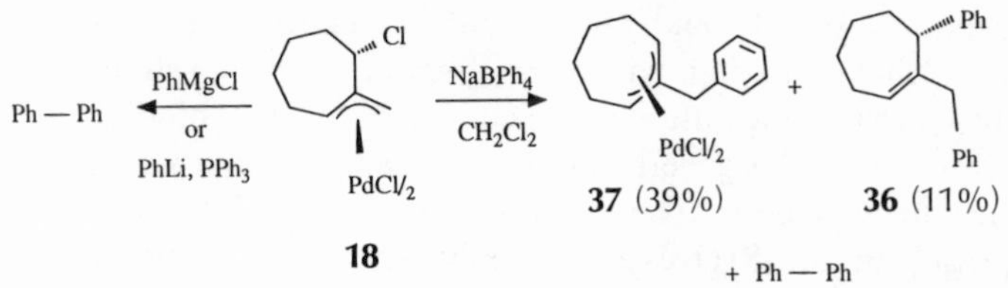

18

37 (39%)   36 (11%)

+ Ph — Ph

It has been imperative to investigate the scope and the reactivity of complexes **12** in order to be able to exploit these in the synthesis of compounds of biological interest. Our results indicate that the two reactive sites of synthon **21** are sequentially generated. Additionally, the intermediacy of the new π-allyl **24** means that isomeric π-allyls (formed in the chloropalladation of unsymmetrically substituted compounds, Scheme 2) will yield the same product(s) upon reaction with nucleophiles since they proceed via a common intermediate. Thus the low regioselectivity involved in step I (*vide supra*) does not pose a synthetic problem, due to the convergent pathway for isomeric starting π-allyls.

IV. SYNTHETIC APPLICATIONS OF COMPLEXES 12 AS SYNTHON 22

As indicated in Scheme 5, the reaction of (3-chloro-2-methylene-cycloheptyl)palladium chloride (**18**) with 1 equivalent of malonate anion affords a cycloheptadiene product (**20**) in good isolated yields. This reactivity might be utilized in a ring homologation–functionalization approach to the tropolonic C ring of colchicine (**38**).

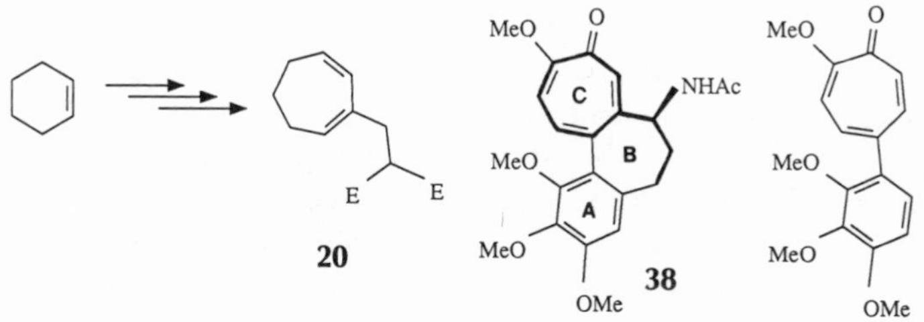

Colchicine is a tricyclic alkaloid isolated from the meadow saffron (*colchicum automnale*). Its antimitotic activity is derived from the slow, irreversible binding of **38** to the protein tubulin.[38] Substantial evidence has established that there are distinct recognition domains for both the A and C rings of colchicine at the binding site on tubulin. The exact contribution of the B ring to tubulin binding is unknown.[39] In fact 2-methoxy-5-(2′,3′,4′-trimethoxyphenyl)tropone **39** will bind rapidly but reversibly to tubulin. Therefore, synthetic colchicine analogs which vary only in the B ring should be of value to a more complete elucidation of the colchicine–tubulin binding mechanism.

Our efforts toward a general route to colchicine and a variety of analogs is outlined in Scheme 6. The chloropalladation of 7-methylene-1-phenyl-bicyclo[4.1.0]heptane (**8**) gives a kinetically controlled 3:2 mixture of isomeric Pd-allyl products **40a** and **40b**. Reaction of either **40a** or **40b**, or a mixture of both, with one equivalent of sodio diethylmalonate affords a mixture of cycloheptadienes **41a** and **41b** (1:1 ratio, 63–75%). Catalytic hydrogenation of the mixture produces a single phenylcycloheptane product, **42**. Saponification–decarboxylation, generation of the labile acid chloride and Friedel-Craft acylation completes the synthesis of the tricyclic Ar-7-7 skeleton of colchicine.[40]

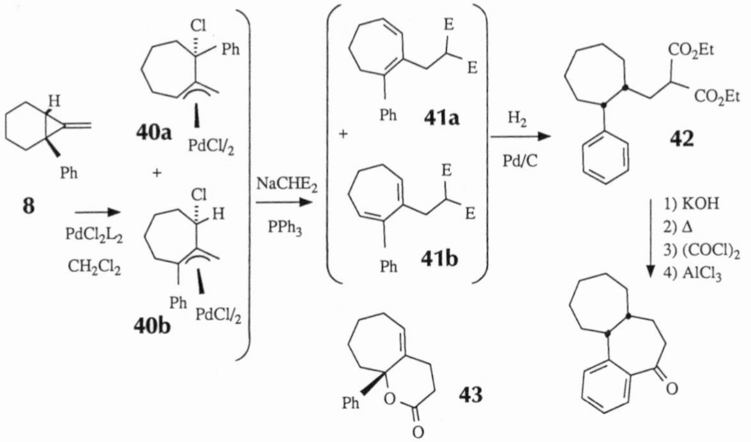

Scheme 6

Reduction of the somewhat oxidized cycloheptadiene ring may seem anti-thetical to a synthesis of colchicine. However previous work had indicated that the saponification–decarboxylation of **41** might result in the formation of the lactone **43**, via addition across the diene. This problem might be avoided if the cycloheptadiene ring is oxidized to a tropone ring prior to saponification–decarboxylation. This has been accomplished as indicated in Scheme 7.

Chloropalladation of **44** gives a mixture of isomeric π-allyls **45** in nearly quantitative yield. Reaction of the mixture with 1 equivalent of sodio di-methylmalonate affords a single cycloheptadiene product **46**. Decarbometh-oxylation and removal of the ketal protecting group yield the cycloheptadie-none product **47**.[40] Aromatization of **47** to the tropone ring **48** might be accomplished by using DDQ,[41] and introduction of the tropolone hydroxyl group has previously been demonstrated by Boger and Brotherton.[42]

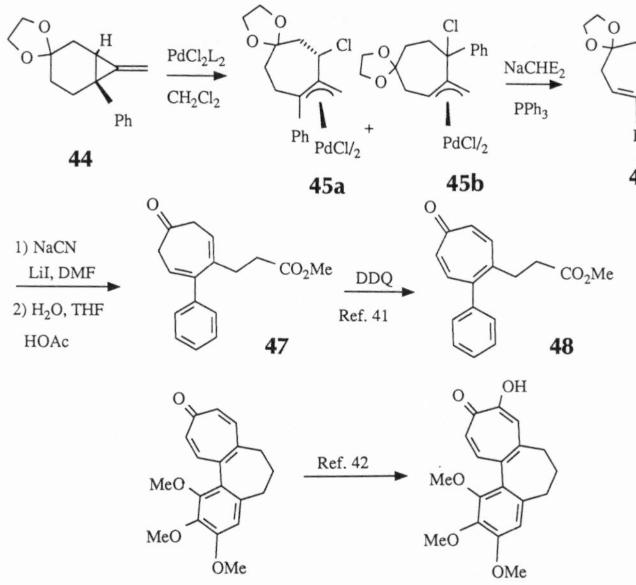

Scheme 7

The preparation of six- and eight-membered B ring colchicine analogs **49** and **50** might be realized from the reaction of **45** with the methylthiomethyl phenylsulfone anion and with the bis(cyclopentadienyl)zirconium(methyl-acryl)chloride, respectively:

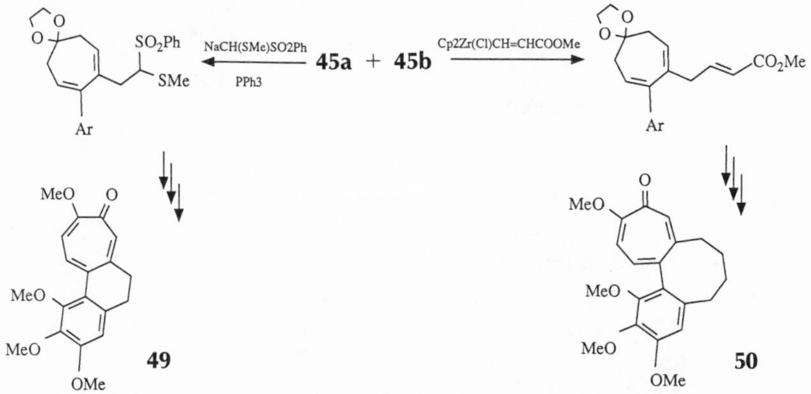

V. SYNTHETIC APPLICATIONS OF COMPLEXES 12 AS SYNTHON 21

As has been demonstrated, the (3-chloro-2-methylenecycloheptyl)palladium chloride dimer (**18**) is reactive with 2 equivalents of a carbon nucleophile as a "trimethylenemethane dication" synthon. Thus complex **12** may be considered as a complimentary synthon to the "zwitterionic trimethylenemethane" synthon[43] (**51**), the "trimethylenemethane dianion" synthon[44] (**52**) and the "trimethylenemethane diradical" synthon[45] (**53**). These latter three synthons have all proven most valuable for cycloannulation reactions. For this reason, we were interested in the reaction of complexes **12** with a *single equivalent of a dinucleophile*.

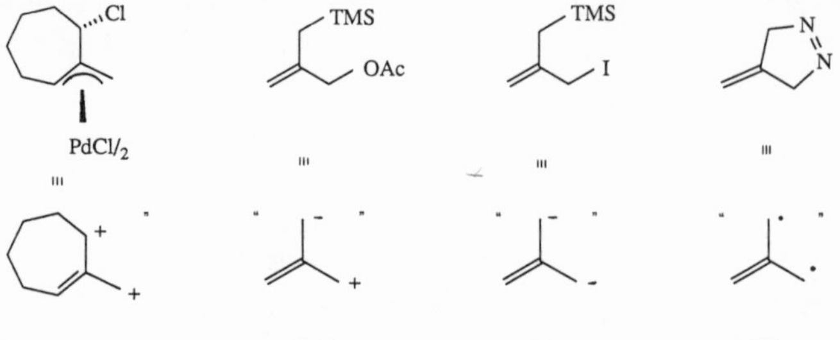

21	**51**	**52**	**53**

The reaction of (3-chloro-2-methylenecycloheptyl)palladium chloride dimer (**18**) with the dianion of dimethyl 3-ketoglutarate[46] in the presence of triphenylphosphine gives the cyclohexanone diester **54** as a mixture of diastereomers.[47] Saponification and decarboxylation affords a single product, 9-ketobicyclo[5.4.0]undec-1-ene (**55**).[48] It is anticipated that the cyclization

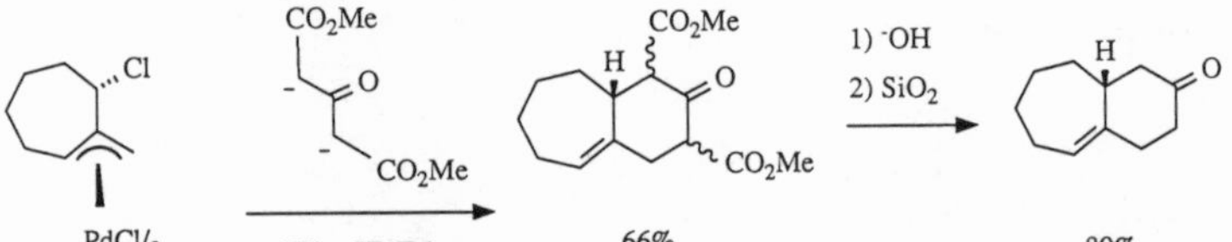

occurs via initial nucleophilic attack at the exocyclic terminus followed by oxidative addition of Pd(0) into the resultant allylic halide to afford the zwitterionic π-allyl intermediate **56**. Subsequent rapid intramolecular attack at either allylic terminus, on the face opposite to the palladium metal, affords the product **54** via a formal "6-*exo*-tet" ring closure.[49] The highly organized nature of this intramolecular attack might allow for the possibility for asymmetric induction. Use of a dinucleophile with chiral ester auxillaries should allow for a high degree of selectivity for one allylic terminus over the other.

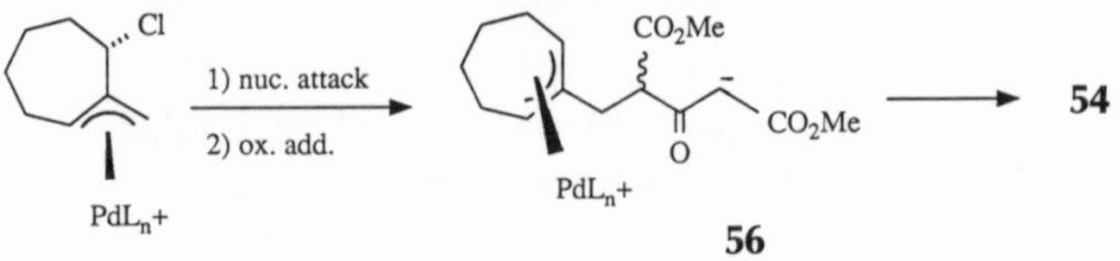

8,10-bis(carbomethoxy)bicyclo[5.4.0]undec-1-en-9-one (**54**). To a solution of dimethyl 3-ketoglutarate (0.16 g, 0.90 mmol) in dry THF (10 mL) at $-78°C$ under N_2 was added, via syringe, t-butyl lithium (1.7 M, 1.06 mL, 1.80 mmol). The dianion solution was stirred for 0.5 h. To the cold dianion solution was added a solution of (3-chloro-2-methylenecycloheptyl)palladium chloride dimer (0.26 g, 0.90 mmol), triphenylphosphine (0.945 g, 3.60 mmol), 18-crown-6 (0.01 g) and HMPA (0.37 g) in dry THF (30 mL). The yellow dimer solution became cloudy upon addition. The reaction mixture was cooled at $-78°C$ for an additional 0.5 h and then allowed to slowly warm to room temperature. The solvent was evaporated and the residue extracted with ether (4 × 30 mL). The resultant extracts were combined, concentrated and "flash" chromatographed[32] using hexane:ethyl acetate (28:1) as eluant. Yield: 0.17 g, 66%.

Further elaboration of the cyclohexanone ring may be accomplished with good stereospecificity. Reduction with $Li(t\text{-}BuO)_3AlH$ affords the equatorial alcohol (**57**) while reduction with K-selectride gives exclusively the axial alcohol product (**58**). The *trans* stereochemistry (**57**) would be desirable for the synthesis of mimics of the tumor promoter, phorbal myristate acetate (**59**).[50]

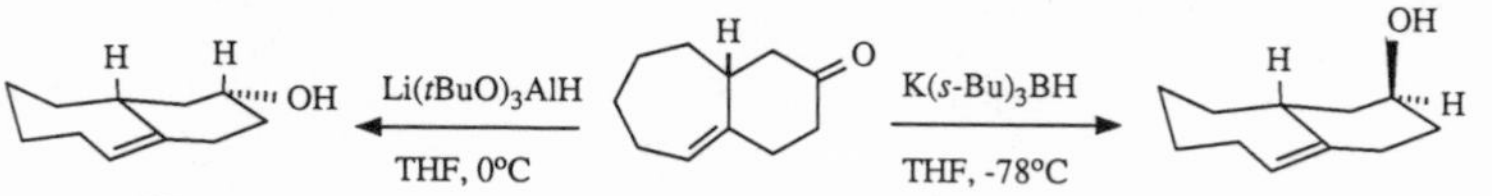

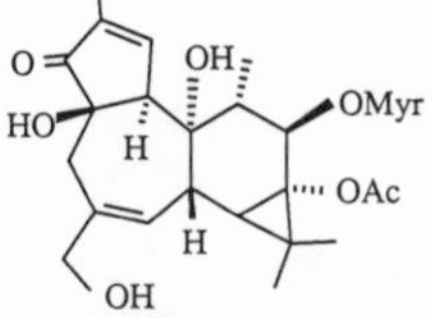

Reaction of the isomeric methyl substituted π-allyls **60a** and/or **60b**[15] with dimethyl 3-ketoglutarate dianion, followed by saponification–decarboxylation, affords a mixture of 7-methyl- and 2-methylbicyclo[5.4.0]undec-1-en-9-ones (**62** and **63**, respectively, 2:3 ratio). Formation of the same product mixture from either **60a** or **60b** may be rationalized on the basis of a common intermediate.

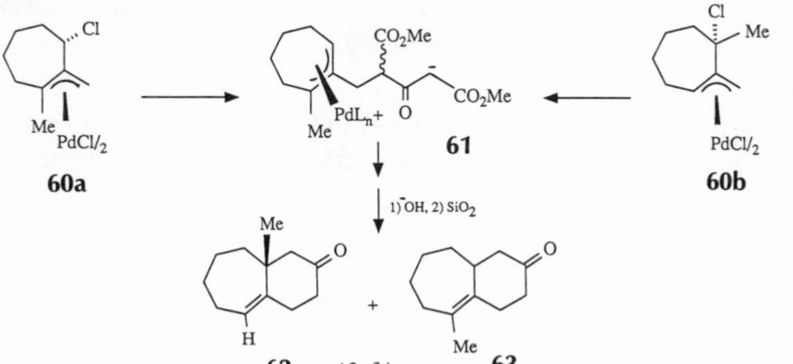

Nucleophilic attack at the exocyclic allylic terminus of either **60a** or **b**, followed by oxidative addition of Pd(0), generates a single π-allyl intermediate **61**. Subsequent intramolecular attack at the more substituted terminus eventually affords the product **62**, while attack at the less hindered terminus gives the product **63**.

In comparison, the reaction of π-allyl complex **18** with tetracarbomethoxyethane dianion gives the octahydroazulene product **64**, albeit in dismally low yield (17% optimized). Due to the wide variety of naturally occurring compounds which contain a hydroazulene skeleton, we explored an alternative approach to the desired molecule **64**. It was anticipated that oxidative coupling of the dianion of tetraester **19**, as an intramolecular process, might be a high yield route. Deprotonation of the tetraester **19** with s-butyl lithium, followed by quenching the dianion with cupric chloride afforded the coupled product **64** in good isolated yield. As indicated previously, the precursor **19** may be obtained in excellent yield from the palladium allyl **18**. Thus, this two step procedure, when in conjunction with the chloropalladation reaction represents a short, novel, efficient route for transformation of cyclohexene into the octahydroazulene skeleton.

$8,8,9,9$-Tetracarbomethoxybicyclo[5.3.0]dec-1-ene (**64**). To a solution of s-butyl lithium (4.2 mmol) in dry THF (60 mL), cooled to 0°C under N_2, was added a solution of **19** (0.22 g, 0.59 mmol) in THF (15 mL). After stirring for 0.5 h, the solution had turned a milky white color. A suspension of cupric chloride (0.65 g, 4.8 mmol) in dry THF (20 mL) was added. The initial green color becomes brown over a period of 1 h. The reaction mixture was washed with H_2O (80 mL) and the aqueous layer extracted with ether (3 × 30 mL). The combined organic layers were dried ($MgSO_4$) and the solvent removed under reduced pressure. Distillation under high vacuum gave the product as a clear oil: 80–110°C/0.1 mm Hg (kugelrohr); 0.16 g, 74%.

There are a myriad of possible target compounds for which this ring homologation–cyclopentannulation methodology would be applicable. We are currently pursuing the asymmetric synthesis of the sesquiterpene bulnesol (**65**)[51] from terpineol.

As a preliminary model study, the ring homologation–cyclopentannulation of 4-t-butylcyclohexene (**66**) was investigated.[36] Chloromethylcyclopropanation of **66** (CH$_3$CHCl$_2$, *n*-BuLi, $-30°C$) followed by dehydrohalogenation (t-BuOK, DMSO, 90°C) afforded 3-t-butyl-7-methylenebicyclo[4.1.0]heptane (**67**) as a mixture of diastereomers (1:1). Reaction of this mixture with palladium chloride gave a mixture of three isomeric π-allyl complexes. High field NMR spectroscopy revealed this mixture to be composed of **68**, **69** and **70** in a 2:2:1 ratio. Since the chloropalladation has been shown to occur with transfer of chloride to the less hindered face of the cyclopropane ring,[15] then **69** must arise from chloropalladation of *trans*-**67** while **68** and **70** must arise from *cis*-**67**. Without further separation, the

mixture of π-allyl complexes was treated with two equivalents of sodio dimethylmalonate in the presence of triphenylphosphine. A mixture of two tetraester products was isolated, **71** and **72** (1:1 ratio), as revealed by high

field NMR spectral analysis. Isolation of only these two products implies that there is remarkable selectivity for nucleophilic attack on the π-allyl intermediates **73** and **74**. While the origin of this selectivity is not readily apparent, these results indicate that the relative stereochemistry at C3 of **69** may be "relayed" into regioselective and stereoselective introduction of the cyclopentane ring. This selectivity is currently being applied to the synthesis of bulnesol.

The ring homologation–cyclopentannulation of 3-methylcyclohexene (**75**) was also investigated. Dichlorocyclopropanation under phase transfer conditions (CHCl₃, 50% aqueous NaOH, CTAB)[14] gave a single product in good yield. By analogy to the reported dibromocyclopropanation,[52] this product (**76**) was assigned the *trans* stereochemistry. Lithium–halogen exchange followed by quenching with iodomethane and subsequent dehydrohalogenation afforded the corresponding methylenebicycloheptane (**77**). Chloropalladation of **77** afforded a mixture of two isomeric π-allyl complexes **78** and **79** (1:1 ratio).

Reaction of the mixture of **78** and **79** with 2 equivalents of sodio dimethylmalonate afforded a separable mixture of diene diesters **80** and a single tetraester product. The tetraester was assigned structure **81** on the basis of NMR spectral data. The vinylic proton of **81** (H$_a$) appears as a doublet, thus indicating the presence of a single adjacent proton. Selectivity for this regioisomer may be rationalized by inspection of the intermediate Pd-allyl species

82. Nucleophilic attack of malonate anion on the intermediate **82**, on the face opposite the palladium metal, is hindered along approach "a" by the axial methyl substituent. For this reason, only attack along the alternative approach "b" is observed. Further oxidative coupling of the dianion of **81** should afford a 3-methyl octahydroazulene (**83**).

While the above intramolecular oxidative coupling reaction is an integral portion of the overall ring homologation–cyclopentannulation scheme, it also exposes the one liability of all of the aforementioned chemistry. All of the reactions presented to this point are stoichiometric in palladium, since Pd(II) is required for the chloropalladation while the product from reactions of the Pd-allyl complexes is Pd(0). Attempts to render the overall transformation catalytic in palladium, by the use of a Pd(0) reoxidant (i.e. $CuCl_2$), will fail. This is due to the rapid oxidative coupling of the more abundant malonate anion instead of oxidation of Pd(0). In the next section is outlined our attempts to use alternative Pd(0) oxidants which would be unreactive toward malonate nucleophiles.

VI. CARBOPALLADAITON OF Ω-METHYLENEBICYCLO[*n*.1.0]ALKANES

The stoichiometric aryl-palladation of methylenecyclopropane (**2**) has been reported to afford the (2-phenylcrotyl)palladium chloride dimer (**84**) in low isolated yield, via cleavage of the C1–C2 bond.[53] This selectivity for C1–C2

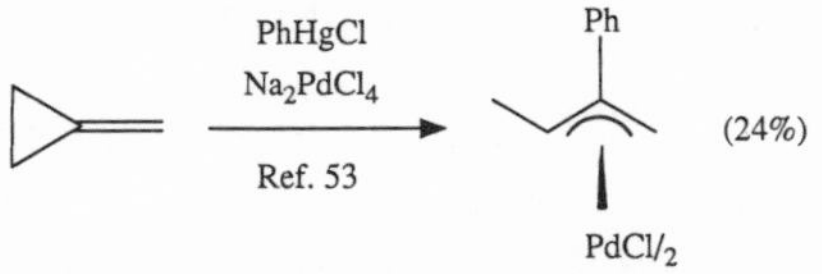

bond cleavage is also observed for the chloropalladation of **2**. The mechanism proposed for the formation of **84** is similar to that proposed for the formation of **4** (*vide supra*). More recently, the catalytic carbopalladation of methylenecyclopropane, in conjunction with nucleophilic addition to the resultant π-allyl, was reported to afford a mixture of styrenes **85a, b** and **86**.[54] The isomeric styrenes result from attack of malonate anion at either of the termini of the intermediate π-allyl **87**.

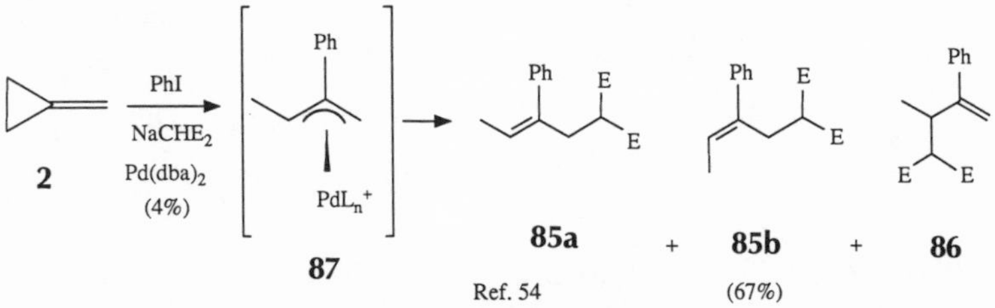

The reaction of 7-methylenebicyclo[4.1.0]heptane (**10**) under similar reaction conditions affords only a mixture of cyclohexenyl styrenes **88a** and

b and recovered malonate. Chemical yields are significantly improved if triethylamine is substituted for malonate anion. These results indicate that compound **10** undergoes cleavage of the C1–C2 cyclopropane bond under carbopalladation conditions.[55]

Carbopalladation of 7-methylenebicyclo[4.1.0]heptane. To a solution of iodobenzene (1.51 g, 7.4 mmol), 7-methylenebicyclo[4.1.0]heptane (0.80 g, 7.4 mmol) and triphenylphosphine (0.1 g, 3.8 mmol) in dry THF (15 mL) under N_2 was added tetrakis(triphenylphosphine)palladium (0.34 g, 0.29 mmol, 0.04 mol equivalent). Triethylamine (2.24 g, 22.2 mmol) was added and the reaction mixture was brought to reflux for 7 days. The reaction mixture was cooled and the solvent removed under reduced pressure. The resultant oil was taken up in CH_2Cl_2 (50 mL) and washed three times with 5% aqueous HCl. The organic layer was concentrated and filtered through a bed of silica gel with petroleum ether elution. The solvent was removed and the product was isolated by distillation under high vacuum, 1.03 g, 76% yield.

We have proposed the following mechanism for this transformation: Initial oxidative addition of the halobenzene generates an arylpalladium species which coordinates to the least hindered face of **10**. Olefin insertion generates the σ-cyclopropylcarbinyl palladium complex **89** which may undergo relative free C–C bond rotation. In rotomer **89′** the σ-cyclopropyl-carbinyl palladium complex may rearrange to the σ-homoallyl palladium complex **90**. The σ-homoallyl complex **90** can only undergo *syn*-β-hydride elimination to afford the (olefin)palladium hydride **91**. Loss of the ligand generates the product **88a** while olefin insertion into the Pd–H bond generates the σ-complex **92**. β-Hydride elimination of **92** may afford either **91** or the isomeric **93**. Loss of the olefin ligand from **93** would give rise to **88b**.

While the C1–C2 bond cleavage observed for carbopalladation of **10** is similar to that of carbo- and chloro-palladation of the parent methylenecyclopropane, it is in *direct contrast* with the C2–C3 bond cleavage observed for chloropalladation of these same substrates (**10**). The most likely cause for this difference in reactivity is due to the presence of lone pair electrons on the chloride ligand. The chlorine atom may act as a "bridging" ligand to attack at the more distant C2 cyclopropane carbon. In fact, Extended Hückel calculations indicated that there is strong interaction between the filled C1 lone pair orbital and an unoccupied antibonding orbital which is concentrated on C2 and C3.[9a] It is anticipated that the aryl ligand present in *carbopalladation* would not have the correct spatial orientation necessary for this type of interaction, and thus that carbopalladation should proceed in a fashion different from chloropalladation.

The regioselectivity for cyclopropane bond cleavage in substituted 7-methylenebicyclo[4.1.0]heptanes has been studied in order to determine the nature of the cyclopropylcarbinyl-to-homoallyl rearrangement.[56] Catalytic carbopalladation of 7-methylene-1-phenylbicyclo[4.1.0]heptane (**8**) affords a single product **94**, via cleavage exclusively at the phenyl substituted cyclopropane carbon. In comparison, catalytic carbopalladation of 1-methyl-7-methylenebicyclo[4.1.0]heptane (**95**) gave a complex mixture of substituted cyclohexenyl styrenes. Analysis of the mixture by gas chromatography/mass spectroscopy (GC/MS) along with comparison to independently synthesized samples, indicates that approximately 84% of the products arise via cleavage at the methyl substituted cyclopropane carbon (**96**) while 16% of the mixture results from cleavage at the less substituted carbon (**97**). As might be expected, reaction of 3-methyl-7-methylenebicyclo[4.1.0]heptane (**98**) under these conditions, also gave a mixture of cyclohexenyl styrene products. Analysis of the mixture indicated that cleavage of both of the cyclopropane bonds occurs with equal probability.

The results on the regioselectivity for cyclopropane bond cleavage are consistent with either a cationic[57] or a radical [58] cyclopropylcarbinyl-to-homoallyl rearrangement mechanism. However the lack of any cyclobutyl rearranged product is relatively strong evidence against a cationic mechanism. The possibility of a concerted cyclopropylcarbinyl-to-homoallyl rearrangement cannot be ruled out.

VII. SUMMARY

In summary, we have in hand an efficient ring homologation methodology for cyclic alkenes. We have investigated the reactivity of the product (3-chloro-2-methyloenecycloalkyl)palladium chloride dimers and have found that they may react as either "isoprenyl monocation" synthons or "trimethylenemethane dication" synthons. In the latter synthon, we have recently shown that two of the same nucleophiles can be added, or one dinucleophile, or two different nucleophiles may be sequently added.

Additionally, we have found that the catalytic carbopalladation of Ω-methylenebicyclo[n.1.0]alkanes proceeds via cleavage of a different cyclopropane bond (C1–C2) than does chloropalladation of these same compounds. We are currently investigating the application of both of these reactivities to the synthesis of biologically interesting compounds.

ACKNOWLEDGEMENTS

The author expresses his deep appreciation to the graduate and undergraduate students who contributed to this work. Their names appear in the references cited. Financial support was provided by the Petroleum Research Fund and by Marquette University. Acknowledgement is due to Johnson–Matthey, Inc. for generous donations of palladium chloride.

REFERENCES

1. H.C. Brown, "Boranes in Organic Chemistry", Cornell University Press: Ithaca, NY, 1972.
2. E. Negishi, Acc. Chem. Res. 1987, 20, 65; R.F. Heck, Acc. Chem. Res. 1979, 12, 146.
3. K. Kaneda, T. Uchiyama, H. Kobayashi, Y. Fuhiwara, T. Ianaka and S. Teranishi,

Tetrahedron Lett. 1977, 2005; P. Mushak and M. A. Battiste, *J. Chem. Soc., Chem. Commun.* 1969, 1146.

4. P. M. Henry, "Palladium Catalyzed Oxidation of Hydrocarbons", Reidel: Dordrecht, 1980; W. Kitching, *Organometallic Reactions* 1972, *3*, 319.

5. (a) R. G. Schultz, *Tetrahedron Lett.* 1964, 301; (b) M. S. Lupin and B. L. Shaw, *Tetrahedron Lett.* 1964, 883; (c) R. G. Schultz, *Tetrahedron* 1964, *20*, 2809.

6. R. Noyori and H. Takaya, *J. Chem. Soc., Chem. Commun.* 1969, 77.

7. (a) M. Green and R. P. Hughes, *J. Chem. Soc., Dalton Trans.* 1976, 1880; (b) R. Goddard, M. Green, R. P. Hughes and P. Woodward, *Ibid.* 1976, 1890.

8. (a) B. K. Dallas and R. P. Hughes, *J. Organometal. Chem.* 1980, *184*, C67; (b) R. P. Hughes, D. E. Hunton and K. Schumann, *Ibid.* 1979, *169*, C37.

9. (a) T. A. Albright, P. R. Clemens, R. P. Hughes, D. E. Hunton and L. D. Margerum, *J. Am. Chem. Soc.* 1982, *104*, 5359; (b) B. K. Dallas, R. P. Hughes and K. Schumann, *Ibid.* 1982, *104*, 5380; (c) P. R. Clemens, R. P. Hughes and L. D. Margerum, *Ibid.* 1981, *103*, 2428.

10. W. A. Donaldson, *J. Organometal. Chem.* 1984, *269*, C25.

11. W. A. Donaldson, J. T. North, J. A. Gruetzmacher, M. Finley and D. J. Stepuszek, *Tetrahedron* 1990, *46*, 2263.

12. B. M. Trost and T. R. Verhoeven, "Comprehensive Organometallic Chemistry", Pergamon Press: New York, 1983, Vol. 8, Chapter 57.

13. S. Arora and P. Binger, *Synthesis* 1974, 801.

14. K. Kitatani, T. Hiyama and H. Nazaki, *Bull. Chem. Soc. Jpn.* 1977, *50*, 3288; K. N. Slessor, A. C. Oehlschlager, B. D. Johnston, H. D. Pierce, Jr., S. K. Grewel and L. K. G. Wickremesinghe, *J. Org. Chem.* 1980, *45*, 2290.

15. W. A. Donaldson, *Organometallics* 1986, *5*, 223.

16. The (3-chloro-2-methylenecyclohexyl)palladium chloride dimer (**12**, *n* = 6) has previously been prepared from the chloropalladation of 1,2,6-heptatriene, albeit in lower yield: J. Powell and N. I. Dowling, *J. Organometal. Chem.* 1984, *264*, 387.

17. P. M. Maitlis, "The Organic Chemistry of Palladium", Academic Press: New York, Vo. 1, pp. 175–252.

18. J. Lukas, J. E. Ramakers-Blom, T. G. Hewitt and J. J. DeBoer, *J. Organometal. Chem.* 1972, *46*, 167.

19. W. C. Still and I. Ealynker, *Tetrahedron* 1981, *37*, 3981.

20. MMX™, Serena Software, Bloomington, IN.

21. R. P. Hughes and C. S. Day, *Organometallics* 1982, *1*, 1221.

22. (a) H. Christ and R. Huttel, *Angew. Chem., Int. Ed. Engl.* 1963, *2*, 626; (b) R. Huttel, H. Dietl and H. Christ, *Chem. Ber.* 1964, *97*, 2037; (c) R. Huttel and P. Kochs, *Ibid.* 1968, *101*, 1043.

23. W. A. Donaldson, *Tetrahedron* 1987, *43*, 2901.

24. (a) R. D. Reike, A. V. Kavaliunas, L. D. Rhyne and D. J. J. Fraser, *J. Am. Chem. Soc.* 1979, *101*, 246; (b) Y. Inone, J. Yamashita and H. Hashimoto, *Synthesis* 1984, 224.

25. H. C. Brown and M. Borkowski, *J. Am. Chem. Soc* 1952, *74*, 1894.

26. (a) K. J. Klabunde and J. S. Roberts, *J. Organometal. Chem.* 1977, *137*, 113; (b) K. J. Klabunde, *Acc. Chem. Res.* 1983, *8*, 393.

27. W. A. Donaldson and B. S. Taylor, *Tetrahedron Lett.* 1985, 4163.

28. R. Kaiser and D. Lamparsky, *Helv. Chim. Acta* 1978, *61*, 2671.
29. B. M. Trost, L. Weber, P. E. Strege, T. J. Fullerton and T. J. Dietche, *J. Am. Chem. Soc.* 1978, *100*, 3416.
30. W. A. Donaldson and V. J. Grief, *Tetrahedron Lett.* 1986, 2345.
31. Another trimethylenemethane dication synthon has recently been reported: Y. Haung and X. Lu, *Tetrahedron Lett.* 1987, 6219.
32. W. C. Still, M. Kahn and A. Mitra, *J. Org. Chem.* 1978, *43*, 2923.
33. F. K. Sheffy, J. P. Godschalx and J. K. Stille, *J. Am. Chem. Soc.* 1984, *106*, 4833.
34. Pd-allyls are less susceptible to β-hydride elimination than Pd-alkyls, however the thermal decomposition of certain Pd-allyls to afford 1,3-dienes is believed to proceed via β-hydride elimination than Pd-alkyls, however the thermal decomposition of certain Pd-allyls to afford 1,3-dienes is believed to proceed via β-hydride elimination: K. Dunne and F. J. McQuillan, *J. Chem. Soc. (C)* 1970, 2200; B. M. Trost, T. R. Verhoeven and J. M. Fortunak, *Tetrahedron Lett.* 1979, 2301.
35. S. D. Ittel, F. A. Van-Catledge and J. P. Jenson, *J. Am. Chem. Soc.* 1979, *101*, 6905; S. D. Ittel, F. A. Van-Catledge, C. A. Tolman and J. P. Jensen, *Ibid.* 1978, *100*, 1317.
36. W. A. Donaldson, J. Wang and V. J. Grief, "Abstracts of the 4th IUPAC Symposium on Organometallic Chemistry Applied to Organic Synthesis," Vancouver, BC, July 26–30, 1987, Paper No. PS2-49.
37. W. A. Donaldson and J. Wang, *J. Organometal. Chem.* 1990, *395*, 113.
38. L. Wilson and I. Meza, *J. Cell. Biol.* 1973, *58*, 709.
39. F. Cortese, B. Bhattacharyya and J. Wolff, *J. Biol. Chem.* 1977, *252*, 1134; K. Ray, B. Bhattacharyya and B. B. Biswas, *J. Biol. Chem.* 1981, *256*, 6241.
40. (a) W. A. Donaldson, D. J. Stepuszek and J. A. Gruetzmacher, *Tetrahedron* 1990, *46*, 2273; (b) J. A. Gruetzmacher, M.S. Thesis, Marquette University, 1987.
41. D. A. Evans, S. P. Tanis and D. J. Hart, *J. Am. Chem. Soc.* 1981, *103*, 5813.
42. D. L. Boger and C. E. Brotherton, *J. Am. Chem. Soc.* 1986, *108*, 6713.
43. B. M. Trost and D. M. T. Chan, *J. Am. Chem. Soc.* 1983, *105*, 2326.
44. G. Molander and D. C. Shubert, *J. Am. Chem. Soc.* 1987, *109*, 576.
45. R. D. Little and G. W. Muller, *J. Am. Chem. Soc.* 1981, *103*, 2744.
46. Dimethyl 3-ketoglutarate has previously been condensed with 1,2-diketones for the formation of the bicyclo[3.3.0]octane skeleton: R. W. Weber and J. M. Cook, *Can. J. Chem.* 1978, *56*, 189; S. H. Bertz, J. M. Cook, A. Gawish and U. Weiss, *Org. Syn.* 1985, *64*, 27.
47. W. A. Donaldson, J. Wang, V. G. Cepa and J. D. Suson, *J. Org. Chem.* 1989, *54*, 6056.
48. The compound 8-keto-bicyclo[4.4.0]dec-1-ene has previously been prepared via a biomimetic cyclization route: W. S. Johnson, W. H. Lunn and W. K. Fitzi, *J. Am. Chem. Soc.* 1964, *86*, 1972.
49. The terminal allylic carbons are hybridized between sp^2 and sp^3, and thus application of Baldwin's rules for ring closure might also suggest a favored "6-*endo*-trig" closure: J. E. Baldwin, *J. Chem. Soc., Chem. Commun.* 1976, 734.
50. "Naturally Occurring Phorbal Esterse"; F. J. Evans, Ed.; CRC Press: Boca Raton, FL, 1986; P. A. Wender, K. F. Koehler, N. A. Sharkey and M. L. Dell'Aquilla, *Proc. Natl. Acad. Sci.* 1986, *83*, 4214.
51. Structure: E. J. Eisenbraun, T. George, B. Rinike and C. Djerassi, *J. Am. Chem. Soc.* 1960, *82*, 3648; Synthesis: J. A. Marshall and J. J. Partridge, *Tetrahedron* 1969, *25*, 2159; C. H. Heathcock and R. Ratcliffe, *J. Am. Chem., Soc.* 1971, *93*, 1746; N. H. Anderson and H. Uh, *Syn. Commun.*, 1973, *3*, 115.
52. R. B. Reinarz and G. J. Fonken, *Tetrahedron Lett.* 1973, 4013.
53. R. C. Larock and S. Varaprath, *J. Org. Chem.* 1984, *40*, 3432.
54. G. Balme, G. Fournet and J. Gore, *Tetrahedron Lett.* 1986, 3855.
55. W. A. Donaldson and C. A. Brodt, *J. Organometal. Chem.* 1987, *330*, C33.
56. C. A. Brodt, M.S. Thesis, Marquette University, 1988.

57. R. H. Mazur, W. N. White, D. A. Senenow, C. C. Lee, M. S. Silver and J. D. Roberts, *J. Am. Chem. Soc.* 1959, *81*, 4390; H. C. Brown "The Nonclassical Ion Problem", Plenum Press: New York, 1977.
58. M. P. Atkins, B. T. Golding and P. J. Sellars, *J. Chem. Soc., Chem. Commun.* 1978, 954; A. Burry, M. R. Ashcroft and M. D. Johnson, *J. Am. Chem. Soc.* 1978, *100*, 3217.

Advances in Metal-Organic Chemistry

Edited by **Lanny S. Liebskind,** *Department of Chemistry, Emory University*

Organometallic chemistry is having a major impact on modern day chemistry in industry and academia. Within the last ten years, the use of transition metal based chemistry to perform reactions with significant potential in organic synthesis has come of age. *Advances in Metal-Organic Chemistry* will contain in-depth accounts of newly emerging synthetic organic methods that emphasize the unique attributes of transition metal chemistry problems in organic synthesis. Each issue contains six to eight articles by leading investigators in the field. Particular emphasis is placed on giving the reader a true feeling of the particular strengths and weaknesses of the new chemistry with ample experimental details for typical procedures.

Volume 1, 1989, 408 pp.
ISBN 0-89232-863-0

JAI PRESS LTD • **JAI PRESS INC.**

118 Pentonville Road
London N1 9JN

55 Old Post Road No. 2
Greenwich, CT 06836-1678

Tel: 071-833 1778
Fax: 071-837 2917

Tel: 203-661 7602
Fax: 203-661 0792